T0213494

True Tales of Medical Physics

Jacob Van Dyk
Editor

True Tales of Medical Physics

Insights into a Life-Saving Specialty

 Springer

Editor
Jacob Van Dyk
Western University
London, ON, Canada

ISBN 978-3-030-91723-4   ISBN 978-3-030-91724-1   (eBook)
https://doi.org/10.1007/978-3-030-91724-1

This Springer imprint is published by the registered company Springer Nature Switzerland AG
The registered company address is: Gewerbestrasse 11, 6330 Cham, Switzerland

To my children: Tonia, Jon, Ben, and Amy
and
to all those children of medical physicists who never quite figured out what their
parents did.

Foreword

In 1895, Wilhelm Roentgen discovered the existence of x-rays, and Henri Becquerel detected spontaneous emissions from nuclei that has become known as radioactivity, which was further researched by Marie Sklodowska Curie. Both of these fundamental physics phenomena were rapidly transformed into medical applications, which are primarily known as the radiology, nuclear medicine and radiotherapy specialties today. All these disciplines harness radiation to perform diagnosis through imaging or treatment of diseases using radiation sources. Complex techniques, technologies and applications evolve continuously, and medical physicists are healthcare professionals that are responsible for ensuring the accuracy of radiation doses, the high quality of imaging techniques and the safe use of radiation technologies to benefit patients.

Medical physics is a global enterprise. The International Atomic Energy Agency (IAEA) has supported the harmonization of radiation dosimetry for more than 50 years through calibration, comparison and audit services, the development of international codes of practice, scientific and professional guidelines and the coordination of research and educational opportunities.

Fellowship training of medical physicists, medical physics expert missions to support centres on the ground and the new IAEA Marie Sklodowska Curie Fellowship Programme have all provided many opportunities for support, training and education of medical physicists and encourage women and young professionals.

While the work of the IAEA provides synergistic support to medical physics activities globally, it has an added impact on low- and middle-income countries. It is also interesting to note that of the 22 award-winning authors of chapters in this book, at least 12 have been involved in IAEA-related activities, hence, providing their support and influence on this global endeavour.

The work of medical physicists is often divided into three major components: clinical service, teaching and research. These three components and the related activities are well described through the stories of the different authors. While the extent of these activities varies from one medical physicist to another, all three are needed to provide continuity and advancements in the world of medical physics.

This book will provide insights into the world of medical physics, serving as an inspiration to the young generation to consider it as a career option and supporting more experienced physicists in their leadership development. In addition, it will expose the public to the life-saving activities that are served by this profession.

As for the IAEA, we recognize the important contribution of medical physics to good quality and safe medical use of radiation and will continue to support the field and medical physicists worldwide through our many activities and programmes.

Vienna, Austria

Rafael M. Grossi
Director General, International
Atomic Energy Agency

Preface

Welcome to the *True Tales of Medical Physics: Insights into a Life-Saving Specialty*. This is not your standard textbook with a structured format to describe a specific medical or scientific topic. While *medical physics* is in the title, this is not a medical physics book; rather, it is a book *about* medical physics. The contributing authors, all of whom are award-winning all-stars in the field of medical physics, have described stories, i.e., *true tales*, from their life experiences. By telling these stories, we hope that you, the reader, will gain some insight into the world of medical physics. These *tales* may be especially relevant for science students, graduate students, medical physics residents or even experienced medical physicists who are trying to establish directions in their careers. Friends and family of medical physicists may wonder what their close connection does at work. Each contributor to the book provides a different perspective with different experiences. In that sense, this book is of interest to anyone who has a general interest in the sciences and especially if they would like to learn more about medical physics. Perhaps this may be even more relevant, if you, or a closely connected person, is undergoing health-related issues which involve diagnostic imaging tests, or if you are being treated with radiation therapy, or if you have a general concern about radiation exposures. By providing these *true tales*, the authors provide in-depth *insights* into the world of medical physics.

The format of the book is very much like that of a novel with a *prologue* and an *epilogue*. The *prologue* provides a detailed introduction to the book. The *epilogue* provides a retrospective overview. In between are all the *true tales* divided into six parts. While the title of each part gives a hint as to some of the emphasis of the contents of the chapters in that part, the divisions are not that clear cut and much of the subject matter overlaps with that of other parts; hence, each part title indicates that medical physics is *"More Than ..."*.

As you read this book, there may be times that you feel this is too much medical or scientific detail. The suggestion is to keep reading since the stories do provide concluding messages that are more universal, sometimes well beyond the world of medical physics. Many chapters have phrases like "the message here is ..." or "the lesson learned is ...". These are experiences from ultrasmart and well-honed individuals. The lessons they learned should be taken seriously!

My hope is that the *insights* gained by the readers will generate a better understanding of the role of medical physicists in the healthcare system, a role that is indeed involved in saving millions of lives.

London, ON, Canada   Jacob Van Dyk
September 2021

Acknowledgments

There are many people who make a production of a book of this nature possible. Some provide direct support, for others it is more indirect. A brief listing provides a sampling of those who have provided some input and to whom I am grateful:

The **contributing authors** have allowed this book to happen. They have given of their time and shed some light on stories from their careers. Without them, there would be no *true tales*.

Jerry Battista read an early draft of the book proposal and provided his usual wise and thoughtful comments.

Shelagh Rogers and **Eleanor Wachtel**, hosts of the Canadian Broadcasting Corporation (CBC) programmes *The Next Chapter* and *Writers and Company*, seeded the concept, through their interviews on the radio, that we medical physicists should share our stories with our colleagues as well as the general public.

Springer for enthusiastically accepting my book proposal.

Zachary Evenson, Springer's Senior Editor, for his strong support and providing open communications and rapid responses to all my comments and questions.

Ravi Vengadachalam, Springer's Project Coordinator, and **Monica Janet Michael**, Production Editor, and their staff for organizing the production of this book.

Terry Peters, **Ravi Menon**, **Joe Gati**, **Trevor Szekeres** and **Oksana Opalevych** who were all involved in my MRI scans used for the front cover.

Susan Cunningham for providing the video and lyrics of Jack Cunningham's song in relation to the story of *The Elevator that Couldn't*.

Rafael Grossi, Director General of the International Atomic Energy Agency, for providing a *Foreword* strongly supporting the work of medical physicists.

Eduardo Zubizarreta and **May Abdel-Wahab** for their positive and encouraging support to my request about the *Foreword* of this book.

Christine Van Dyk for constant and endearing support for the various "crazy" ventures that I undertake.

Prologue

In 1978, I was summoned to jury duty in Peel County—a county that contains the Toronto Pearson International Airport. The charges against the accused, who were from South America, included drug smuggling. At the first day of the trial, I was part of a large jury pool of about 140 from which names were selected. I was number 127. The court officer randomly selected numbers, after which the name, the address and the person's occupation were called out and the lawyers for both the defence and the prosecution teams then decided as to whether to accept this individual as a juror or not. Lo and behold, my name was called, "Jacob Van Dyk, 2315 Bromsgrove Rd, Mississauga, medical physicist". Even from the back of the room, I could read the lips of one of the lawyers for the accused lean over to his colleague and whisper, "What's a medical physicist?" I was chosen as juror number 6.

This brief episode is a metaphor of my life. When I am asked about my occupation and I answer, "I am a medical physicist", the general response is, "Oh! What's that?" or "What exactly do you do?" When my children in their younger school years were asked, "What does your Dad do?", their general answer was, "He works in a hospital." "He wears a lab coat." "He sees patients.". If pushed, they might say, "He works in a cancer hospital", or "He works with radiation machines" but that was about the extent of it. My eldest daughter, Tonia, points out that they even had a hard time pronouncing the word "physicist". Often the responses get further questions, like, "Oh, so your Dad is a doctor?" When I am asked and proceed to explain in a little more depth as to what I do, often the eyes of the listener glaze over and I realize that it is time to move on to another topic.

This book is not a logically organized description of what medical physicists do. Rather, it is a series of snippets from the lives of medical physicists who have made their mark in the field. These stories give real-life experiences within their careers. They represent illustrations and examples of the work of medical physicists told from a light perspective in easy-to-understand language. By telling these stories, they provide a glimpse of the world of medical physics, a profession that is indeed involved in saving lives.

The Boring Answer

When I asked one of the Internet search engines, "What do medical physicists do?", I found the following answer from Wikipedia, "*A medical physicist is a **professional who applies the principles and methods of both physics and medicine**. They focus on the areas of prevention, diagnosis, and treatment, as well as ensuring quality services and prevention of risks to the patients, and members of the public in general. A medical physicist plays a fundamental role in applying physics to medicine, but particularly in the diagnosis and treatment of cancer. The scientific and technological progress in medical physics has led to a variety of skills that must be integrated into the role of a medical physicist in order for them to perform their job. The "medical services" provided to patients undergoing diagnostic and therapeutic treatments must, therefore, be the result of different but complementary skills.*" There, that is the boring answer!

When I ask the search engine to "Tell me stories about medical physics", it provides links to a series of websites, including *Medical Physics—Goodreads*, which leads me to a series of textbooks that are barely readable, even by medical physicists, or to a series of websites that lead to various medical physics organizations. It also leads to other questions, like "How do I become a medical physicist?" The answer provided by the web, "*A prospective medical physicist should first have an honors degree in physics. Courses in computing, electronics, and mathematics are advantageous. They may then undertake graduate work in medical physics or another area of physics followed by a one or two year training program in medical physics*".

The web automatically comes up with the statement, "People also ask": "*What is the job description of a medical physicist?*". The answer is "*Most medical physicists work in one or more of the following areas: The responsibilities of a clinical medical physicist lie predominantly in the areas of radiotherapy and diagnostic imaging. The roles of a medical physicist in radiotherapy include treatment planning and radiotherapy machine design, testing, calibration, and troubleshooting".*

"What are the four areas of Medical Physics?" "Certification is offered in four areas of medical physics, including therapeutic radiological physics, diagnostic radiological physics, medical nuclear physics and medical health physics".

The latter is not entirely complete since another area of specialization, at least in some jurisdictions, is magnetic resonance imaging (MRI) or sub-specialization in mammography. Furthermore, there are areas of medical physics that are not covered by certification procedures, examples being, photodynamic therapy or laser biophysics. Thus, even on the web, the answer as to what medical physics is and what medical physicists do is rather pedantic. So, it is no wonder that the general public is confused.

It is interesting that historically people often think of the beginning of medical physics occurring with the discovery of x-rays by Röntgen in 1895 and the discovery of radioactivity by Becquerel in 1896, or the work of Marie Sklodowska Curie for her research on radium and polonium and her application of the principles of physics in the field of medicine with a focus on diagnosis and treatment of diseases. However, there was already *"A Text-book on Medical Physics"* published in 1885 (see Fig. 1). This book has 774 pages along with 377 illustrations. To quote from the Preface, *"Broadly speaking, this work aims to impart a knowledge of the relations existing between Physics and Medicine in their latest state of development, and to embody in the pursuit of this objective whatever experience the author has gained during a long period of teaching this special branch of applied science".* It touched on topics like the properties of matter, potential and kinetic energy, properties of sound and light, the microscope and telescope, the eye and vision, heat, ventilation, electricity, electromagnetism, electrobiology and electrotherapy.

One should not forget the work by Antoni van Leeuwenhoek (1632–1723), one of the first individuals to develop and use the microscope for looking at microbial life. Clearly, he needed to use the principles of optics (a branch of physics) to develop and construct his own microscopes. His observations then led to the foundations of the science of microbiology and into the world of medicine, i.e., physics as applied to medicine. A leading cancer therapy hospital in the Netherlands has been named the *Antoni van Leeuwenhoek Ziekenhuis* and has been very much involved in groundbreaking medical physics activities, especially in the last 50 years (see Chapter 12 by Marcel van Herk). In 2004, a public poll in the Netherlands to determine the greatest Dutchman ("De Grootste Nederlander") named van Leeuwenhoek as the fourth-greatest Dutchman of all time. Note that Chapter 1 by

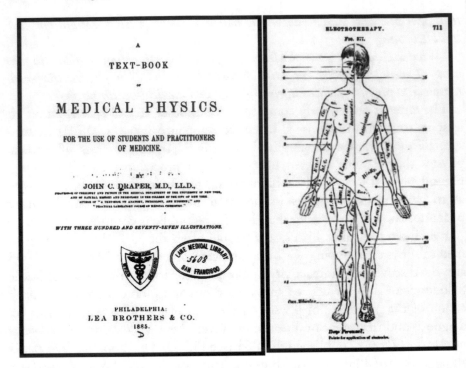

Fig. 1 One of the first textbooks on medical physics by John C. Draper published in 1885. The figure on the left shows the title page and the one on the right is the last figure in the book showing the points of application of electrodes for electrotherapy. (Draper, John C., A Text-Book of Medical Physics. For the Use of Students and Practitioners of Medicine, Lea Brothers & Company, Philadelphia, 1885)

David Thwaites gives a much more detailed description of the origins and history of medical physics.

Of course, a lot has changed in the meantime with new discoveries and new applications of physics in medicine.

Why?

As physicists, we often ask questions. In fact, in my early years of marriage, my wife was often perturbed by the incessant number of questions I asked, about anything, but especially about the things we were about to purchase, like, "Why do we need it?", "Do we really need it?", "If we really need it, then which is the best one?". This again leads to a series of other questions. (By the way, she has adapted over the years and tolerates it now; sometimes, she even acknowledges the benefit of the questions!) Note that the above

Table 1 Some comparative statistics for different vocations in Canada and the USA

Occupation	Canada Total	Canada Number per 100,000	USA Total	USA Number per 100,000
Physician	91,400	241	1,085,800	331
Lawyers	105,000	276	1,329,000	408
Engineers*	305,300	803	1,630,000	500
Police Officers	65,300	172	697,200	211
Medical Physicists	420	1.1	5400	1.6

*The large differences in engineers between Canada and the USA likely relate to the definition of an engineer. The Canadian information is based on Engineers Canada membership data. The US data are based on a congressional report and appear to be more general. Either way, the numbers are large.

list of questions began with "why?". "Why?" is a question often asked by physicists in their professional or academic lives. Why is it so that the general public knows so little about medical physics? Perhaps the first answer to that question is the "boring answer" given in the above section. However, there are other reasons as well.

One can begin by simply looking at the statistics. In Canada, in 2019, there were over 91,000 physicians corresponding to over 240 doctors per 100,000 population (see Table 1). At the same, there were a total of about 420 medical physicists in Canada corresponding to 1.1 medical physicists per 100,000 population. Similarly, in the USA, there are about 330 physicians per 100,000 population, 410 lawyers per 100,000 and only about 1.6 medical physicists per 100,000. Comparable data exist for engineers and police officers. Clearly, in terms of numbers, there are very few medical physicists compared to other occupations.

Another issue that relates to the lack of knowledge about medical physicists is their visibility to the public. Medical doctors, for example, have direct patient contact all the time. Lawyers connect directly with their clients. Medical physicists generally are involved in "behind the scenes" activities related to diagnostic or treatment technologies and the radiation treatment process but do not usually interact directly with the patient. They only see patients directly when the patient situation is rather unusual or more complex than normal, and, therefore, they are consulted by the physician and brought in to see the patient. Similarly, for other common professions such as engineers and police officers, there tends to be more of an interaction between the general public and the professionals—hence, a much higher visibility. There is presently an initiative underway that is evaluating the benefits of routine and direct patient interactions with medical physicists, spearheaded at the

University of California in San Diego. This study relates to the benefits of educating the patient and the patient's family regarding the radiation therapy treatment process, the technologies involved and quality assurance and radiation safety issues. So far, this evaluation has received a positive response and, thus, could lead to a more direct and more frequent interactions of patients and medical physicists.

Furthermore, how many medical physics toys do children have? My children had doctor's kits as kids, fire trucks, police cars and Meccano® sets. They did not have any Geiger counters, radon detectors or water phantoms. Maybe medical physicists should become creative and develop some toys that can be used by children, or kits with which they can do medical physics experiments. However, until that happens, we will try to use another approach.

True Tales of Medical Physics

This book takes a completely different approach in looking at medical physics. To my knowledge, there is no equivalent book. The contributors to this book were invited *"to communicate what medical physics is and what medical physicists do to a broad audience including science students, graduate students and residents, experienced medical physicists and their family members, and the general public who are wondering about medical physics.*

The book will consist of a series of short stories written by award-winning medical physicists—stories that are of personal interest as it relates to their careers. Each story will be unique to the author and could serve any one or more of the following purposes:

1. *Be an inspiration to young people searching for career directions, as well as more experienced physicists who are seeking direction on leadership development.*
2. *Provide an overview of what medical physicists do with a level of description that is understandable by the non-medical physicist.*
3. *Provide lessons on life's experiences from high-profile medical physicists who have significant experience and who are clearly at the top of the field as shown by the awards that they have won.*
4. *Be entertaining for those working in the field as well as others."*

The emphasis is on "tales" from high-profile, award-winning medical physicists—stories that normally do not get published in scientific or medical journals—but stories that inevitably involve medical physics and medical

Fig. 2 Authors of four of the chapters of this book from Western University, London, Ontario Canada, showing off their Gold Medals, the highest award from the Canadian Organization of Medical Physicists. From left to right: Terry Peters, Ph.D., Jerry Battista, Ph.D., Jacob Van Dyk, D.Sc. and Aaron Fenster, Ph.D. Picture courtesy of Western University's Schulich School of Medicine and Dentistry

physicists. Based on these stories, you, the reader, will find out about the realities of medical physics and what medical physicists do.

So, who are these high-profile, award-winning medical physicists? They are an all-star cast of medical physics legends each of whom has received the top awards from their respective national or international medical physics organizations, or the highest government awards. For the American Association of Physicist in Medicine (the largest medical physics organization in the world), the top award is the *Coolidge Award*, which is granted to one awardee per year. For the Canadian Organization of Medical Physicists, it is the *Gold Medal*. For the Canadian government, it is the *Order of Canada*; for the Australian government, it is the *Order of Australia Medal*. Also included are award winners from the International Organization for Medical Physics (IOMP). In other words, they are the *crème-de-la crème* of the world of medical physics. They represent nine countries; thus, they provide a very global perspective. The end of each chapter includes a brief summary of their careers along with a listing of their top awards, up to a maximum of five.

Anecdotally, of the 22 contributors to this book, four are from Western University, London, Ontario Canada—the greatest number from any one institution although representation from the MD Anderson Cancer Center in Houston, Texas, USA, runs a close second. The four from Western are pictured in Fig. 2 with their Gold Medals—the top award from the Canadian Organization of Medical Physicists (COMP). There was a fifth Gold

Medallist from Western who died in 2018, Dr. John C.F. MacDonald. John actually supervised the master's degree graduate studies of Jerry Battista and Jacob Van Dyk. No other university or institution has had this kind of representation since the inauguration of the COMP Gold Medal in 2006. This picture was taken as part of an eNewsletter article published by Western's Schulich School of Medicine and Dentistry in March of 2020. The number of Gold Medal winners from Western is evidence that the quality of medical physics and imaging research at Western is outstanding. Western also represents one of the largest graduate schools in medical physics in Canada, if not North America. More award winners from Western University could have been included but the editor did not want to make this a Western dominated book.

In view of the top-tier nature of the contributors, many of the chapters contain stories about superlatives, such as the "first", the "best", the "biggest", the "clearest", the "finest", the "smallest", …. In addition to the superlative nature of these stories, they provide background information that is not found in standard scientific or medical journals, making these *tales* truly outstanding. Indeed, many of the contributors provide words of wisdom that are useful to medical physicists at any stage in their career, be it at the beginning of their graduate studies or later in their more senior years. Furthermore, these words of wisdom are often of a general nature, being relevant well beyond medical physics.

How to Read This Book

There are several approaches to reading this book. For those who are totally new to the concept of medical physics, it is perhaps best to begin with the prologue and then to read the book from the beginning to the end. For those with specific connections to medical physicists, perhaps your brother, sister, uncle, aunt, cousin, colleague, friend or acquaintance, you might be interested in specific topics that relate to them. To help with that, I have generated a table with a very high-level overview of the chapters including some topic categories (Table 2). You may want to go straight to the relevant chapter and see what tales your particular acquaintance has to tell and then read the rest for comparison. For those interested in specific subtopics within medical physics, it is again worth looking at Table 2 to see if that topic is listed; otherwise, check the index.

Table 2 Overview of chapter contents

Chapter Number	Author	Country	Radiation Therapy	Nuclear Medicine Imaging	X-ray/CT Imaging	Ultrasound Imaging	Magnetic Resonance Imaging	Mammography	Industrial or Commercial Connections	Research	Clinical Service	Teaching	Professional Activities	Health Physics/ Radiation Protection
-	Jacob Van Dyk (Prologue)	Canada	-	-	-	-	-	-	-	-	-	-	-	-
Part I. Medical Physics: More than History														
1	David Thwaites	Australia/UK	**	*	**	-	-	-	-	**	**	-	**	**
2	Jacob Van Dyk	Canada	***	-	**	-	-	-	-	***	***	*	**	-
3	Peter R. Almond	US	***	-	-	-	-	-	-	***	***	***	**	-
4	Gary T. Barnes	US	-	-	***	-	-	***	**	***	***	***	**	-
Part II. Medical Physics: More than Clinical Service														
5	Arthur L. Boyer	US	***	-	-	*	-	-	-	**	***	**	-	*
6	James A. Purdy	US	***	-	-	-	-	-	-	***	***	**	***	-
7	John Wong	US/Canada	***	*	*	-	-	-	**	***	***	-	**	-
Part III. Medical Physics: More than Research														
8	Paul L. Carson	US	*	*	*	***	-	**	*	***	*	*	-	-
9	C. Clifton Ling	US	***	*	*	-	*	-	*	***	***	-	-	*
10	Terry M. Peters	Canada/New Zealand	**	**	***	**	***	-	*	***	*	*	*	-
11	Stephen R. Thomas	US	***	***	-	-	***	-	*	***	*	*	**	-
12	Marcel van Herk	UK/Netherlands	***	-	***	-	-	-	**	***	*	-	**	-
Part IV. Medical Physics: More than Protection of the Public														
13	Caridad Borras	US/Spain	-	-	-	-	-	-	-	-	-	-	**	***
14	Carlos de Almeida	Brazil	***	-	-	-	-	-	-	**	***	*	-	***
15	Arun Chougule	India	***	*	-	-	-	-	-	**	***	***	***	**
Part V. Medical Physics: More than Teaching														
16	Jerry J. Battista	Canada	***	**	***	*	***	-	-	**	*	*	-	-
17	Tomas Kron	Australia/Canada	***	-	-	*	*	-	-	**	***	***	***	*
18	Martin Yaffe	Canada	-	-	***	-	-	***	-	***	*	**	-	-
Part VI. Medical Physics: More than Commercial Developments														
19	Aaron Fenster	Canada	*	-	***	***	*	-	***	***	-	-	-	-
20	Maryellen L. Giger	US	-	-	*	*	*	***	***	***	-	***	***	-
21	Thomas R. Mackie	US/Canada	***	*	***	-	-	-	***	***	**	**	-	-
22	Radhe Mohan	US	***	-	*	-	-	-	**	***	*	-	-	-
-	Jacob Van Dyk (Epilogue)	Canada	-	-	-	-	-	-	-	-	-	-	-	-

Broad topic categories associated with each chapter:

*** Significant emphasis   ** Moderate emphasis   * Mild emphasis   - No emphasis

How Did I Get to Edit This Book?

My eldest daughter, Tonia, lives about a three-and-a-half-hour drive from our home. On the drive home from one of our visits to her place, we were listening to the Canadian Broadcasting Corporation (CBC) on the radio. One of the programmes we listened to was Shelagh Rogers' *The Next Chapter*. Shelagh has travelled the length and breadth of Canada, interviewing thousands of Canadians and collecting their stories. That is her passion, and she believes that sharing our stories enlarges our understanding of each other. In her program, Shelagh interviews authors of recently published books. On this occasion, the author who was interviewed summarized his book which consisted of a series of short stories. The ultimate message of the discussion was that it is important for all of us to tell our stories. Her programme was followed by another which also had an emphasis on telling our stories. This led me to think about medical physicists and how we tell *our* stories. First, my conclusion was that we are not very good at telling our stories. And then it hit me, we *should* tell *our* stories, not so much from a technical/physics perspective but more from a human-interest perspective. From that, the concept of this book evolved.

The Cover Story

Two of the great discoveries in medicine include computerized tomography (CT) scanning and magnetic resonance imaging (MRI). These discoveries resulted in four Nobel Prize winners. The Nobel Prize in Physiology or Medicine in 1979 was awarded jointly to Allan M. Cormack and Godfrey N. Hounsfield "for the development of computer-assisted tomography". The Nobel Prize in Physiology or Medicine in 2003 was awarded jointly to Paul C. Lauterbur and Sir Peter Mansfield "for their discoveries concerning magnetic resonance imaging". Thus, what better image to display on the cover of this book than either a CT scan or an MR image. However, CT scanning involves x-rays which has a very minor risk associated with it in terms of potential long-term secondary cancers. In general, this very minor risk needs to be balanced by the clinical benefit of the scan when it is performed. For a cover picture, there is no clinical benefit; hence, it is not a good idea to CT scan a volunteer without a clinical need. On the other hand, a magnetic resonance (MR) scan involves the use of magnetic fields which have no clinical risk, and therefore, it is acceptable to scan a volunteer. (See Chapter 11 by Stephen Thomas for a brief tutorial on how MRI works.)

When I thought of generating some MR images of me for the cover, the question was how I was going to accomplish that. I contacted one of my medical physics colleagues here in London, Ontario. He is also the author of Chapter 10 of this book, Terry Peters. Terry in turn referred me to Ravi Menon, another very accomplished medical physicist associated with Western University's Robarts Research Institute. He has appointments in the Departments of Medical Biophysics, Medical Imaging, Neuroscience and Psychiatry. In addition, he is the co-scientific director of BrainsCAN and a Canada research chair in Functional and Molecular Imaging. When I contacted him, his immediate response was "Sounds like a great project" and then went on to describe how the scanning might be done. He then continued, "Joe Gati is the Managing Director of the Centre and you can follow up with him on logistics and art aspects." After further communications with Joe and his senior MR technologist, Trevor Szekeres, I had to fill in an MRI screening form to assess whether I was "MRI compatible". This is to confirm that I have no metal objects in my body that may be affected by the strong magnetic fields. The form basically asks questions about the types of surgeries I have had, whether I have had any injuries involving metallic objects, whether I am pregnant or whether I have had any previous reactions to contrast agents. And then, a whole list of "Please indicate if you have any of the following:" The only thing that I had to say "Yes" to was "Cardiac pacemaker, pacemaker wires, or stents" since I had two cardiac stents inserted during a working trip to Vienna, Austria, in 2006. So, this generated some concern for me as to whether I was MRI compatible. However, when I responded to Trevor's request about the makes and models of the stents, he assured me that there were no concerns about me going into the magnet.

Having agreed upon an appointment time, I asked Trevor about advice for the actual scanning process. His response, "Just wear anything comfortable without metal (i.e., zippers, snaps etc.). If you don't have anything like that, we do have gowns and hospital pants here for you to wear." So, on the day of the scanning, I went in my biking pants and sport shirt. In a nutshell, I was going to the Centre for Functional and Metabolic Mapping (CFMM) at Western's Robarts Research Institute which houses Canada's only collection of high-field (3T human) and ultra-high-field (7T human and 9.4T animal) MR systems. The centre is dedicated to establishing the anatomical, metabolic and functional characteristics of normal brain development and healthy ageing across the lifespan as well as establishing the brain basis of developmental, neuropsychiatric and neurodegenerative deficits. Wow! And they were scanning me for a book cover! What a privilege! Note that the T stands for Tesla the unit of magnetic field strength. The most common

MRI scanners have a field strength of 1.5T. I was being scanned on the 3T machine. 3T provide exceptional detail, lower scan times and increased signal, which leads to an improved signal-to-noise ratio (SNR). The 3T represents a magnetic field pull that is 60,000 times stronger than the pull of the earth's magnetic field. (See Section 3.3 of Chapter 11 for a more detailed tutorial on how MRI works.)

Upon arrival at the Robarts Research Institute, I was greeted by Trevor and Joe. When the previous appointment was completed, I was taken to the scanning suite where I also met Oksana Opalevych, who is the registered MR technologist who operated the scanner for my scans. My requests were very unusual since I was looking for images of my internal anatomy as they would represent someone who is telling *True Tales of Medical Physics*. Thus, we had some discussion about this. Should it include an open mouth? Did we want to have just a head or a head and torso? Did we want any hand and arm gestures like someone in animated story telling? Well, the net result was that we tried to do all of those things, all to be done within the scheduled one-hour appointment since the day schedule was fully booked. I was led into the scanning room with shoes off out of fear that there might be metallic objects built into the shoes. My glasses were also left in the control area. I was positioned on the scanner couch and given a bulb in my hand to be pressed by me if I felt uncomfortable and needed to abort the scanning process. I was also given some earplugs to muffle the noise of the scanner during its scanning procedures. A head coil was positioned over my face and a body coil over my chest. These are parts of the technology to provide high-quality images. Figure 3 is a picture of me on the couch of the 3T Siemens scanner and shows both the head and body coils. I was gradually moved into the scanner tunnel. There was a mirror just above my head (also visible in Fig. 3) which allowed me to see the scanner operators who were behind the window in the scanner control area. After Joe and Oksana had some discussion in the control area about what settings to use for my scans, the machine started its scanning procedures. I could tell by the various loud sounds that the scanner made. The nature of the sounds varied significantly depending on what scanning procedures were being used. Some sounds were continuous, and others were pulsed. Various words have been used to describe these sounds including banging, clicking, whirring, clanging and beeping. Some of the scans were quick, in the neighbourhood of 90s, others were slow, perhaps around 5–10 min. Figure 4 shows two versions of one of the MR images of my head/brain and part of my chest. Such MR scans are now used for a variety of circumstances including the diagnosis of cancer, brain injury, assessing the impact of concussions on brain function, multiple sclerosis,

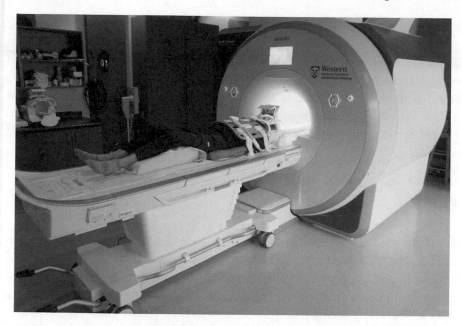

Fig. 3 This is me on the couch of the 3T Siemens MR scanner. Special coils for the head and body are shown over my face and chest

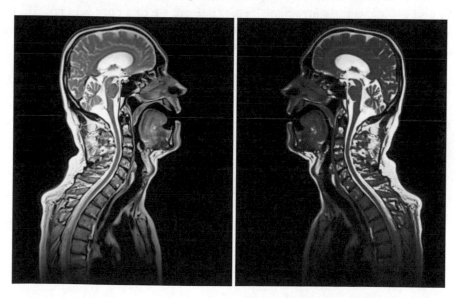

Fig. 4 (*Left*) Picture of the MRI scan of part of my head and torso. This is known as a sagittal slice. It is clear where my nose and mouth are and the tongue inside my mouth. Clearly, I am telling a "true tale". In addition, you can see details of anatomy such as my brain, spinal cord and vertebral bodies. The picture on the right is a mirror image of the one on the left only the colour setting is for "VR Muscles and Bones". It looks like these images are telling true tales to each other

spinal cord injuries, stroke, eye problems, inner ear problems, and the list goes on. As far as I know, nothing is wrong with my head or body. Medical physicists have been heavily involved in the developments associated with magnetic resonance imaging (MRI) right from its early formation to today's research for improving and optimizing its use for various diagnoses and as a tool for real-time image-guided radiation therapy. Furthermore, medical physicists are involved in ensuring its best implementation in daily clinical practice.

Enjoy

Enjoy the *tales* as they unfold. These stories are clear demonstrations by each of these individuals of their fulfilment in their careers, their dedication and hard work and the tremendous results of their labours. You will be impressed by the variety and creativity of their activities. Hopefully, the next time you meet a medical physicist, you will not have to say "What's that? Or "What do you do?" Rather, you can ask a question like "So, are you in radiation therapy physics or in diagnostic imaging physics?" or "What's your area of research?" We also get asked questions like, "How close are we to a cure for cancer?" or "How does magnetic resonance imaging help resolve concussion issues?". In summary, enjoy your brief glimpse into the world of medical physics!

Jacob Van Dyk
Western University
London, ON, Canada

Contents

Contributors

Peter R. Almond The University of Texas, MD Anderson Cancer Center, Houston, TX, USA

Gary T. Barnes University of Alabama Birmingham, Birmingham, AL, USA

Jerry J. Battista Western University, London, ON, Canada

Cari Borrás Radiological Physics and Health Services, Washington, D.C., USA

Arthur L. Boyer Belton, TX, USA

Paul L. Carson Departments of Radiology and Biomedical Engineering, University of Michigan, Ann Arbor, MI, USA

Arun Chougule Sawai Man Singh (SMS) Medical College and Hospital, Jaipur, India

C. Clifton Ling Varian Medical Systems, Palo Alto, CA, USA

Carlos E. de Almeida Universidade do Estado do Rio de Janeiro, Rio de Janeiro, Brazil

Aaron Fenster Robarts Research Institute, Western University, London, ON, Canada

Maryellen L. Giger University of Chicago, Chicago, IL, USA

Tomas Kron Peter MacCallum Cancer Centre, Melbourne, Australia

Radhe Mohan MD Anderson Cancer Center, Houston, TX, USA

Terry M. Peters Robarts Research Institute, Western University, London, Ontario, Canada

James A. Purdy Washington University School of Medicine, Saint Louis, MO, USA

Thomas "Rock" Mackie University of Wisconsin, Madison, WI, USA

Stephen R. Thomas University of Cincinnati Medical Center, Cincinnati, Ohio, USA

David Thwaites The University of Sydney, Sydney, Australia; The University of Leeds, Leeds, UK

Jacob Van Dyk Western University, London, ON, Canada

Marcel van Herk University of Manchester and Christie NHS Foundation Trust, Manchester, UK

John W. Wong Johns Hopkins University School of Medicine, Baltimore, MD, USA

Martin Yaffe Sunnybrook Research Institute and University of Toronto, Toronto, ON, Canada

Part I

Medical Physics: More than History

"I was taught that the way of progress is neither swift nor easy."

Marie Sklodowska-Curie, a pioneer of research on radioactivity and winner of two Nobel prizes in Physics (1903) and Chemistry (1911). She was the first woman to win a Nobel prize and the only women to win the Nobel prize twice.

1

What *Is* Medical Physics? And Other Related Questions

David Thwaites

1 What *Is* Medical Physics?

In my experience, medical physicists, when asked what they do and saying they are a medical physicist, then usually attract the next question, 'but what *exactly* do you do?' This comes from family, friends, neighbours, adjacent passengers in planes, even other medical workers or from the same questioner again! It seems people can see the place of the obvious: doctors of lots of types and specialties, nurses and porters. And also, radiographers, receptionists, admin staff, therapists of various descriptions (physio-, radiation, occupational, art, play, speech…). "Oh and lab technicians" and many more. "BUT physicists? In hospitals? Engineers? Yes, I can see the need for them with all that high-tech equipment, BUT physicists?"

So, we, the medical physicist, try to explain, with our emphasis probably depending on what branch of medical physics we are in. We might talk about any combination of the following:

- medical uses of ionising radiation and radioactivity, safety and protection of staff and patients and even about radiobiology;

D. Thwaites (✉)
The University of Sydney, Sydney, Australia
e-mail: david.thwaites@sydney.edu.au

The University of Leeds, Leeds, UK

J. Van Dyk (ed.), *True Tales of Medical Physics*,
https://doi.org/10.1007/978-3-030-91724-1_1

- radiotherapy (radiation therapy, radiation oncology) for cancer treatment: either using radiation beams, or 'brachytherapy' using sealed radioactive sources inserted directly into or beside tumours;
- radiotherapy beam delivery systems, most commonly using x-rays produced by increasingly sophisticated clinical linear electron accelerators (linacs), but also using gamma rays from radioactive source units or a variety of particle beams;
- medical imaging based on lots of physics, such as x-rays, computerised tomography (CT) scanning, ultrasound imaging, nuclear medicine imaging methods including single photon emission computed tomography (SPECT), and positron emission tomography (PET), magnetic resonance imaging (MRI) and the evolution from anatomical imaging to functional imaging, i.e., imaging of biological functions and processes;
- therapeutic applications of nuclear medicine, using unsealed radioactive isotopes;
- medical applications of non-ionising radiation, e.g., terahertz and microwave radiation, the use of ultraviolet and infrared radiation, optical imaging methods, laser applications and photodynamic therapy; and we might leave the intriguing acronym HIFU, to be googled by the questioner, with the hint: 'ultrasound';
- instrumentation, equipment, large and small machines, and the wide range of techniques using them for imaging, diagnosis, analysis and treatment;
- electric and magnetic signals from the body and other measurable characteristics, or more widely physiological measurement;
- physics-based instrumentation, methods or imaging giving involvement in or impact on a wide range of less obvious medical specialties besides radiology, nuclear medicine and radiotherapy; e.g., anaesthetics, audiology, cardiology, dermatology, neurology, neurophysiology, obstetrics, other oncology areas, ophthalmology, orthopaedics, physical medicine, surgery, urology and more,
- quality assurance and accuracy; and
- research and development.

We might also mention bio-engineering, for people/patients, as well as for equipment. This might lead to discussing the interface between clinical engineering or biomedical engineering and medical physics. And that departments might be called Medical Physics plus some form of Engineering, (e.g., Medical, Clinical, Bio-Engineering), depending on country and healthcare system.

We'd also likely refer to computer science applications, including modelling (creating models of the real world being a key physics passion!), other information technology (IT) applications and bio-informatics; "Big Data", machine learning and artificial intelligence; and even 'radiomics', extracting and learning quantitative features from medical images.

Or the 'physics of/in medicine', or biophysics, and talk about understanding biological and human structures and functions from a physics point of view, down to the very small scale, with a famous now-historical example having been the determination of DNA structure using x-ray diffraction; and about biomaterials and bio-nanoscience.

We might at some point discuss troubleshooting, picking up novel problems in hospitals and solving them! … we're scientists after all!

And we might mention how we fit in with other medical professionals, teams and specialties; or that medical physics has proportionately more female members of the profession than in many other science, technology, engineering and math/s (STEM) occupations.

Eventually (as we pause for breath?) there's the possibility that the questioner wishes they hadn't asked! Or, in my experience, if they're school students or university physics students, or simply others who are interested, they might become enthused too and ask still more questions!

The bottom-line message I hope they take away is that medical physics is as broad as 'physics applied to medicine and healthcare', with a wide range of applications and roles. It impacts essentially any medical problem or specialty that can benefit from physics or engineering knowledge, theories, tools or methods. Also, that medical physicists are enthusiastic about what they do and want to share that!

A memorable version of the 'what is' question came from an elderly family acquaintance…

2   A Brief History of *My* Time

However, before I get to that, let me first outline how I got into medical physics. I was brought up in a rural community in north-west England just south of the border with Scotland. My grandfather was a farmer who was great with machinery and at making and fixing things and I was raised wanting to know how things worked and were put together. I was torn between studying physics or medicine at university. I chose a degree in Natural Sciences, majoring in physics and math/s (with the 's' reflecting my UK English!). That decision was influenced by the exciting fact that physics is the basic science that underpins all the others. Also, it holds the promise of understanding how the world and the universe work, and of where we and other 'stuff' come from, at least within a materialistic framework. To quote Sheldon Cooper, the obsessive theoretical physicist from the *Big Bang Theory* TV series: "Physics encompasses the entire universe, from quantum particles to supernovas, from spinning electrons to spinning galaxies". For me, that's enough to spin the mind.

Towards the end of my studies, I began considering what might be next and came across medical physics. A 'wow moment'… you could use physics on a daily basis, apply it directly to help people and also have opportunities to do research at the same time. In the mid-1970s, I began a Medical Physics M.Sc. programme at Aberdeen University in north-east Scotland. This was in the joint clinical and academic department founded by Professor John Mallard, a pioneer in nuclear medicine scanning methods including PET, and in MRI. The course had a research project at the end. My chosen topic was on a luminescence-based radiation dosimetry method using organic materials as the sensitive detector material. I discovered that research was fun, fascinating, rewarding … and moreish (I hesitate to say addictive, but there *is* always a next step!). Therefore, I began a Ph.D. and then a postdoctoral fellowship at Dundee University, also in eastern Scotland, and also where I met Catherine, my life partner. My calmly encouraging supervisor was David Watt, who had been involved in nuclear research, reactor physics, microdosimetry, radiation and environmental safety applications and more. The topic was in radiation interactions and energy loss of charged particles (stopping power) in biological and dosimetric materials, with applications in radiation safety, dosimetry, and particle therapy. I spent short periods with our collaborators at the Argonne National Laboratory and Fermilab, near Chicago, IL, and in three Danish universities' physics research groups, the Niels Bohr Institute, Copenhagen and in Aarhus and Odense. I was set for a university-based career, but UK public sector funding cuts around 1980 reduced the number of purely academic posts. Instead, I looked for a clinical hospital-based post, but in a teaching hospital to be linked with a university to provide research and teaching opportunities as well as the clinical service role.

In 1980, I found myself in a radiotherapy physics job in Edinburgh in the Medical Physics and Medical Engineering department. This was headed by Prof John Greening, an international expert in, among other things, radiation dosimetry and also with both theoretical and practical interest in the physics of golf! My clinical training and research enthusiasm were nurtured by my three immediate seniors. John Law was a painstakingly thorough dosimetry and radiation protection specialist. Tony Redpath was an innovative and inspirational pioneer of computerised treatment planning. David Bottrill was a widely knowledgeable and patient teacher and mentor, as well as a linac 'whisperer' in his feeling for the temperamental 1970s machines then in Edinburgh. I had received some advice not to go into radiotherapy (RT) physics, as it had reached a plateau, but to go into diagnostic imaging physics where the exciting developments would be. How wrong that advice was, as the concept of intensity-modulated radiotherapy (IMRT) was just about to emerge. Major advances in the complexity and flexibility of computer-controlled radiotherapy methods followed. The overall aim was to better match radiation dose distributions to the irregular tumour volumes that need to receive radiation and yet to spare nearby healthy tissues that don't. It required, and generated, inter-linked and evolving advances in:

- 3D imaging for RT to accurately determine the patient tumour/target and normal tissue shapes and positional relationships; at that time using CT scanners, developed not long before by Godfrey Hounsfield with the head scanner in 1971 and the body scanner in 1975, and more recently using 4D (i.e., considering time and corresponding motion) methods and other imaging such as MRI and PET;
- increasingly complex 3D computer modelling (treatment planning) to optimise radiation dose distributions to the individual patient anatomy and geometry to achieve the clinical dose requirements;
- hardware and software control of delivery equipment and methods, evolving from static to dynamic methods of IMRT, such as tomotherapy and volumetric modulated arc therapy (VMAT);
- dosimetric and geometric/positioning accuracy and precision,
- quality assurance and quality control of systems and techniques;
- image-based patient and beam positional verification methods.

The last one led to imaging systems directly on-board radiotherapy treatment machines and image-guided radiotherapy of increasing sophistication. This laid the foundations for adaptive radiotherapy to account for changes during a treatment course or individual treatment session (fraction) and motion management methods to deal with organ and target motion.

Thus, RT and RT physics were at the threshold of a series of continuing and overlapping waves of significant paradigm shifts in equipment, methods and precision. In addition, image-based methods began to be increasingly integrated into RT.

From then, I have had a 40-plus year career, based in Edinburgh and Leeds in the UK and in Sydney in Australia. It has covered a wide spectrum of clinical and research areas of medical physics in radiation oncology. Broad research objectives have consistently been to evaluate, optimise and improve the accuracy, safety, quality and effectiveness of RT at whatever stage it is at that time. Thus, this has evolved with advancing RT technology, tools, systems and techniques. My earliest research publications were in basic and applied radiation dosimetry, with applications in conventional x-ray RT and particle therapy. Most recently they have included dosimetry (still!), dosimetric and geometric accuracy, treatment verification and quality audit, but now for IMRT and VMAT; MRI-guided treatment planning for external beam RT and brachytherapy; treatment planning study standards; particle therapy use; and machine learning and datamining applications in oncology. I have always held roles across the interface of clinical and academic medical physics, with the strong belief that research and development need to be firmly rooted in clinical service needs and those of the patient.

Incidentally, imaging medical physics would also have been a good area to enter in 1980, as that too has had an exciting explosion of research, development, technology, and methods over the last 40 years!

Meanwhile, maybe you remember my elderly family acquaintance: somewhere in my early period of medical physics study, research and job-hunting when she asked me … 'but what ARE medical physics?'.

3   Words, Meanings, and Origins

This leads me to consider some relevant words, meanings and origins. Scientists and doctors need to be precise in language, to achieve clarity. Words can mutate to take different meanings. For example, they can evolve with time or application; bend to fit social or political orthodoxy; escape academic use into the public sphere and change (as a physics-originated example, 'quantum leap'); and be taken from the public sphere and used in science (e.g., quark descriptors). To consider the terms *medical physics* and *medical physicist*, we should first consider *physics* and *physicist*.

The 's' on the end of physics led my questioner to think it was a plural word. However, physics is singular but in English sounds plural. In fact, it did arise as a plural, of the old English physic in the fifteenth century, linked to a 'physical thing'. That was from the Latin physica and the ancient Greek phusika (φυσική), around the study of natural things and of nature itself. It meant the natural sciences generally, particularly empirical and practical, based on observation and experience. Natural philosophy was a wider term implying a deeper quest to explain the nature of things in the physical world. This later narrowed to denote what we now call physics and it continued to be used in English until the early twentieth century and is still used in some course and post titles in older universities. 'Scientist' was coined in English in 1833 and 'physicist' in 1840.

Because of the root word *physic*, words for the physical sciences have intertwined with words related to medicine and the body, since many of those arise from the same root. Physic (or sometimes physick) as a noun in old English was also used for a medicine or drug and as a verb for the practice of medicine. Physician is the older established word for a medical practitioner, not doctor. The latter is historically an academic term for the holder of a doctoral degree. There are also parallels across languages, e.g., *la physique* is French for physics, *le physique* for physical appearance, close to the English usage of the word physique. *Physician* in English is a medical doctor (*médecin* in French), *physicien* in French is a physicist.

I now come back to my family acquaintance, who was elderly in the late 1970s, likely born in the very late 1800s. I speculate that her question may have arisen from hearing her parents or grandparents referring to physics (plural) meaning medicines, so implying to her that medical physics were medical medicines. No wonder she was confused as to what I might do!

Besides words, it is also worth considering who carried out science and why, before the more modern concept beginning in the 1800s of professional and academic scientists being employed more widely to 'do science'. This has some relevance for the origins and development of medical physics. In general terms, official science was conducted by relatively small numbers of people, e.g., in universities, or to provide state-of-the-art technology support and scientific development for rich and powerful patrons. Alternatively, some needed science to understand or support other activity, e.g., doctors to help their medical practice, or by ministers of religion to better understand God as expressed in nature. Lastly, science was the hobby of enthusiasts of independent financial means, i.e., with time and resources to follow their curiosity about the world around them. Conversely, small-scale science and technology development, of course, has occurred everywhere throughout history, since fire and flint tools, to solve problems or improve work and life; and carried out by human artisans in the widest sense!

So, what then *is* medical physics? It is physics applied to medicine and healthcare, broadly encompassing any applications of physics concepts, knowledge, theories and methods. And who did it first? Often doctors! Or more accurately, physician-scientists.

4   The Origins of Medical Physics: Back to the Future

4.1   Back to the Future?

The title of this section is part-inspired by the 1980s three-film/movie series *Back to the Future* (BttF). These starred Michael J Fox and Christopher Lloyd as Marty McFly and Emmett 'Doc' Brown, the latter an eccentric scientist (a physicist, of course!). They used a time-travelling DeLorean sports car to move back and forward in time and explored the intricate twistings of time and events that might occur if this was possible.

I have used BttF a few times in discussing medical physics. Firstly, for a review article on the history of the clinical linear accelerator in the July 2006 50th anniversary issue of *Physics in Medicine in Biology* (PMB). PMB is the oldest modern medical physics journal (but not the oldest medical physics journal, as we shall see). It is the journal of the oldest professional medical physics body, founded in the UK in 1943. That was then called the Hospital Physicists Association, now the Institute of Physics and Engineering in Medicine. In that paper, my Leeds colleague, John Tuohy, and I indicated that many innovations on clinical linacs, the workhorse machine for radiotherapy treatments, had been proposed much earlier than they'd appeared. Hence, we'd gone back to find the origins of those futures. However, they needed technology, including computing, to catch up in order for them to be implemented clinically. I used it again in an ESTRO (European SocieTy for Radiation Oncology) award lecture in 2014, on '*BttF, synergies between physics and medicine from history to horizon*'. This considered the history and development of medical physics to then speculate on how things might evolve into the future. I modified it for the third usage as a 'Christmas lecture' in Sydney University, Australia, that became a recurring talk for students and then for school students and science teachers, Lastly, '*BttF4: Medical Physics Roots, Role, Soul and Goal*' was the title of a talk to celebrate the International Day of Medical Physics (7th Nov., Marie Curie's birthdate) at the Australasian medical physics conference in Wellington, NZ, in 2015. There was no full BttF4 movie (although there was an online short, travelling to 2045), but Nov. 2015 was beyond the future time travel date of 21 Oct. 2015 in the second BttF movie. Thus, we were already just into that future then!

The idea of turning to look back, so standing with our back to the future, is appealing. It lets us consider where we have come from, helps to understand where we are and from that we can turn again to consider where we might be going and to shape that future. This is an interesting activity in all areas of society, including medical physics. We need to know and understand history, as often quoted, so that we don't make the same mistakes. Also, so that we efficiently use what's already there and build appropriately on it. Looking at history in a realistic and measured way implies not sanitising or re-writing it. Instead, we need to try to understand it in context, realising that we can't necessarily easily apply modern beliefs and values retrospectively to the past. Nevertheless, whilst the past might be 'another country', the present *is* intrinsically linked to the past and therefore histories and horizons are similarly linked.

4.2   Where Did Medical Physics Originate? And When?

Wilhelm Conrad Röntgen (1845–1923) (Fig. 1a), Professor of Physics in Würzburg, discovered x-rays on the 8th of Nov. 1895, for which he received the 1901 Nobel prize in physics. He chose not to patent his discovery so that society should benefit freely from it. Medical physics in modern terms is often dated to this event, because of the huge direct impact of physics on medicine that it promptly produced. For the first time ever, doctors could routinely visualise body structures inside living patients. This translated into clinical imaging and diagnostic use almost immediately. Radiotherapy also began shortly afterwards, using radiation to treat various diseases, particularly cancer. Further, in 1896, Antoine Henri Becquerel (1852–1908) (Fig. 1b) discovered "radiation activity" (radioactivity) of uranium salts. Following directly from this, in 1898, Marie and Pierre Curie (Fig. 1c) identified new radioactive elements, polonium and radium. All three worked in Paris and together shared the 1903 Nobel physics prize for their work, with Marie also receiving the 1911 Nobel prize for chemistry. Radioactive materials were destined to play a large part in medicine, initially in radiotherapy applications, then later in nuclear medicine for diagnosis and treatment. The latter applications were developed and expanded only after it became possible to produce new radioisotopes artificially. That was demonstrated by the Curies' daughter Irène and son-in-law Frédéric Joliot-Curie in 1934, for which they jointly received the 1935 Nobel chemistry prize. The needs of radiotherapy led to the first physicists being directly employed in hospitals to harness and quantify radium and x-rays, to ensure their safety and to develop and apply an ever-expanding range of physics, engineering, and technology. Therefore, many medical physics histories and websites discuss medical physics only in terms of the three main areas of radiation medicine arising from those discoveries, i.e., radiation therapy, diagnostic imaging, and nuclear medicine. This

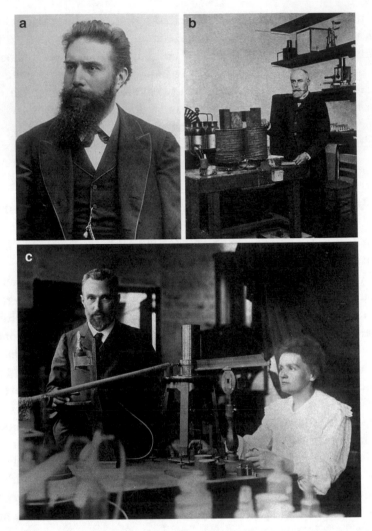

Fig. 1 a Wilhelm Conrad Röntgen, 1900 (LIFE photo archive). **b** Henri Becquerel (circa 1900). In his lab at the Museum of Natural History, Paris. Studies identified three types of radioactivity, from their deflection in magnetic fields. **c** Pierre and Marie Curie in their physics measurement room; the balance electrometer apparatus is used for measuring small amounts of charge produced by radioactivity from an air ionisation chamber. This picture was on the front cover of the first issue of the journal, *Le Radium* (Jan 1904)

approach implies or even directly states that medical physics began with x-rays in 1895.

However, physics in and applied to medicine, i.e., the use of physical techniques involving, for example, heat, light, and magnetism, goes back with the recorded history of medicine itself. The Edwin Smith papyrus, was discovered in the nineteenth century and was written around 1600 BCE, but is believed to be a copy of works from around 3000 BCE and so likely the oldest medical text known. It discusses surgery and describes cauterisation of breast cancer using 'fire drills'. The 'father of medicine', Hippocrates (circa 460–370 BCE), described the use of a thermographic method to map skin temperatures and potentially to determine body temperature. It used the differential drying rates of wet clay applied to the skin. So immediately my physicist sensibilities think of controlling experimental methods and conditions. This needs uniform wetness of the clay and uniform thickness of its spreading onto the body area, in a short time. It also needs environment control to ensure no variable external factors, such as the sun, other heat sources or air movement, can affect the observed drying. It requires setting benchmarks and calibrations of the method for different body areas and sizes, repeating measurements, and more. I am applying the now well-accepted scientific method.

If we also require such approaches, similar to those we'd recognise today, then we can select many individuals as the forerunners of modern medical physics and different observers' selections will vary. I give just a few examples here. It has been argued that recorded medical physics in its broadest sense can be traced back more than 1000 years to Alhazen (Hasan Ibn al-Haythem, 965–1039) of Basra and Cairo. He is known as 'the father of modern optics' for his work on the physics of the eye and determining that vision is due to light being received by the eye. For that, he relied on systematic approaches to experimental method and controlled testing. Others have argued for the beginnings of medical physics to be more than 500 years ago with Leonardo da Vinci (1452–1519), for his work on the mechanics of the body as well as on human optics. Francis Duck is a retired imaging/ultrasound medical physicist, a noted medical physics historian and the author of a book[1] which was an inspirational source for this chapter. He selects the work of Santorio Santorio (or Sanctorius, 1561–1636), whilst Professor of Theoretical Medicine in Padua, Italy as one significant starting point of medical physics. Sanctorius carried out very careful physics- and measurement-based experiments, weighing and comparing his bodily intakes and outputs over long periods, thus beginning the quantitative study of physiology. Besides body mass, he measured other related parameters, including temperature and pulse rate to develop wider understanding of body function. He viewed the body as a machine, a mechanistic approach that became widely adopted, with early attempts to support it by mathematical models. Another of Duck's significant selections is Giovanni Borelli (1608–79) who

---

[1] Duck FA, *Physicists and Physicians: A History of Medical Physics from the Renaissance to Röntgen* (Institute of Physics and Engineering in Medicine, York UK, 2013 ISBN 978-1-903,613-55-9).

developed systematic approaches to this, whilst Professor of Mathematics in Pisa. These ideas were later termed iatrophysics, (Fig. 2a), meaning 'the medical application of physics'. However, this was only in the limited sense of applying mechanical principles and the laws of physics to the workings of the body and more generally to the nature of life, but not to medical diagnosis and treatment. Many others through the seventeenth century and into the 18th contributed to the mathematical-physical-mechanical-materialistic model of the human body and its functions. This included well-known names in physics and mathematics, such as Boyle, Hooke, Bernoulli and Euler, as well as the wide supporting influence of Newton's work. Rationalism and rigorous scientific methods had been becoming more firmly established over a similar period. Thus, wider concepts of mechanistic philosophy, e.g., from Descartes, Hobbes and others, were also supportive, stating that all of nature could be characterised and described in mechanistic terms and by physical laws.

In the second half of the eighteenth century, ideas on the role of electricity in the human body and in life itself began to develop. Experiments were enabled by the discovery of capacitor storage of electric charge, which arose from various workers' inter-related studies over 1745–1746. Notable amongst these were Ewald Georg von Kleist, a German cleric and amateur scientist, and Dutch scientist Pieter van Musschenbroek, the Leiden Professor of Mathematics and (natural) Philosophy. A significant part of their discovery and development resulted from accidentally receiving severe electric shocks whilst conducting experiments! Another accidental observation in 1780 also contributed. Luigi Galvani, the Bologna Professor of Anatomy noticed that electricity could stimulate muscle contraction, initially in a partly dissected frog's leg. This effect was later termed galvanism. Medical use of 'electrotherapy' developed rapidly in the second half of the 1700s and through the nineteenth century, with a growing number of hospitals establishing departments of medical electricity. They used 'Leyden jar' (or Kleistian jar) capacitors and electrical friction machines for shocking and charging, including charging up patients in 'electric baths' (Fig. 2b). The latter process was termed Franklinisation in the US. The name was from Benjamin Franklin, US statesman, scientist, and polymath, who briefly conducted experiments with it in 1757, attempting to treat symptoms of patients suffering from paralysis. Electrotherapy was further taken forward by the availability of voltaic piles, or electrochemical batteries, to provide continuous current. These were first reported in 1800 by Alessandra Volta, the Pavia Professor of Experimental Physics. Then in 1895, the clinical use of high-frequency alternating

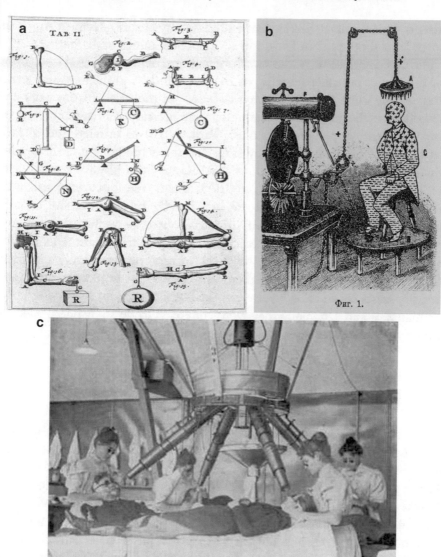

Fig. 2 a Limbs modelled as machines (page illustrating biomechanics principles, from Borelli, *De Motu Animalium*, 1680, Movement of Animals, a title previously used by Aristotle (384-322 BC)). b High voltage 'electric bath'; a frictional electrical source charges an insulated patient, used for a range of conditions (from *Brockhaus and Efron Encyclopedic Dictionary*, St Petersburg, Russia, 1890–1907). c The Finsen hospital lamp, 1890s. The adjustable tubes focus various types of light through hollow water-cooled lenses. Multiple tubes treat multiple patients simultaneously for up to an hour, with a nurse for each. Staff and patients wear eye protection (Wellcome Library, London)

currents for deep tissue heating was established by Jacques-Arsène d'Arsonval, Professor of Biophysics in Paris, forming the beginnings of diathermy.

The nineteenth century also saw the expansion of physics-based modelling, measurement and understanding of the body and its physiology and structures. At the same time many physics-based technologies, instruments and methods for diagnosis and treatment were developed. Examples ranged from the stethoscope (René Laennec, Paris, 1816) to infrared and ultraviolet phototherapy or light therapy (Fig. 2c), for debilitating and disfiguring skin conditions (Niels Finsen, the Faroe Islands-born physician and scientist, Copenhagen, 1890s). Then, in 1895, x-rays were discovered and were immediately adopted for clinical use in a rapidly growing number of radiology facilities. These were often initially housed in well-established and long-standing departments of medical electricity. Such departments already employed 'medical electricians', some of whom were medically qualified, others not, but many of whom then also became 'x-ray technicians'. Soon after, physicists began to be involved in the hospital use of x-rays and radioactive materials and eventually they were employed to do so. Modern medical physics was established.

4.3   When Was the Term 'Medical Physics' First Formally Used?

One of the first clear published recognitions of a wider linkage of physics to medicine, rather than only for specific disconnected problems and situations, was in 1719. In that year the British scientist-apothecary-physician John Quincy produced his Lexicon Physico-Medicum: A New Physical Dictionary. However, the term 'Medical Physics', or more precisely 'Physique Médicale' as its first use was in French, was not formally recorded until 1779 and that was almost immediately in a journal! The Parisian physician-scientist, and last physician to Queen Marie-Antoinette, Félix Vicq d'Azyr (or d'Azir, 1748–1794) (Fig. 3a), was the founder and general secretary of the Société Royale de Médicine. He established two Society scientific publications. L'Histoire de la Société Royale de Médicine contained a section on Physique Médicale. The other was the first journal, or indeed publication of any type, with medical physics directly in its title, Les Mémoires de Médecine et de Physique Médicale. Among many other things, these contained some accounts of general physics applied to medicine. Examples included therapy with physical agents, reports on medical electricity and magnetism, and applications in physiology and public health. The journals were published erratically and infrequently in their later years, given the intense events unfolding during and after the

Fig. 3 a Félix Vicq-d'Azyr, circa 1790 (Académie Nationale de Médecine, Paris). **b** Jean-Noël Hallé, circa 1815, (Duck FA, *Physicists and Physicians: A History of Medical Physics from the Renaissance to Röntgen* (IPEM: York) 2013; and Wellcome Library, London)

1789 French revolution. The last issues appeared in 1798, now having no royal reference in the title!

Vicq d'Azyr died in 1794, but had developed a comprehensive plan for reforming French medical education and training. This stipulated knowledge, practical hospital training and formal examinations. Medical physics was listed as one of the relevant basic sciences. It was specified as the body of knowledge in physics that doctors needed to know 'for the mastery of medicine' and 'through (that)…to make progress'. However, it would be sufficient to include '*only* those discoveries of physics that are truly applicable to practical medicine'.

The wider recognition that the physical sciences should be taught to medical students meant more European medical schools began providing it. Teaching was often initially given by clinician-scientists or from linked university physics or natural philosophy departments, but later increasingly by directly employing physicists in medical schools. In addition, in the nineteenth century, books on medical physics by physician-scientists began to appear, to summarise the required knowledge. Thus, the earliest use of Physique Médicale in a book title was in Pierre Pelletan's first ever physics textbook for medical students (Paris, 1824). This was followed by the specific terms Medical Physics (the Scot, Neil Arnott, London, 1827) and Medizinischen Physik (Friedrich Heidenreich, Ansbach, 1843) being also first formally recorded in book titles. These aimed to distil and report knowledge of the physics that was by then increasingly required to support medical practice. There were also other similar first texts in Spanish (José Maria López, Cadiz,

1835) and Italian (Carlo Matteucci, Pisa, 1844), but with not exactly the use of 'medical physics' in the title. However, of these earliest books, the latter has been recognised to have the strongest emphasis towards medical and physiological applications of physics. Medical Physics took a little longer to become established in North America. The first 'Textbook of Medical Physics' was published there in 1885, by John C Draper, professor of chemistry and physics in the Medical Department of New York University (see Fig. 1 in the *Prologue*).

4.4   When Was 'Medical Physics' First Used in a Position's Title?

The answer may be surprising. The first person with medical physics in his title was appointed in December 1794. The French needed to re-start medical training following the closure of medical schools and other centres of learning during the Revolution and the subsequent societal upheaval. The need was urgent, since doctors were required to support the army in the French Revolutionary Wars. Since Vicq d'Azyr's plan for reformed medical education was ready and available, it was used. Its recommendation for medical physics teaching was adopted and so too the suggestion that this topic could be combined with 'hygiene'. So, in that year, the Paris École de Santé (School of Health) was established with 12 professors, one of whom was Jean-Noel Hallé (1754–1822) (Fig. 3b), the first Professeur de Physique Médicale et d'Hygiène. Translating, this becomes Professor of Medical Physics and Hygiene, although at times the first part is translated incorrectly into English as physical medicine. He held this post until his death. His successor was Pierre Pelletan, whose professorship was of medical physics only, a role that has continued in the Paris Faculty of Medicine.

Hallé had previously practiced medicine to a community including poorer patients, and had developed wider health concerns from his observations of their lives, health and needs. He had surveyed pollution in industrial parts of Paris in 1789/90 and its potential links to disease and had given practical recommendations to alleviate the observed problems. He had outlined a syllabus for a course in hygiene, or preservation of health. Topics included environmental and physical factors with potential for impact on health, e.g., temperature, light, atmospheric electricity and magnetism, meteorology. It also included subjects such as climate, housing and occupational factors. A similar wording is still met in modern radiation and medical physics, where we use health physics, radiation protection, or radiation hygiene. More widely, we consider occupational, environmental and public health or safety, i.e., having the aim to ensure health.

Hallé and his associate professor (Philippe Pinel, physician and mathematician) rapidly developed a physics syllabus for medical students. It included mechanics of the body, properties of liquids and gases, acoustics, optics, microscopy, medical electricity, magnetism, animal experimentation (for physiological studies) and the general principles of applying physics to medicine. In 1814, Pierre-Hubert Nysten, French physiologist and physician, compiled a dictionary of 'medicine, surgery, physics, chemistry and natural history'. In it, Hallé provided a succinct definition of medical physics: "Physique appliquée à la connaisance du corps humain, à son conservasion et à la guerison de ses maladies". This definition, "Physics applied to the knowledge of the human body, to its preservation and to the cure of its illnesses", is wide-ranging and comprehensive and has stood the test of time. It is still valid! Physics applied to the knowledge of the human body covers understanding form, function, physiology and processes of and in the body. This began, as above, from bio-mechanics and physiological physics and its current areas include biophysics and physics in medicine at the cellular, sub-cellular, molecular biology and nano-science scales. Physics applied to the preservation of the human body covers those 'hygiene' or health physics aspects of occupational, environmental and personal safety and protection, as well as prevention of disease. Thus, it incorporates physics methods applied to maintain and optimise health and quality of life. Physics applied to the cure of the body's illnesses is clear and includes most of the current mainstream medical physics applications and techniques in hospitals, as given in section 1 in response to the 'what is?' question.

4.5   Who Was the First 'Modern' Medical Physicist?

'Modern' medical physics began with Röntgen's Nov. 1895 discovery of x-rays. The discovery was announced to the world at the end of December in a paper titled '*On a new type of ray*'. After his first observations of the effects of these unknown rays, he had spent seven weeks virtually locked in his laboratory carrying out systematic experiments to investigate their properties and behaviour. This included x-raying his hand and, famously, his wife's. Her reaction, 'I have seen my own death', reflected the consternation, even horror, that x-rays initially produced in many people. The discovery of invisible rays that could 'see inside' people, and through walls and boxes, caused intense worldwide interest and both negative and positive speculation from the media and the public. It also caused strong reactions in the scientific and medical communities. The diagnostic possibilities were immediately exploited. Hospital and independent radiology/radiography departments and practices were set up within weeks in early 1896 and x-rays were used in battlefield situations soon after. Therapeutic x-ray use is also reported almost as fast, later in 1896 in a number of cities and centres. This was administered for a range of benign and malignant conditions, even though the x-ray energies available were still quite low, likely to be from a few tens of kilovolts applied to the x-ray tube. X-rays were also rapidly used for public amusement and demonstration, including by Edison, the US inventor who developed efficient fluoroscopic materials for x-ray imaging use. A manufacturing industry for x-ray apparatus grew rapidly. By the late 1890s, accumulated experience from all these activities and applications led to the negative effects of radiation becoming increasingly recognised. This prompted the development of radiation protection recommendations for safety design and operational requirements.

The history of x-rays is well-documented, as is the early development of medical imaging and radiotherapy (Fig. 4). So too is their evolution and growth into the current sophisticated specialties of Diagnostic Radiology (or Medical Imaging) and Radiation Oncology respectively. The role of medical physics has been vital in these clinical specialties, which along with nuclear medicine, employ the large majority of medical physicists in modern hospitals. Nuclear medicine involves the use of unsealed radioisotopes introduced into the body for imaging and therapy purposes. It dates from the growing availability of selected radionuclides made possible by the development of nuclear reactors in the 1940s. Besides direct involvement in the clinical services, a key role for medical physicists has been in the necessary radiation protection of staff, patients and the public to ensure the safe operation of clinical facilities.

The need for direct physics support of medical radiation applications and radiation protection was increasingly seen from 1895 onwards. At first this was ad hoc involving local university and other physicists, but likely not directly participating in clinical work. However, there is at least one known exception in those early days. Charles E S Phillips (1871–1945) (Fig. 5a) was an amateur, but gifted, scientist appointed by the Royal Cancer Hospital in London, UK (now the Royal Marsden Hospital) as their honorary physicist from around 1892 until 1927, when he retired. Thus, he was already embedded in the hospital when x-rays were announced and by Feb. 1896 was working on them. Among other things relevant to this particular story, he was

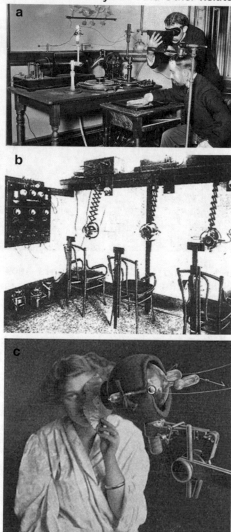

Fig. 4 **a** Early radiology facility: William J Morton's New York x-ray laboratory in early 1896, demonstrating simultaneous directly viewed fluoroscopy (of the standing subject's hand) and radiography (seated subject with hand on a photographic plate). An unprotected Crookes tube is irradiating both, powered by an induction coil (left) with a primary circuit interruptor to its right and a rheostat (flat disk in the centre of the table) to adjust the coil primary current. Spare tubes are in the wall rack (William J. Morton and Edwin W. Hammer, *The x-ray, or Photography of the Invisible and its Value in Surgery*, American Technical Book Co: New York USA, 1890). Morton was a physician and Hammer an electrical engineer. **b** Early radiotherapy facility, The London Hospital, UK, 1903. Multiple patients, radiation and electrical hazards, rudimentary patient support and radiation beam positioning (Richard F Mould, *Radiation Protection in Hospitals*, Adam Hilger: Bristol, UK, 1985). **c** 'Roentgenotherapy' for epithelioma of the face. x-ray tube in a localizing shield; perforated sheet of x-ray shielding metal securely fastened to the surface by adhesive plaster (Sinclair Tousey, *Medical Electricity and Röntgen rays, with chapters on Phototherapy and Radium*, W.B Saunders Co: Philadelphia, USA, 1910.)

Fig. 5 a Major Charles Edmund Stanley Phillips, circa 1909 (British Institute of Radiology Archives). b Sidney Russ, circa 1905 (*History of the Hospital Physicists' Association, 1943–83*, IPEM: York, 1983)

an early proposer (1907) of an ionisation-based unit of measurement for radiation. Also, in 1897, he compiled a bibliography of x-ray literature. This was no mean feat, as it is estimated that there were around 1000 papers on x-rays in their first year. He has been called the first medical physicist in the modern sense, since he was hospital appointed. However, since he had independent wealth, he was likely not paid for his work! Phillips was President of the UK Röntgen Society in 1909–1910. This was the oldest radiological society in the world, founded in 1897, and is the forerunner of the present British Institute of Radiology. Its first President was the physicist Silvanus P Thompson (1851–1916). Whilst the society's main impetus came from doctors, around half of its early membership was non-medical, including physicists, engineers, and photographers.

By around 1910, x-ray imaging and radiotherapy had become more firmly established, with the latter either using x-rays or gamma-rays emitted from radium sources. The need for full-time physicists in these clinical departments was recognised, to support the technology, metrology, standardisation and clinical methods and techniques. The first such full-time medical physicist placed directly in a hospital appears to be Sidney Russ (1879–1963) (Fig. 5b). He had been working on radioactivity research with Rutherford in Manchester, UK. However, his interest in the potential of medical radiation led him to work at the Middlesex Hospital, London, UK, in the general area of radium use and cancer. He was funded by a three-year fellowship covering 1910–1912. When this ended at the end of 1912, he was directly appointed physicist to the hospital. Later, in 1920, he became the first professor of physics at a London medical school. Among many other achievements, he contributed significantly to radiation dosimetry and protection and developed an early radium gamma-ray radiotherapy machine. He was also a prime mover and founding Chairman of the previously mentioned first medical physics organisation, the UK Hospital Physicists Association, hospital physics being Russ's preferred term rather than medical physics. However, since he was initially funded by a fellowship, he was not the first directly hospital-employed *and* hospital-paid full-time medical physicist. F Volz has been reported to fulfil that role, appointed just before in 1912 to the radiation therapy department of a Munich hospital.

4.6   Wasn't Marie Curie the First Medical Physicist? And Didn't She Set the Foundations for Women in All of Science?

Marie Curie (1867–1934) has been proposed as a proto- medical physicist, from the range of her work in, and impacting on, healthcare. Marie/Maria Sklodowska-Curie, to recognize her original surname and her Polish origins as well as her working life in France and her French partner, was extraordinary and inspirational in many ways. Not least, this is for overcoming the barriers she experienced, as a woman in research at that time, from the science establishment and officialdom. Her work was formative, not only in its own right but also in providing an incentive and a tool for the work of others to develop understanding of atomic and nuclear structure and, hence, a foundation for nuclear physics. She began her doctoral research in 1896 on the rays emitted by uranium, shortly after their discovery by Becquerel. The research proposal and topic were entirely hers. Becquerel was a mentor and supervisor, along with Pierre, her husband (1859–1906), and the work used a sensitive piezo-electric quartz electrometer developed some years before by Pierre's brother Jacques, based on some previous work of both brothers. This could measure small amounts of ionizing radiation-induced electric charge. Pierre soon moved from his magnetism research to join Marie's ground-breaking project on radioactivity, a word that they coined. From 1898 to 1902, they published more than 30 papers on radioactivity, including the discovery of new radioactive elements, polonium and radium, in 1898. One paper reported that the radiation from radium would damage and destroy tumour cells more effectively than healthy cells. Arising from this, Marie was later instrumental in establishing the first work on investigating cancer treatment using radioactive isotopes, particularly using radium. After long and painstaking work, they isolated radium in April 1902, one tenth of a gram of pure radium chloride from several tons of pitchblende waste. After Pierre's accidental death in 1906, her research program continued, isolating metallic radium in 1910. As with Röntgen and x-rays, she chose to take no patents on her radium-isolation process so that the scientific and wider community could carry out and benefit from such research. The Radium Institute, later becoming the Curie Institute, was established in Pierre's memory in 1909 between the University of Paris and the Pasteur Institute. However, its buildings and labs reached completion only in July 1914, just before World War 1. Thus, its work program did not develop and mature until after that. Marie Curie directed one part concentrating on physics and chemistry research. The radiobiology and radiotherapy pioneer, Claudius Regaud, directed the other. He focussed on investigating biological and medical effects and applications of radium and radioactivity, particularly for cancer. As noted above, Curie received two Nobel prizes, one joint with Pierre and Becquerel and one alone; the first woman winner, the first two-prize winner and only one of two to receive the Nobel prize in two fields. The Curie family total was 5!

During WWI, Marie Curie promoted and organised mobile radiography vans (Fig. 6) to provide x-ray services to field hospitals close to the battle-fronts. She raised funding for this, initially including from the French Union for Women's Suffrage. She set up around 20 vans in the first year, as well as many radiological units in field hospitals. She worked in these herself and trained other women as radiology helpers. She also developed sealed radioactive sources to sterilise wounded soldiers' infected tissues. These contained gamma-ray emitting radioactive radon gas, a short half-life (3.8 days) product of radium decay.

At the same time as Curie was promoting battlefront radiology units at the outset of WWI, so too were the Dublin-born sisters Edith and Florence Stoney. They were from a family of scientists and engineers. Their father, George Johnstone Stoney, was an Irish mathematical physicist, who among other work had introduced the concept of the electron in the 1870s (which he first called the electrine, only coining the name electron in 1891). He was a strong advocate for women's right to higher education and instrumental in women being able to take medical qualifications in Britain. Edith (1869–1938) studied maths and physics at Cambridge University. Florence (1870–1932) studied medicine in London and became a 'medical electrician' and the first woman radiologist in the UK. In 1898, Edith was appointed to set up the first physics department in the London (Royal Free) School of Medicine for Women. This had been established soon after a UK law change to ensure women could enter medical training. Her post was mainly to

Fig. 6 **a** A French mobile radiology car (a 'petite Curie') and equipment, Argonne, 1915 (Société Française de Radiologie). **b** Marie Curie driving a petite Curie during WW1 (Eve Curie: *Madame Curie*, Heinemann: London 1938)

support physics teaching to medical students and not involvement in clinical services. However, in 1914, the sisters offered to provide radiological services in military hospitals. Like Marie Curie, they were initially refused because they were women. However, they simply found ways around this and set up their own units, also in collaboration with women's organizations (and with some support from Madame Curie). These included hospital and mobile radiology units provided by the Scottish Women's Hospitals (SWH). SWH was established in 1914 by Dr. Elsie Inglis, an Edinburgh doctor and suffragette activist. It was founded to provide women-run and staffed hospitals close to the battlefront and supported 14 of those. Among other achievements, Edith Stoney developed a radiographic method to rapidly diagnose gas gangrene for fast decisions on life-or-death amputations. She too has been identified by some as 'the first woman medical physicist'.

Given the topic of this section, it may be noted in passing that at the 1943 formation of the UK Hospital Physicists Association, there were 53 founding members, reflecting the number of medical physicists in the UK. This had increased from 10 or 12 in the early 1930s and grew to become 1400 or so in the 1980s and approximately 2000 in 2010. Of the 53 HPA founding members, four were women medical physicists. In 2018, the proportion in the UK was reported to have grown to be around 40%.

I would like to mention one other woman medical physicist. In 1953, Andrée Dutreix took up a post in the Institut Gustave Roussy (IGR), Ville-juif, outside Paris to provide dosimetry calculations and measurements for a relatively recently installed betatron and for brachytherapy, likely the first French physicist working on this. Her husband, Jean, was a radiation oncol-ogist at IGR. She was advised by physics colleagues not to work in a hospital, for fear of not doing any scientific work, but this contrary advice helped to persuade her to take the post! From that beginning, she became a key figure in the development of medical physics in France! In addition, she was instrumental in securing the place of medical physicists in ESTRO on its formation in 1980. Along with others, notably Hans Svennson (Umea, Sweden) and Ben Mijnheer (Amsterdam, Netherlands), she developed the medical physics activities of ESTRO and many international collaborations. Interestingly, Andrée had been taught by Irène Joliot-Curie. Another Curie link was that Jean's father had trained as an electrician and became an x-ray 'manipulator' and mechanic in the mobile x-ray vans in WWI. After that he set up a radiological supply business and that eventually led Jean into radiation oncology. I attended the first ESTRO scientific conference, which was held in London, UK in 1982, i.e., not long after I had taken up my Edinburgh post. I was encouraged to become involved in ESTRO activi-ties by Andrée and Hans in the late 1980s, initially from their interest in radiotherapy dosimetry accuracy, since I had set up the UK's first national dosimetric intercomparison at that time. From that I became increasingly involved in ESTRO and its scientific journal, *Radiotherapy and Oncology*, as well as in wider developing international dosimetry audit work.

4.7   Who's the 'Genius' Who Has Worked (Loosely) in Medical Physics?

Well, clearly there are some great minds in the people mentioned already in this chapter… and also in the authors of this book! However, I refer to Ludwig Wittgenstein (1889–1951). The Austrian-born philosopher is widely viewed as one of the greatest thinkers of the twentieth century, perhaps even its greatest philosopher. He worked mainly on logic and the philosophy of mathematics and also of language and the mind. He began studying mechanical engineering in Germany, coming to Manchester, UK, in 1908 to begin doctoral studies in aeronautics. This considered the behaviour of balloons and kites in the upper atmosphere, for ionization and meteorology studies. He also studied, and patented, plane propeller designs using gas jet concepts. For this, he needed complex mathematical modelling and became interested, obsessed even, with the foundations and philosophy of mathematics and logic. In 1911, he moved to Cambridge, UK to study mathematical logic under Bertrand Russell, eventually gaining his Ph.D., and by 1939 a professorship of philosophy. As an aside, Alan Turing also had a mathematics fellowship in Cambridge and attended his lectures. They are reported to have had regular argumentative discussions, for example, on how computational logic and conventional ideas of truth are related. His work has been described as 'genius'; so, he fits the role in that respect.

So, where's the medical physics? During WWII, Wittgenstein felt he should do something more practical than philosophy and became a pharmacy porter in a London hospital from 1941 to 1942, delivering drugs to wards and patients. He became friendly with a doctor, Basil Reeve, who was interested in philosophy. Reeve and a colleague were working on radical approaches to shock in bombing casualties. When the bombing of London reduced in 1942, they moved to Newcastle-upon-Tyne, UK to continue similar studies on road and industrial accident victims. They offered Wittgenstein a laboratory assistant post to support their research. In this role, and using his engineering background, he built equipment and carried out physiological measurement experiments to study the characteristics of, and links between, breathing and pulse rate. When Reeve went to Italy later in the war to study battlefield casualties, Wittgenstein returned, in Feb 1944, to being a philosophy professor in Cambridge!

5   Last Thoughts and Any Other Questions?

I mentioned a talk on 'roots, role, soul and goal' of medical physics. The questions above have covered, roots, roles and goals, although 'goals' also implies direction and aims. The aims are already embedded in Hallé's definition, to use physics towards optimising each person's, each patient's, health and, also where possible, that of society.

Direction? I don't intend to discuss the future, but simply draw attention to a few currently developing areas that will involve physicists very directly. For example:

- the increasing use of functional imaging, i.e., imaging of biological function and process, or molecular imaging, i.e., imaging molecules of medical interest within living patients, to be used to enhance current clinical paradigms, for better diagnosis, targeting and therapy;
- the increasing applications of imaging for use in rapidly evolving precision, or personalised, treatment and prevention, i.e., more closely tailored to the individual variability in biological characteristics (genes, environment and lifestyle) and, hence, individual response or risk; and taking into account a growing list of '–omics', e.g., genomics, proteomics, image-based radiomics; or in theragnostics, the combination of diagnostics and therapy;
- data-mining or data-farming approaches using existing clinical data on patients to learn and model personalised treatment, and outcome prediction using machine learning and also expanding into wider artificial intelligence approaches; and requiring robust data handling, protection and privacy frameworks;
- machine learning also linking to automation of other clinical tasks, to assist human skills, along with robotics;
- the increasing impact of biophysics, physics of medicine, using physics tools and methods to probe human structure and function at increasingly small scales, fusing biology and physics for the understanding of biological heterogeneity and of disease and its causes, evolution and prevention;
- the continuing and continuous development of imaging, nuclear medicine and radiotherapy tools and methods;
- and, lastly, unknown unknowns! New applications of physics in biology, medicine and healthcare.

And soul? Soul also encompasses aims, but motivation too. That can be multi-factorial. Firstly, as physicists, to use our physics knowledge and skills in the clinical medical physics service, in collaboration with our clinician colleagues to help diagnose and treat individual patients and alleviate illness. Secondly, though, there is the exciting challenge of research and development to improve the clinical possibilities. The rationale of embedding scientists into clinical structures is, yes, to provide scientific services and advice, but also to carry out research. This generates and develops new solutions and knowledge to be translated into clinical use and disseminated for wider benefit. It is a necessary part of the medical physicist's work, realisable at whatever level is appropriate in each role. Often, for clinical medical physicists, research is applied, focussed, translational or developmental, close to clinical applications. However, many scientific developments that eventually come to practical clinical use have emerged from basic curiosity-driven research that was conducted without necessarily a prior expectation of possible clinical applications. An obvious example is the discovery of x-rays! Others arise from lateral thinking across assumed, perceived, or defined research field boundaries. Medical physicists must be aware of the development history of their field, as well as of other potentially interacting or impacting research fields. This underpins the vision, basis, and direction to look ahead to formulate next questions and then to further horizons.

The past is 'another country'? I was teaching radiotherapy physics on the University of Sydney's Medical Physics Master's degree course until relatively recently. I ran 'problem-solving sessions', where the students could turn up with any problem or topic in the course. That was fun, but could be challenging. One time the students were discussing the pros and cons of having lectures about equipment before seeing it, or alternatively seeing equipment in hospital visits before knowing about it. I mentioned that whilst on my Aberdeen Master's course around 1975, we were taught about linear accelerators for radiotherapy treatment, but the local centre only had cobalt-60 gamma ray machines. Therefore, we couldn't see linacs without going to another city. One enthusiastic, but not fully critically-thinking, student asked, 'but couldn't you google them'! It's a small example of how easy it is to not see or understand that times, situations and mindsets can be completely different looking back. Hence, they are not necessarily easily evaluable from here and now without pause and thought. It made me reflect that, as seen by my circa 2020 students, 1975 was as far back and different as 1930 was to me when I was doing my M.Sc. in 1975. Reflecting on the changes in medical physics in those two time spans made me wonder how medical physics, and medicine and health generally, will develop over the *next* 45 years. I'd like to see it!

In this chapter I've selected some examples to answer the questions I chose to pose; there are many more... people and discoveries, developments and applications, as well as questions! So, I take the blame for anything that others think I should have included! The take-home messages I hope to have given include that we need to know our history and our subject. At the very least, we should be ready to answer the questions that we will meet, and to explain medical physics and what we do! That might be to students, colleagues, or the public, but also to funders and decision makers, including hospital managers and politicians. I also hope I have conveyed the enthusiasm I still hold for medical physics towards the end of my career. I have found it an exciting and worthwhile branch of physics to be in and hopefully to have contributed to! I recommend it to others!

And finally ... any other questions?

Acknowledgements I have drawn heavily on some sources, the main one of which is Francis Duck's 2013 book (footnote 1). Other references were editorially discouraged, but I particularly acknowledge use of medical physics history papers by Francis himself (Physica Medica June 2014), Steve Keevil (Lancet April 21, 2012) and two PMB papers, by Richard Mould (Nov 1995) and Freidrich-Ernst Stieve (a radiologist, June 1991). I have also multiply consulted English and French Wikipedia to check information.

David Thwaites is honorary Professor of Medical Physics, formerly Director of the Institute of Medical Physics, at The University of Sydney, Australia; and honorary Professor of Oncology Physics at The University of Leeds, UK, formerly Head of the Department of Medical Physics and Engineering in Leeds Teaching Hospitals. He has had a 40+ years career in clinical and academic medical physics, mainly for radiation oncology applications, in the Edinburgh and Leeds Cancer Centres, as well as in Edinburgh, Leeds and Sydney Universities; and has honorary appointments in Liverpool and Westmead hospitals' cancer centres in Sydney. He has been privileged to be involved in developments at local, national and international levels in scientific, clinical, educational and professional areas of medical physics, including around 30 years advising International Atomic Energy Agency (IAEA) programs for low- and middle-income countries, as well as government advisory roles in both northern and southern hemispheres. He was the lead physics editor for the European SocieTy for Radiation Oncology (ESTRO) journal *Radiotherapy and Oncology* for 25+ years until 2020, now having a Physics Editor Emeritus role. His research and other work have produced more than 250 publications and involvement as author or editor in around a dozen books.

Awards

2018. European SocieTy for Radiation Oncology (ESTRO) lifetime achievement award, for contributions to radiation oncology and to ESTRO.

2014. ESTRO's Emmanuel van der Schueren award, for scientific excellence and contributions to education and practice of radiation oncology.

2013. International Organization for Medical Physics (IOMP) 50th anniversary award, as one of 50 selected medical physicists "*who have made an outstanding contribution to the advancement of medical physics over the last 50 years.*"

2011. Fellowship of the Australasian College of Physical Scientists and Engineers in Medicine (FACPSEM), for outstanding contribution to engineering or physical science applied to medicine.

2008. Honorary Fellowship of the Royal College of Radiologists (FRCR) for contributions to UK clinical oncology education and practice.

2

The Elevator that Couldn't and the Laundry Cart that Could

Jacob Van Dyk

1 The Elevator that Couldn't[1]

In 1971, I was hired by Professor Harold Elford Johns to work as a medical physicist at the Princess Margaret Hospital (PMH) in Toronto. Professor Jack Cunningham was my immediate boss. Professor Johns was considered the guru of Medical Physics with a world-renowned reputation for being a great scientist, a feared graduate student supervisor, and humanitarian. Over the years, he received multiple awards including five honorary doctorate degrees, and Officer of the Order of Canada. He was the first medical physicist to be inducted into the Canadian Medical Hall of Fame. Jack Cunningham also received the Officer of the Order of Canada along with multiple other awards, largely for his work on software development for computerized radiation treatment planning systems. Johns and Cunningham were the authors of *The Physics of Radiology*, the textbook which gave me my medical physics grounding as it did for all other young, aspiring medical physicists at that time. Professor Johns is generally considered to be the main driving force for

J. Van Dyk (✉)
Western University, London, ON, Canada
e-mail: vandyk@uwo.ca

[1] "The Elevator that Couldn't" is based on a true story described by J. R. (Jack) Cunningham in his self-published memoir "*And I Thought I Came from a Cabbage Patch.*"

the development of cobalt-60 radiation therapy, with the first patients being treated towards the end of 1951.

The introduction of these great minds is relevant background to this story. Further background is required on cobalt-60 radiation therapy. This mode of treating cancer patients is largely outdated now and has been replaced with the use of x-rays from linear accelerators. Cobalt-60 is a radioactive source emitting high-energy gamma rays continuously. The number of nuclear decays that occur, thus, the number of gamma rays emitted, is described by the source's "activity." The source decays in such a way that it is reduced to half of its activity in about five years. This means that the time required to give cancer patients a specified dose has to be increased by almost a factor of two after five years of use. Generally, this requires the need for a source replacement with a newer, higher-activity source, every five years.

Since cobalt-60 emits high energy gamma rays all the time, the source needs to be housed in a shielded container to protect the staff and the patients. Thus, the "head" of a cobalt-60 machine has a large quantity of high-density material in it, primarily lead which is encapsulated in a steel housing. The high density is important since this means that the shielding material takes up relatively little volume to absorb virtually all the radiation being emitted. The "beam" is turned on by moving the source electronically or pneumatically from the shielded part of the head to a variable opening that allows a beam to be incident on the patient. Because of all the shielding, these cobalt-60 radiation therapy machines are very heavy, with the entire machine weighing up to six to seven tons.

The sources can be replenished by removing the old, low-activity source and replacing it with a higher activity source. These sources are produced in nuclear reactors by bombarding a stable cobalt-59 with neutrons to generate radioactive cobalt-60. The stable form of cobalt is generally mined along with other metals but in only a few locations in the world. The neutron bombardment occurs in the same nuclear reactors that are used to generate electricity. These bombardments can take several years to get high enough activities.

Safety is a major consideration when sources are replaced in the cobalt-60 treatment machines. Both the old and new sources are transported in "source containers" that also weigh two or three tons so that the gamma rays are shielded to protect the staff and the general public during transport and the source exchange process.

This story involves a cobalt machine that was designed by Johns and Cunningham and built in the Princess Margaret Hospital's machine shop in 1959. It was known as the X-otron because it was the first cobalt-60 machine that also had a diagnostic x-ray source for taking images of the treatment volume in the patient. Since the X-otron was a "homemade" device, the cobalt-60 source had to be installed by its makers, i.e., the local staff.

At the time of loading the source into the X-otron, a suitable source container available from the Picker x-ray Company was used. The process involves removing the collimator (the beam shaping device) from the treatment machine and jacking up the source container such that it was flush with the bottom of the head of the cobalt machine and then bolted to it. In this configuration, the new source was installed. Since this was a brand-new machine, there was no old source to be removed.

The four people involved in this process were Professor Harold Johns, Professor Jack Cunningham, Professor John Hunt, who was another high-profile researcher at the Princess Margaret Hospital, and Jan Cederland from Sweden who spent two-years in Toronto as a visiting scientist. Figure 1 is a picture of these four individuals with the source transfer container under the head of the X-otron unit.

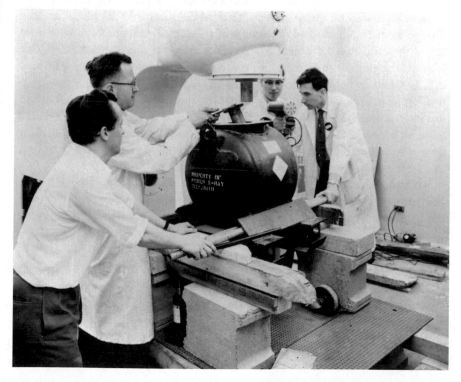

Fig. 1 The source transfer container under the head of the X-otron cobalt-60 radiation therapy machine. From left to right are Professors Jack Cunningham, Harold Johns, John Hunt and visiting scientist Jan Cederland. Courtesy of Anthony Olsen, Visual services, University Health Network

The following is quoted directly from Professor Jack Cunningham's memoir. The story starts after the source was installed in the head of the therapy machine.

> Our next task was to take the (now empty) source container to the basement for temporary storage. We used a (laundry) cart to carry the gear that we had used to jack up the container, and it was noticed that the wheels of the source container were just about level with the top of the laundry cart. If we could just transfer the container to the cart, it would save us a lot of work. This was done, and the cart, now carrying its two ton burden (much, much heavier than its normal load of laundry) was wheeled by the other three along the hall to the elevator. I had stayed in the room to clean up, but in a short time Jan returned with a rather enigmatic look on his face, saying the container had fallen off the cart onto the floor of the elevator, and that I should come. Jan was a bit of a practical joker, so my first response was to disbelieve this strange story, but I soon realized he was serious. What happened was that when the front wheels of the cart went onto the elevator, the great weight caused the elevator to descend a short distance. This, of course, made the cart slant, and though we had well blocked the wheels of the source container to secure it on top of the cart, it began to slide off and toppled over on the elevator floor. This lowered the elevator further and, as one might anticipate, produced a marked depression in its floor. The first task was to stop the elevator door alarm from making its incessant noise during the evening quiet of the hospital, the second was to lever the source container to an upright position and remove it from the elevator. This all took considerable time and effort. The damage to the elevator floor was embarrassing (especially for Doctor Johns) to explain the next day, to a rather unsympathetic director of the hospital.
>
> The story continued in various forms as a legend of the early days of PMH. Sometimes the story says there was a cobalt source in the container, but that is not true. There were numerous subsequent source loadings but we never again tried to use either the elevator or a laundry cart to carry a source container.

The moral of the story? … Even great minds do not always know their limits.

As a follow-up to this episode, I was recently made aware of a song that was written about this event. In fact, the song was sung by Jack Cunningham and recorded by his daughter, Susan Cunningham, in his 93rd year, during the last month of his life, in December 2020. The following are the lyrics:

> When Friday night after all had gone,
>   some boys were playing with a cobalt bomb.
>   They dropped it through the elevator floor,
>   and Harry's pride it was no more.

Of course, "Harry" refers to Harold Johns.
Now, on to the laundry cart that could ...

2   The Laundry Cart that Could

Fast forward from 1959 to 1977. Professor Johns was still designing cobalt-60 machines. By now, I was on staff at the Princess Margaret Hospital. In those days, the Princess Margaret Hospital was probably the largest cancer therapy institution in North America with a correspondingly high patient load. Because of the high patient load, special treatment techniques could occur with great frequency. One of these special procedures involved very large-field radiation therapy. These were used either for total body irradiation for patients requiring bone marrow transplants or half-body radiation therapy for patients who needed relief from their pain symptoms due to wide-spread metastatic disease. The frequency of half-body treatments had become so large that a special time was set aside for these patients on Saturday mornings. Because of the complexity and high doses involved, these half-body or total-body procedures were very time consuming, usually between one to three hours per patient and always required the presence of a medical physicist.

One day, probably in 1976, as we were having lunch in the Princess Margaret Hospital cafeteria with Professor Johns, Dr. Bill Rider, who was the head of the Radiation Oncology Department, Dr. Phil Leung, a fellow medical physicist, and myself, the discussion came up about developing a cobalt-60 machine specifically designed for large-field radiation therapy. The basic machine was designed on paper napkins—a commodity often used in the Princess Margaret Hospital cafeteria to generate new ideas or to move research projects into their next phases. Coffee and lunch times were extremely useful for the development of creative next steps. The result of this conversation was that a new cobalt machine was designed and built in the Princess Margaret Hospital machine shop that was capable of providing field sizes that were 160 cm by 50 cm (63 in by 20 in) at 90 cm (35 in) from the cobalt-60 source, and nearly twice that size for patients who were positioned on a special trolley on the floor. Thus, this machine was capable of both half-body and total-body irradiation; it became known as the *Hemitron*. Its picture is shown in Fig. 2.

Medical physicists are known for performing radiation measurements in water tanks, generally termed as "water phantoms". Water is known to be very close to human tissues with respect to radiation beam interactions. However, most water phantoms are limited in size. At that time, the maximum field size

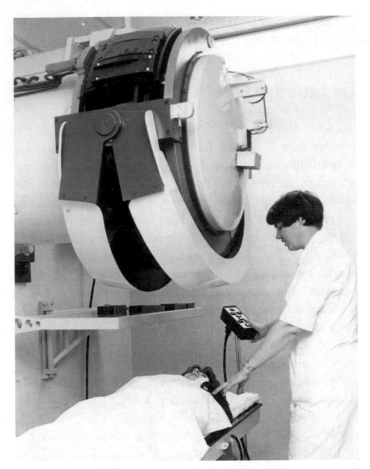

Fig. 2 The Hemitron with a patient in treatment position and a radiation therapist at the machine's hand controls. Note the collimators located just in front of, and slightly below, the radiation therapist's face. These contain lead inside a stainless-steel housing and are used to define the field size. This cobalt-60 unit became affectionately known by staff members as "Jaws"

on a regular therapy machine was 35 cm by 35 cm (14 in by 14 in) although some of the linear accelerators were moving up to 40 cm by 40 cm (16 in by 16 in) fields. Thus, commercial water phantoms with remote controls for the movement of a radiation detector in three dimensions was not going to work for total-body or half-body fields.

My solution … build a 117 cm × 117 cm × 48 cm (46 in × 46 in × 19 in) wood enclosure with a waterproof liner to hold approximately 600 kg (1320 lb) of water. However, after some experimentation and not being able to use the full 48 cm (19 in) depth, I concluded that this phantom depth was not deep enough to provide full radiation scattering conditions as required for these large fields, especially for the measurements at larger depths. The next solution … a 90 cm × 64 cm × 75 cm (35 in × 25 in × 30 in) laundry cart [approximately 430 kg (950 lb) of water]. This was quite a different type of laundry cart from that used earlier for the cobalt-60 source change. I used this laundry cart to generate the world's largest field tissue-air ratio measurements for cobalt-60 gamma rays. Tissue-air ratios are quantities that are important for radiation dose calculations used for planning the patient's treatment. The first tissue-air ratio measurements were performed in the 1960s by Professors Johns and Cunningham. These were done up to field sizes of 35 cm × 35 cm (14 in by 14 in).

While I measured tissue-air ratios out to the equivalent of 75 cm × 75 cm (30 in × 30 in), I also measured them for smaller field sizes to compare these data to what had been measured before. The interesting thing was that when I showed Professor Johns my tissue-air ratio measurements and he saw that they did not completely agree with the data that had been published under his name (differences generally being within 1–2%, up to a maximum of 5%), his first question was "Jake, what's wrong with your data?" Eventually, my data were incorporated into the standard data sets that were published in the supplements of the British Journal of Radiology and are still in use today.

The morals of this story? …

1. Social interactions, such as those over coffee or lunch, provide a great venue for generating new ideas.
2. Take advantage of the circumstances.
3. Argue with (or respond to) your boss as appropriate. Professor Johns showed leadership by asking difficult and tough questions; however, if you provided appropriate responses, he was ready to accept your arguments. He definitely provided leadership training to those who observed his actions and decisions.

3   How Research Evolves

The development of the *Hemitron* and the deliberations with Professor Johns regarding the differences in tissue-air ratio measurements resulted in some other consequences. Up until that time, patient dose calculations were performed assuming that patients consisted of water-like densities, primarily because there was no way of determining the "real" densities inside the patient. Computerized tomography (CT) scanners [or computerized axial

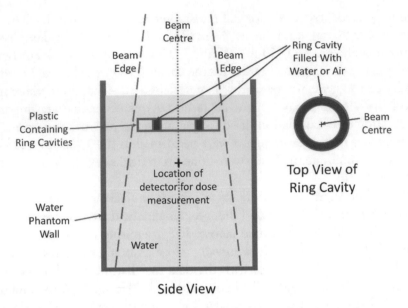

Side View

Fig. 3 Schematic of ring-shaped cavities that could be filled with air or water. In reality, there were a number of rings in the plastic holder

tomography (CAT) scanners] were developed in the 1970s and were clinically implemented for radiation treatment planning in the late 1970s. A Ph.D. student, Marc Sontag, was working on a new patient dose calculation methodology which accounted for "real" patient tissue densities based on patient data provided by CT scans. The method became known as the *equivalent-tissue air ratio* method. It was the first method that accounted for the three-dimensional scattering effects of the radiation beam interacting in the patient using "real" patient densities. However, the method required the use of a set of *weighting factors* that determined the importance of a scattering element in the patient based on its location and distance from the point at which the dose was being calculated. Marc calculated these weighting factors theoretically. Professor Johns believed that these factors could be determined experimentally. With the recent discussions about the cobalt-60 tissue-air ratio measurements, Professor Johns and I spent hours manipulating the measurements that I had made to see if we could determine the weighting factors from these measured data. It became clear that additional measurements would be required.

To make a long story short, the net result was a new experimental configuration using a water tank along with a specially designed, and somewhat complex, set of plastic rings from which water could be pumped leaving behind an annular ring of air inside the water tank (see Fig. 3). Thus, by measuring the dose at a point with a ring of air and again with that air replaced by water, I could get a direct measure of the impact of that scattering ring to the point of measurement. Because of the number of rings and the number of depths involved, these became very detailed and tedious experiments. Furthermore, since the *Hemitron* was used for patient treatments during the day, the experiments had to be performed later in the day and during the night.

There are two brief anecdotes related to these experiments. Both anecdotes demonstrate the unexpected risks of performing experiments that are new, using methodologies not used before.

Because the *Hemitron* was a radiation therapy machine designed to provide very large radiation fields, it had correspondingly large beam defining collimators, i.e., shielding blocks that were made of lead held inside a stainless steel sheath that could be moved electronically by the push of a button on a hand control. The collimators defining the width of the field were very bulky and to some extent looked like a large set of jaws (see Fig. 2). The suspense movie, *Jaws*, came out in 1975. These experiments were performed in 1978. Because of the intimidating look of this machine, some people referred to it facetiously as *Jaws*. The bottom of the lower jaws was less than 5 feet (~1.5 m) from the floor. At that time, I was 6 ft 2 inches (1.88 m) in height.

The experiments I performed involved setting up a fairly large water tank underneath the head of this machine. Inside the water tank was the plastic device containing the set of annular rings from which water could be pumped resulting in ring-shaped air cavities (see Fig. 3). The setting up of this experiment required a fair amount of careful positioning of the plastic device inside the water tank. To do that, I would have to bend over the edge of the top of the plastic wall of the water tank and reach down into the water to position the plastic device containing the rings. It took a bit of body contortion to get access to these rings in the water. On one occasion, I stood up too straight, too quickly, and banged my head full force against the lower jaws of the *Hemitron*. These jaws were much heavier and denser than my head. While I did not pass out (at least as far as I know), I was not far from it. It took me some time to recover; however, I was eventually able to continue with the experiment that evening.

The message here is … that you can never be too careful and that a clear and alert mind is always required during any experimental procedures.

The other anecdote relates to the experimental equipment. To pump water in and out of the plastic, annular ring cavities, I used a vacuum pump along with a reservoir system consisting of a series of tubes and two Erlenmeyer flasks, one being an air reservoir and the other allowing the collection of the water extracted from the plastic ring cavities. This way the ring cavities could be filled with water or air. The pump was always on and the direction of water flow was dependent on how a specific valve was turned. As indicated above, these experiments tended to be performed later into the evening when virtually nobody else was around. Hence, it was quiet as a mouse. Suddenly, there was a horrendous noise, sounding like a gun shot in a very small room, and glass shards flew all over the place. Being in this very quiet evening environment, it scared the life out of me! After some investigation, I discovered that as a result of the vacuum pump, which was pumping out air, one of the connecting tubes had collapsed and completely closed-in on itself, resulting in extremely low pressure in one of the glass Erlenmeyer flasks. Since the flask was not able to handle this pressure difference, it imploded, causing pieces of glass to fly all over.

The resulting data from these experiments created a lot of discussion between Professor Johns and me and led to more experiments. (This is typical of just about all research projects.) Between my day job as a medical physicist working in the clinic on patient-related activities and this research activity, it became clear that another set of hands would be required. The result was that a post-doctoral fellow, John Andrew, took on this role. With his involvement, the experimental methodology evolved so that the water pumping system to generate air cavities was replaced by conical Styrofoam rings that could be moved up and down in the water tank. Since Styrofoam has a very low density (less than 0.03 g/cc), it effectively behaved like an air cavity. Following John Andrew's work, John Wong took on this project for his Ph.D. research and brought it to fruition. In the meantime, Professor Harold Johns, who had supervised this project, retired in 1980 and Professor Mark Henkelman became John Wong's Ph.D. supervisor. John Wong, as well as John Andrew, became very successful medical physicists, John Wong in the US and John Andrew in Canada. (See John Wong's Chap. 7.)

As indicated in the beginning of this section, the purpose of this research related to providing accurate dose calculations using "real" patient tissue densities obtained from CT scanning patients for treatment planning purposes. I was heavily involved in the early implementation of CT scanning for treatment planning in the late 1970s. As indicated in Sect. 2 of this chapter, we were also heavily involved in total-body irradiation for bone marrow transplants and half-body irradiation for palliative purposes (i.e., alleviation of symptoms). In this context, we performed a study of a series of patients who had been planned for thoracic treatments with the new CT planning processes. Using data from these patients, we were able to establish a method of retrospectively determining the radiation dose to lung tissues for

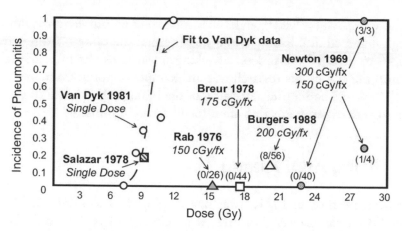

Fig. 4 The incidence of radiation pneumonitis versus dose as obtained from various reports published over the years. The open circles are the Van Dyk data for single fraction, whole lung irradiation. The dashed line is the mathematical fit to these data. Note that this represents a full dose-response curve which is very steep indicating that a small increase in dose can have a significant increase in lung complications; hence, these data are very useful for knowing how to limit lung doses. The figure is reprinted with permission from Marks et al. Int J Radiat Oncol Biol Phys 76(3): Suppl, S70-S76; 2010

patients that had been treated with total-body and half-body procedures in the previous decade. From this retrospective analysis, in 1981, we were able to generate the first clinical dose-response data for lung complications (radiation pneumonitis) following single treatment, whole lung irradiation, the results of which have subsequently been used around the world as a guide for lung dose limits for these large field radiation therapy procedures. Such data are important for being able to optimize our treatment procedures by giving the maximum dose to the target tissues and minimizing the effects on lung tissues. Such guidance data have had an impact on the treatments of many patients. These data were still presented as part of the Quantitative Analysis of Normal Tissue Effects in the Clinic (QUANTEC) review in 2010 (Fig. 4). Further research evolved into animal experimentation in the early 1990s (600 rat experiment, conducted by Ph.D. student, Chris H. Newcomb, with additional input from my radiobiology colleague, Richard P. Hill) to assess isoeffect formulae that allowed the comparison of different treatment regimens (including the effects of dose per treatment fraction, i.e., the dose given on each treatment day, the number of daily fractions given in a treatment course, and the total number of days over which the total treatment course is given). Later in the 1990s, lung-response experiments in rats with partial volume lung irradiation instead of irradiating the total lung, demonstrated, for the first time, unexpected out-of-field effects. It is interesting to note how research evolved from basic medical physics problems to addressing biological and clinical issues. While we often use the phrase "from bench to bedside" to describe the process of taking research results from the laboratory into the clinic, this research was an example of going from "bedside to bench" to gain a better understanding of the biological and clinical processes involved with experiments that are difficult to do in humans.

4   Being Appreciated

As indicated earlier, in 1971, I was first employed by the Princess Margaret Hospital/Ontario Cancer Institute as a "clinical physicist" and Professor Harold Johns was my "upper" boss and Professor Jack Cunningham was my immediate boss. The letter of offer that I received is shown in Fig. 5.

There are several things of note in this letter. First, "the philosophy of the Physics Division", which is actually Professor Johns' philosophy, is that "None of the people in our Division is on a fixed salary schedule, and all promotions depend on the individual's performance." Second, my salary offer was CAD$8500, not a high salary even at that time. The other Johns' philosophy was that, in view of its reputation, it was an honour to work at the Princess Margaret Hospital/Ontario Cancer Institute and, thus, salary should not be a major determining factor for taking on a position there. Third, "your major responsibility will be in the clinical area, but we hope you will tackle this area in such a way that something new would periodically come out of it." Thus, while my major responsibility was "clinical service", there was a research expectation as well. In those days, the relative amount of time spent on research versus clinical service was never well-defined. Whereas today that is generally clearly quantified even though the distinction between what is clinical service and what is research can often be very nebulous. The other point to note is that when I was hired, I had a Master of Science degree but not a Ph.D. That was actually an overt decision on my part since my initial main goal with respect to my career was to provide the best clinical service possible.

By 1979, I had developed eight years of wonderful experience because of the great group of people that I worked with in Toronto. This included a one-year leave-of-absence as the acting head of Medical Physics at the Centre de Radiothérapie, Hôpital Cantonal de Génève, in Geneva, Switzerland during 1974–75. By 1979, I had published approximately 11 papers in medical physics-related academic journals. It was during this year that I received a request from Professor Ervin Podgorsak to consider an academic, medical

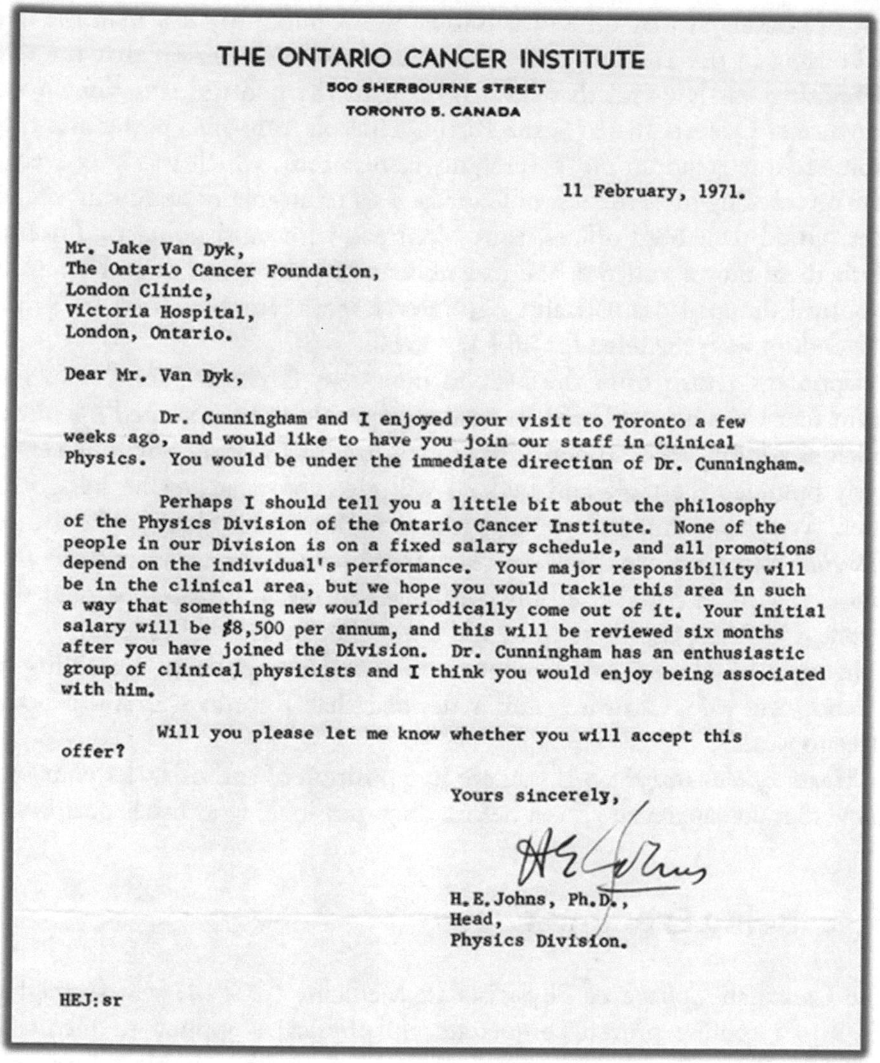

THE ONTARIO CANCER INSTITUTE
500 SHERBOURNE STREET
TORONTO 5. CANADA

11 February, 1971.

Mr. Jake Van Dyk,
The Ontario Cancer Foundation,
London Clinic,
Victoria Hospital,
London, Ontario.

Dear Mr. Van Dyk,

Dr. Cunningham and I enjoyed your visit to Toronto a few weeks ago, and would like to have you join our staff in Clinical Physics. You would be under the immediate direction of Dr. Cunningham.

Perhaps I should tell you a little bit about the philosophy of the Physics Division of the Ontario Cancer Institute. None of the people in our Division is on a fixed salary schedule, and all promotions depend on the individual's performance. Your major responsibility will be in the clinical area, but we hope you would tackle this area in such a way that something new would periodically come out of it. Your initial salary will be $8,500 per annum, and this will be reviewed six months after you have joined the Division. Dr. Cunningham has an enthusiastic group of clinical physicists and I think you would enjoy being associated with him.

Will you please let me know whether you will accept this offer?

Yours sincerely,

H. E. Johns, Ph.D.,
Head,
Physics Division.

HEJ: sr

Fig. 5 Letter of offer for employment as a clinical physicist at the Princess Margaret Hospital/Ontario Cancer Institute in Toronto in 1971

physics position at McGill University in Montreal. Ervin had worked at the Princess Margaret Hospital earlier in the 1970s as a post-doctoral fellow and we became well-acquainted during that time. For someone like me without a Ph.D., this was a very attractive consideration, especially at a prestigious institution like McGill University. I went to Montreal to give a seminar and to undergo the job interview process. I received a job offer which was very attractive from a variety of perspectives, not the least of which was that the cost of housing at that time in Montreal was significantly less than the cost of housing in the Toronto area. However, there was a reason that the cost of housing was low and that had to do with the political situation in the province of Quebec. In 1976, the Parti Québécois came into power and they promised a referendum on "sovereignty-association", which was effectively a vote on seceding from the rest of Canada. As a result, many significant industries moved their head offices from Montreal, with most going to Toronto. With these moves and the political uncertainty, the value of real estate in Montreal dropped dramatically. My interview was in the fall of 1979; the referendum was scheduled for 20 May 1980.

Upon my return from the McGill interview, Professor Johns, who had heard that I was being offered an academic position there, invited me out to lunch at a local "greasy spoon." To summarize briefly, Professor Johns effectively pounded the table and said, "I will give you whatever he has got to offer." Professor Johns always "kept his cards close to his chest." *This was the first time that I realized that I was very much valued at the Princess Margaret Hospital.* While I appreciated his confidence in me, I was disappointed that it took a job offer elsewhere to let me know that I was valued there.

In view of Harold Johns' counteroffer and the political instability in Quebec, my wife, Christine, and I decided that Toronto was still a better place to work.

Moral of this story? ... If you are in a position of authority, let your staff know that you appreciate them before they start looking at other positions.

5   Being Examined

The Canadian College of Physicists in Medicine (CCPM) was formed in 1979 to recognize proven competence in physics as applied to medicine. As such, it is a "certifying body." "Candidates with suitable educational background and experience become members of the College by passing examinations." One of the main founding members of the College was Professor Johns. He along with Professor Cunningham, i.e., my bosses, were on the examination committee. The first written certifying exams were to be held on 29 March 1980. To get the College started, a number of medical physicists had been "grandfathered" into the college without formally undergoing the certifying exams. By that time, I had worked as a medical physicist for nine years. It was not clear to me (nor anyone else) as to what the criteria were for being grandfathered into the College. Of course, my perspective was that I had enough experience that I should have been grandfathered as well, especially when I compared myself to some of the individuals who had been grandfathered. However, that was not to be. I had to write these certification exams—exams that had never been given before, which meant that there was no previous experience as to what kinds of questions might be posed.

Thus, in January 1980, I gave my wife, Christine, a compendium of multiple novels by Neville Shute for her to read during the evenings while I would be spending time studying for the certification exams. My big concern was that I was going to be examined by my bosses. They would find out what I knew and what I didn't know. Every deficiency in my knowledge would be exposed. I felt that my career was on the line!

At that time, Professors Johns and Cunningham were working on a draft of the fourth edition of their book, *The Physics of Radiology*. In January, I asked Jack Cunningham, if I could review the available draft of this fourth edition. Considering that they would be contributors to the certification examination questions, my guess was that some, if not all, of the questions could be answered if I knew everything in this new edition. So, I went through this draft of the book from cover to cover and I solved (at least I worked on) every problem that was posed at the end of each chapter. As part of this process, I provided Jack with some comments on some questionable things that I found in the draft. As a result, my name was listed, along with others, in the acknowledgments when the book was published in 1983. Figure 6 shows inscribed autographs of Johns and Cunningham in my copy of *The Physics of Radiology*.

On 29 March 1980, I wrote six hours of CCPM exams—a three-hour closed book exam in the morning, and a three-hour open book exam in the

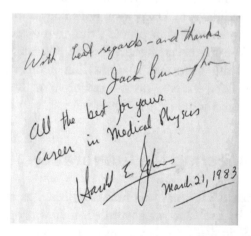

Fig. 6 Autographs of Professors Harold Johns and Jack Cunningham inscribed in the front of my copy of the Johns and Cunningham "The Physics of Radiology"

afternoon. By this stage, I was 35 years old, well past my university years and certainly not used to writing exams anymore. My fingers on my right hand were indented and sore from the pressure of the pen. I was totally wiped and exhausted at the end of the afternoon. While I felt that I had answered most questions reasonably well, there were some that I was concerned about. I had no idea what the accepted standards would be for these exams since they had never been done before. Thus, I was tentative about the possible outcome. That evening, I went home and had a quiet supper with Christine and our three children, Tonia, Jon, and Ben. After they went to bed, Christine and I sat down in our recreation room to watch a movie. I do not remember which movie it was, but I do remember that after the movie was over, Christine announced that she was in labour pains. She had patiently refused to acknowledge this during the movie since she wanted me to relax, at least while the movie was on. The next morning, at 8:30 am, our youngest daughter, Amanda Marguerite (Amy), was born. Because of her birthdate, I have an easy recollection of the date of the first Canadian certification exams.

As it turns out, I did pass the written exams; however, this meant that now I qualified for the oral exams to be held later in May, just prior to the meeting of the Division of Medical and Biological Physics of the Canadian Association of Physicists to be held in Montreal. Thus, the trauma was not over yet, since again I would be placed face-to-face in an oral exam setting with my bosses, Johns and Cunningham. The exams existed of a presentation to be made by the candidate, questioning on that presentation, and then questioning specific to my area of specialization in Medical Physics, which in my case was Radiation Therapy and for others it was Diagnostic Radiology or Nuclear Medicine. Since these were the first exams for the CCPM and since there were fewer candidates in Diagnostic Radiology and Nuclear Medicine, to cross calibrate the examination process, the examiners of the other disciplines decided to join my exam; hence, further pressure on me. While very stressful, I did pass the exam and it seems that it did not hurt my career. In fact, a few years later, I served on the executive of the CCPM, including being its President for 5 years (1991–1995).

6   Re-engineering the Corporation

The Princess Margaret Hospital was first established as the Ontario Cancer Institute in 1952 by the Ontario government. It was originally located on 500 Sherbourne St in Toronto. It was officially opened on 1 May 1958 by Her Royal Highness The Princess Margaret, Countess of Snowdon, and it was renamed Princess Margaret Hospital at that time. Expansions to the Princess Margaret Hospital had been discussed at length ever since I joined the staff in 1971. Finally, in the 1980s and early 1990s, a firm decision was made to move the Princess Margaret Hospital from the Sherbourne St site to University Ave to be close to the University of Toronto and a number of other major hospitals, with the Toronto General Hospital being across the road and Mount Sinai Hospital being a physically connected neighbour. The move of the Princess Margaret Hospital began by constructing four bunkers in the basement of Mount Sinai Hospital in a configuration that would eventually link in with the new radiation therapy department of the relocated Princess Margaret Hospital once it was built. The plan was for a total of 16 high-energy radiation therapy machines with the 12 additional machines being on the same floor and interlinked to the previously installed four machines in the Mount Sinai Hospital.

There are several historical issues that are relevant for subsequent activities that took place in the 1990s. The first was that there was an economic recession affecting much of the Western world. The impacts included the resignation of Canadian Prime Minister Brian Mulroney. I was told that when the new Princess Margaret Hospital was under construction, it was the largest construction project in Canada—largely because the rest of the construction industry had mostly closed down as a result of the economic circumstances. Because the Princess Margert Hospital was moving to a new facility that was at least three times the size of the Sherbourne St site (from a 7 floor building to a 19 floor building), there was a concern about the affordability of the increased operating budget in the new facility.

"Re-engineering" was a new term that became prominent in the 1990s. Many consultants refer to the book by Michael Hammer and James Champy, entitled *Reengineering the Corporation* published in 1993 as the impetus for the re-engineering movement. Practically, re-engineering consists of a total review of any organization's mode of operation and transforming how it operates to be more efficient, more effective and at a lower cost. To quote Hammer and Champy, re-engineering served to "achieve dramatic improvements in critical contemporary measures of performance, such as cost, quality, service, and speed." As a result of the economic circumstances, upper level management of the Princess Margaret Hospital first performed an "operational review" and then, based on the advice of the consultants, implemented a total re-engineering process. This occurred in 1995.

The following outlines some anecdotes related to my involvement in this re-engineering process. When this re-engineering process was initiated, the word spread quickly throughout North America, and probably the entire world. With its positive international reputation, the large scope of its operations, and its well-known history of high-level medical physicists like Johns and Cunningham, I started getting telephone calls from medical physicists around North America asking me what was going on at the Princess Margaret Hospital. While the calls were partially out of concern as to the impact of re-engineering on the practice of medical physics at the Princess Margaret Hospital, the main concerns related to their own circumstances should a re-engineering process be implemented in their setting. They really wanted to know how to address re-engineering should it hit their instituion.

When, in one of my meetings with the consultants in charge of this re-engineering process, I explained to them that I was getting telephone calls from across North America asking what was happening at the Princess Margaret Hospital, their question was, "Why are *you* getting these calls?" When I explained that the Princess Margaret Hospital has a world-class reputation, especially in medical physics, because of pioneers like Johns and Cunningham, they had no idea of the history of the Hospital, nor did they have any idea about who Johns and Cunningham were ... nor did they seem to care.

At the time that the re-engineering process unfolded, I was in charge of the medical physics-related treatment planning aspects of the radiation treatment program. Since the total treatment delivery program was under review, one of the significant considerations was the restructuring and reorganization of the radiation treatment planning process. As part of the restructuring, I was asked to be the *Treatment Planning Team Leader.* The process was that I would get a new letter with a job offer (after having worked at the hospital for 24 years)! It should be noted that medical physicists often have three major components to their activities including clinical service, teaching and research, in addition to other activities related to administration and involvement with professional organizations. The re-engineering consulting company had decided to address the clinical service aspects of the radiation therapy activities since that kept the process focussed. In my letter of offer, there was no indication about research and teaching. So, when I asked them about research and teaching and how that would fold in with my responsibilities for clinical service, I was told that they would let me know within a week. I never did receive an answer to that question. Since I had not accepted the position as a *Treatment Planning Team Leader*, I became the *Treatment Planning Interim Team Leader*. Being in this position, I received a visitor in my office from the re-engineering consulting company on a Friday afternoon. With him, he had a thick binder with several inches of paper describing the transition process into the new organizational structures. The only thing that was missing was the actual timelines of implementation. His request, could I fill in the timelines by the next meeting to be held the following Thursday? I explained to him that I was leaving on the weekend for meetings of the Canadian Organization of Medical Physicists and the Canadian College of Physicists in Medicine (of which I was the president at that time) in Montreal and that I was returning on the following Wednesday; being busy at the meetings, I would not have a chance to look at these documents in any detail by then. Upon my return to work, the following Thursday, I was catching up on my mail and by about 10 AM I listened to my telephone messages. A message from the re-engineering consultant sent on the Tuesday of that week reiterated his request to have the timelines available for the meeting on Thursday morning at 11 am. Clearly, this was not to be.

I arrived at the meeting at 11 AM. There were several people at the meeting including the consultant who had approached me about providing the time-lines, as well as the top person of this re-engineering consultants' group for this entire re-engineering project, in addition to an administrative person from the Princess Margaret Hospital. It should be noted that Hammer and Champy, in their book, describe three groups of attitudes within an orga-nization undergoing re-engineering. Paraphrasing roughly, as I remember it, the upper group (perhaps 10–20%) are those who "buy into" the process (possibly called "ambassadors") and actively participate in its implementation. A middle group (perhaps 50–70%) who will follow along with the process (perhaps called "accommodators"). Finally, the bottom group (perhaps 10–30%) who are resistant to change and perhaps somewhat rebellious about the process (possibly called "rebels"). In summary, the undercurrents at the meeting were clear; I was being considered as one of the "rebels" in the re-engineering process. The discussion at the meeting began with the leader of the group talking about the change process and, then, going around the table with each person describing how they had "bought into" and supported the process. And then, it was my turn. While I was quite intimidated by the discussion, I decided to expound on my decision-making theory. This was a theory that I had developed based on my 24-years of experience as a working medical physicist and interacting with "doers" and "shakers" within my field—the likes of Harold Johns and others. I indicated that the decision-making process was like a radiation therapy dose-response curve. If you plot a graph of tumour or normal tissue response, it generally has an "S"-shaped curve, i.e., if you give a little dose, there is virtually no response. If you give a little more, then there is some response after which the response increases rapidly as the radiation dose is increased. Eventually, when the dose gets quite

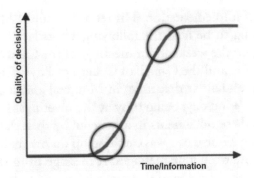

Fig. 7 Jake Van Dyk's decision making theory showing how the quality of a decision depends on the time expended and the information obtained in analogy with a radiation therapy dose-response curve. We should be making decisions once we are in the upper 80–95% range and not in the lower 10–20% range (as shown by the black ovals)

high, then there is 100% response and adding more dose makes no difference since one cannot go beyond 100%. Figure 4 shows an example of a dose-response curve for radiation pneumonitis. If instead, we consider quality of decision making on the vertical axis (compared to tumour or normal tissue response on a dose-response curve) and time expended or information gained to make that decision on the horizontal axis, then we should spend enough time and get enough information such that we are working in the 80–95% range for quality of the decisions and not in the 10–20% range. Figure 7 shows this concept graphically. The re-engineering team seemed to be somewhat stunned by my analogy—I am not sure whether they did not understand my theory or whether they did not know how to respond to it. However, for the next couple of months, I did show my "buy in" to the process by being the *Treatment Planning Interim Team Leader* and organizing meetings with all the relevant "players" including the medical physicists, radiation oncologists and radiation therapists. Once the re-engineering team observed my proactive involvement, there appeared to be a change of attitude. One of the things I learned in the process was that the attitude of the consulting company was that you define the desired changes and determine implementation timelines as quickly as possible and then, if you find that the proposed changes do not yield the desired results, you adjust the process accordingly. This attitude of "if it doesn't work, we'll change it" is extremely frustrating for the staff involved. Perhaps a little more thought and input information would help the decision-making process.

My original career plan was to gain my experience in the "big city" for three years and then to move to a place like London, Ontario, where I had done my graduate studies—a city that Christine and I enjoyed immensely. However, because I really appreciated the group that I was working with in Toronto, it actually took us 24 years! While there were multiple reasons for the move, re-engineering added to the contributing factors. In July of 1995, I made the announcement that I had taken on a new position as the Head of Clinical Physics at the London Regional Cancer Centre, starting in the beginning of September.

The take home message ... major organizational changes need to be addressed with caution and sensitivity. The consulting company's attitude about "if it doesn't work, we'll change it" is potentially fraught with longer term problems. Furthermore, major changes should be based on good information.

7   Summary

A few quotes from Albert Einstein provide a nice summary of this chapter:

1. "Anyone who has never made a mistake has never tried anything new,"
2. "If we knew what it was we were doing, it would not be called research, would it?"
3. "Genius is 1% talent and 99% hard work."
4. "Everything that can be counted does not necessarily count; everything that counts cannot necessarily be counted."

Finally, in my home office, I have a picture of Einstein sitting behind a desk quite cluttered with books and papers and another one of his quotes, "If a cluttered desk is a sign of a cluttered mind, of what, then, is an empty desk a sign?"

Jacob (Jake) Van Dyk is Professor Emeritus of Oncology and Medical Biophysics at Western University, London, Ontario, Canada, and former Manager of Physics and Engineering at the London Regional Cancer Program in London, Ontario, Canada. He has more than 40 years of experience in the practical facets of radiation oncology physics with 24 years at the Princess Margaret Hospital (PMH) in Toronto and 15 years at the London Regional Cancer Program (LRCP). With a leave-of-absence in 1974–75, he worked as the acting head of physics at the Centre de Radiothérapy, Hôpital Cantonal de Génève, in Geneva, Switzerland. From 2009 to 2011, he was employed as a professional expert and consultant at the International Atomic Energy Agency (IAEA) in Vienna, Austria. His research has yielded over 200 publications along with four unique Volumes of *The Modern Technology of Radiation Oncology: A Compendium for Medical Physicists and Radiation Oncologists*. He has lectured in over 41 countries. He was the main founder of *Medical Physics for World Benefit* (www.MPWB.org), an organization devoted to supporting medical physics activities, largely by training and mentoring, especially for lower income settings.

Awards

2022. William D. Coolidge Gold Medal Award, American Association of Physicists in Medicine (AAPM). This is the highest award given by the AAPM and recognizes an eminent career in Medical Physics.

2019. International Day of Medical Physics (IDMP) Award for "*promoting medical physics to a larger audience and highlighting the contributions medical physicists make for patient care.*" Awarded by the International Organization for Medical Physics.

2014. Honorary Doctor of Science (honoris causa) degree granted at Western University's MD Convocation, London, Ontario, Canada.

2013. Selected by the International Organization for Medical Physics (IOMP) as one out of 50 medical physicists "*who have made an outstanding contribution to the advancement of medical physics over the last 50 years.*" This recognition was given as part of IOMP's 50th anniversary.

2011. *Canadian Organization of Medical Physicists (COMP) Gold Medal*. This is the highest honour that COMP bestows on one of its members in recognition of an outstanding career as a medical physicist who has worked mainly in Canada.

3

The Decades of Change

Peter R. Almond

1 Introduction

My career in medical physics started in 1958 when I entered a one-year post graduate course at Bristol University in England. It was a time when radiation oncology was primarily carried out with orthovoltage x-ray equipment (in the 100–300 kV range) and manually handled radioactive radium for interstitial implants or gynecological applications. Diagnostic imaging was done with flat cassettes which were manually loaded, and the film developed in a dark room; nuclear medicine with radioactive isotopes was in its infancy; ultrasound was a curiosity; lasers would be invented in December of that year; and computers were nowhere to be found. But things were changing and the next two decades would bring about significant developments.

P. R. Almond (✉)
The University of Texas, MD Anderson Cancer Center, Houston, TX, USA
e-mail: palmond@reagan.com

© The Author(s), under exclusive license to Springer Nature
Switzerland AG 2022
J. Van Dyk (ed.), *True Tales of Medical Physics*,
https://doi.org/10.1007/978-3-030-91724-1_3

2   As Things Were

2.1   Of Betatrons

I had come to the University of Texas, M D Anderson Hospital and Tumor Institute (MDAH) in September 1959 on a one-year fellowship after completing a post graduate course in medical physics at Bristol University in England. My intention was to return to England at the end of the fellowship. However, in March 1960, Warren Sinclair (then chairman of the physics department at MDAH) called me into his office. He told me that he thought that the real advances in cancer treatment would be made in basic research and he was not sure that there was much of a future in radiation therapy physics. He then offered me a full scholarship in a biophysics Ph.D. program through the University of Texas. It required going to Austin to take the necessary course work on the main campus but returning to MDAH in Houston to do my research. As kind as the offer was, I told him I really wanted to be a clinical medical physicist involved in treating cancer patients. I told him I was sure there were people alive at that time who in the course of their lives would be diagnosed with cancer and many of them would need to be treated with radiation, and that radiation treatments would be around for the rest of my career. In which case he said he had another offer. MDAH had two scholarships for graduate students in the physics department at Rice University; one was available, and he was pleased to offer it to me, but I should realize that the research would be in nuclear physics. The condition for the scholarship was that the student work part-time in the physics department at MDAH. This was not an onerous requirement since Rice University is across Main Street form the Texas Medical Center. This appealed to me since I wanted to continue with the research that I had been involved with during my one-year fellowship. Shortly after my meeting with him, Sinclair resigned from his position at MDAH to take up a post at the Argonne National Laboratory near Chicago, Illinois, and Bob Shalek was appointed chairman.

MD Anderson Hospital was a relatively young institution. In February 1954, before the new hospital building was fully completed, the radiation therapy basement was opened so that the Grimmett cobalt-60 unit could be installed, and patient treatments initiated. The basement had been designed by Dr. Grimmett, the first chairman of the physics department at MDAH, to house his cobalt-60 unit and a 22 MV Allis Chalmer's betatron. A betatron is a particle accelerator that uses the electric field induced by a varying magnetic field to accelerate electrons. Both machines were in use when I arrived and during my fellowship year, I became very familiar with both.

They were located in the basement. There was a central area shared by both machines. On one side was the control panel for the cobalt unit and on the other the control desk for the betatron. Neither room had an entrance maze. Radiotherapy rooms are often known as bunkers because of the thick shielding walls. Bunkers often incorporate a maze of some sort to reduce the need for large and heavy entrance doors. The entrance doors of these rooms were of laminated wood and lead on heavy duty hinges. For the cobalt room there was a short passage from the door into the treatment room. The cobalt unit was ceiling mounted in the center of the room. For the betatron the door was the same but the passage into the room much longer and the room itself was much bigger, with the Allis Chalmer's betatron, floor mounted, at the back of the room. Grimmett had designed the rooms with thick walls made with heavy concrete with ilmenite being added to the concrete to increase its density. Since there were no closed-circuit cameras at that time, Grimmett had designed viewing portals through the walls placed by the control desks for both machines. These portals were in the shape of truncated pyramids with the larger base of approximately 40 cm square on the outside, tapering to an opening of about 30 cm square inside the treatment room. In 1950, Grimmett had come across glass which contained 55% by weight of lead and he recognized its potential as shielding material. He did experiments and found they had excellent optical quality and forwarded the information to the architect designing the building, at the same time strongly urging the administration to place an order for a betatron. Several layers of the glass were put at the treatment end of the viewing port. These treatment rooms met all the appropriate shielding regulations for photons and the patients could be visually observed during their treatments. However, several years later, concern was raised about the neutrons produced by gamma-neutron production at the higher photon energies, which has a threshold around 10 MeV and increases to a maximum at approximately 20 MeV. This was certainly true for the 22 MV betatron beam. As a general rule, treatment rooms that have adequate shielding, with concrete, for the highest energy photon beam, will be adequately shielded for neutrons. However, the viewing portal for the Allis Chalmers' room with only leaded glass as the barrier was transparent to the neutrons. Subsequently, tightly fitting Perspex tanks were installed in the viewing ports in the space from the glass to the control room side of the wall and filled with triply distilled water. This proved sufficient to reduce the neutron leakage to acceptable levels.

There were serious limitations to the use of the betatron. The maximum output was low, 0.2 Gy per minute. (Today's treatment machines typically operate at 6 Gy/min.) The beam was limited to a vertical beam from the top, and only rotated to a horizontal beam from one direction only to a few degrees beyond the horizontal. Field size was set by removeable cones. There was no incorporated treatment couch; the patients were treated laying on a gurney. No isocentric setups (where beams from all directions have a common focus point at the isocentre) were possible, all treatments were carried out using a fixed source-to-surface distance (SSD) of 80 cm. Never-the-less, successful treatment techniques were developed, especially with a four-field box technique for the pelvic region that involved parallel-opposed anterior and posterior fields and parallel-opposed right and left lateral fields. Because of the limitations with the betatron such treatments took quite a lot of time.

Research was ongoing on how to correctly calibrate the 22 MV x-ray beam of the betatron. In practice the beam was calibrated using a Victoreen dosimeter and applying the chamber's cobalt-60 exposure calibration factor to derive, what was called, 'Exposure Dose'.

Sinclair understood that what was really needed was the determination of 'Absorbed Dose' and had designed and built an aluminum calorimeter to do so. This calorimeter was quite large, about the size of a 2 lb coffee can, with a thick aluminum top to accommodate the 22 MV x-rays and was not suitable for the cobalt unit. I designed and built a much smaller pocket watch sized aluminum calorimeter which could be used for cesium and cobalt units, high energy photons and eventually electron beams (Fig. 1).

2.2   Of Cobalt-60

My one major recollection of the cobalt unit was of the day the source leaked cobalt-60. The therapists noted that when the machine was turned on or off, a fine white powder fell from the collimator. This was quickly determined to be radioactive but not before it had been tracked on peoples' shoes all around the hospital. Although the actual activity lost was quite small the US Atomic Energy Commission, under whose license we had the cobalt-60, required that we account for all of it. I spent the whole of one night, in surgical scrubs on my hands and knees on the floor of the cobalt room with a Geiger counter locating the cobalt and scrubbing it up, carefully keeping all the water used, for storage until it had decayed. Surprisingly we accounted for nearly all the leaked cobalt.

Fig. 1 My small calorimeter in place under the Allis Chalmers' Betatron

3   Something New

In the spring of 1964, as I was completing my Ph.D. in nuclear physics at Rice University, Bob Shalek offered me a position at MDAH, first as a postdoc but after a year to be on the faculty. In particular, he wanted me to be the physicist in charge of initiating electron beam radiation treatments at the hospital.

As commercial cobalt-60 machines became available, MDAH began to install Atomic Energy of Canada Limited (AECL) units, and the Grimmett cobalt unit was retired. In its place, a small prototype 18 MV Siemens beta-tron was installed, primarily for the electron beam capabilities of the machine. The fact that the treatment room was adequately shielded for 18 MV x-rays

was a testament to Grimmett's conservative approach to shielding calculations. Ralph Worsnop was on the staff of the physics department at the time
and had overseen the installation of the Siemens betatron but had moved to
California, hence the offer to me to come and help initiate electron beam
radiotherapy at MDAH.

There was only a handful of institutions world-wide beginning to look
at electron beam therapy for cancer and there were many questions to be
answered. Some of which were:

Could electron beams be used to treat cancer?

What cancers could be treated?

What energies should be used?

What was the radiation distribution in the body?

How much radiation was required?

How could that radiation be measured?

What was the optimal equipment to be used?

I would spend the next 50 years answering some of these questions.

There were several institutions in the US, and some in Germany, France
and Sweden looking into these questions at that time. We became a
fairly close-knit group exchanging information, ideas and the results of our
research, and getting together for meetings as often as possible. In 1967, I
published my first major paper on the use of electron beams for the treatment
of cancer which began answering the questions with which I had started out.

Dr Gilbert Fletcher, chairman of the Department of Radiation Oncology
had hired Dr Norah Tapley, a radiation oncologist, to head up the clinical development of electron beam therapy and I worked closely with her.
In 1976, we published: '*Clinical Applications of Electron Beam Therapy*', for
which I wrote the opening chapter on the physics of electron beam therapy.
This was done before the widespread use of computers and empirical relationships had to be determined, from experiments, on the effects that different
tissue densities had on dose distributions. It was one of the first, if not the
first, textbook on electron beam therapy and remained so for many years.

3.1   Think Before You Speak

The Siemens betatron had some interesting features. The scattering foils, to
spread and flatten the electron beams were mounted on a chain mechanism
and were positioned into the beam manually, by a knob. There were about
nine foils consisting of various thickness of aluminum, gold, or lead. The
selection of the foil depended upon the energy of the electron beam and
the field size which was determined by electron beam collimators of various

sizes, manually attached to the machine (Fig. 2). There were no interlocks for energy, scattering foil, and collimator and no automatic readout. The monitor chamber, which was used to control the length of time the beam is on, was a sealed cylindrical ionization chamber off to one side that monitored scattered radiation, which was different for each energy, scattering foil and field size. All of this information had to be manually entered into the patient's therapy chart for each day's treatment along with the treatment time. The radiation therapists, who treated the patients, were instructed to double check everything carefully and if anything unusual showed up, to contact me.

One week when Dr Tapley was out of town, I was called to the betatron on a Thursday. The therapists had noted that the treatment times for one or two of the patients were significantly longer than their previous times. The therapists assured me they had double checked the settings and had recorded

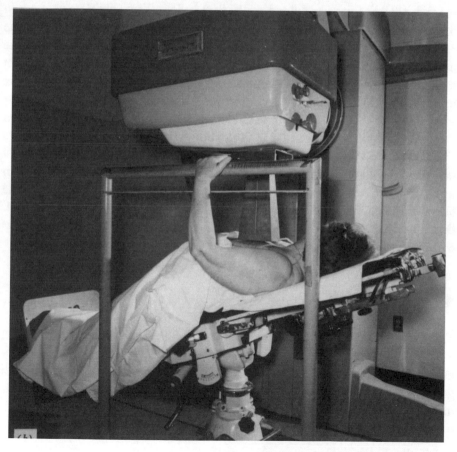

Fig. 2 The Siemens 18 MeV betatron. The top knob in the photo is the one used to move the electron beam scattering foils into place

them correctly. A check of the charts for all the patients treated that day indicated that it had happened for several patients but not all. I immediately took the machine out of clinical service. I brought in a film phantom, had the therapist set up the machine for the patient in question, and took an edge on film (film parallel to the direction of the beam). I did the same for a treatment that showed no significant change in the treatment time. The films told the story. For the latter case, the film showed the normal electron beam dose distribution. The former film showed all the dose concentrated in the middle of the field. It had not been scattered properly, and it took longer for the monitor chamber to accumulate the set number of units. When we took the cover off the machine, we immediately saw that some of the scattering foils were partially detached from the positioning chain. The foils were glued to the chain and I wondered if the high radiation environment had weakened the glue. It was not a serious clinical problem; we reattached the foils with the strongest glue I could find and made appropriate changes in the treatments of the patients involved. It wasn't much since it was only one or two treatment fractions in the course of many. By the end of Friday, the problem had been corrected and the necessary corrections to the treatments made. All that remained was for me to tell Dr. Tapley when she returned the following week. On Monday morning as I made my way to the treatment area, I saw Dr. Tapley ahead of me waiting for the basement elevator when her resident (not one of our brightest), showed up and said in a loud voice: "That was some kind of accident on your treatment machine last week!" I saw the color drain from her face, and I thought she was going to faint. I'm not sure what would have happened if I had not been there, but I was able to reassure her that it wasn't that serious, and it had all been taken care of. But it was a good reminder that we should not make outlandish statements without understanding or knowing all the facts.

4   Lost[1]

There were a handful of institutions in the US, Germany, France, and Sweden looking into electron beam therapy at that time. We became a fairly close-knit group exchanging information, ideas and the results of our research, and getting together for meetings as often as possible. One of those meetings was in Munich, Germany in the early 1970s. My wife, June, made the trip with me and went sightseeing while I worked!

---

[1] Adapted from my memoir, "Here, There, Everywhere."

The Siemens Medical group is headquartered in Erlangen which is only 117 miles (about 190 km) north of Munich and June and I were invited to visit the factory, to discuss my progress with their machine.

My contact person with the company was a man named Richard. He was a regular visitor in Houston, so I got to know him quite well. Richard made all the arrangements. After the meeting was over a Siemens Company chauffeur-driven limousine would come to our hotel in Munich, at 5 o'clock in the afternoon. We were to be packed, checked out of the hotel, and we would be taken to a hotel in Erlangen and then to dinner. Five o'clock came and no limousine, by 5:30 I was concerned enough that I called Richard to let him know. He advised waiting till 6 o'clock and to call him back if the limo had not arrived by then. By six, there was still no car, so Richard talked to the hotel manager and asked him to get a cab to take us to the hotel we were to stay at in Erlangen. When the cab arrived, we got in,

"Erlangen," we said to the driver, our German being very limited.

"Ja, Ja," he replied, his English more limited than our German.

Off we took. After half an hour we were out of Munich and driving through the surrounding countryside. We began to feel that something was wrong, especially when he stopped the car in the middle of nowhere and got a map out!

"Erlangen," we said.

"Ja, Ja," he replied.

We took off again; we could see an autobahn in the distance; but it soon disappeared, and we drove further into the countryside, through small villages. It was now pitch-black outside, and another hour had passed. In the next village, every place looked closed-up but the driver found a tavern that was open and he went inside; to get directions, we assumed.

"Erlangen," we again said.

"Ja, Ja,"

We drove for another hour, wandering through the dark countryside and we were now both very concerned. Again, the driver stopped in a dark and shuttered hamlet and banged on the door of a closed inn. This time I was ready. No one spoke English but I indicated I wanted to make a phone call and I was able to get Richard, who was now very concerned about us since we were now several hours overdue.

"Let me talk with the driver," Richard asked.

When we got back in the car the driver said,

"Erlangen?"

"Ja, Ja," we said.

We later learned the source of the problem. The driver was from Czechoslovakia and spoke very little German and did not know the area at all, but Richard was also from Czechoslovakia and they both spoke Czech and could communicate. The driver was completely lost! We finally arrived at the hotel in Erlangen sometime after 11 o'clock, tired and hungry. The hotel had instructions for us: I would be picked-up at 9 o'clock in the morning and taken to the factory. June would be picked-up at 11:30 to join me at Siemens headquarters for lunch. Just as we made it to the elevator to go to our room, the driver stopped me, he wanted to be paid. The bill was much more than the deutschmarks I had on me, but I was in no mood to argue with him and I refused, and I told the hotel manager to settle with him and to charge Siemens.

Next day started out well. After a good breakfast I was picked up and taken to the factory. At 11:30 AM Richard and I went to the lobby of the Siemens administrative building and we waited and waited. Once again, the Siemens driver had failed to show up. Poor Richard was beside himself, never in the history of the Siemens Company had their driver failed to show up, not once and now it was two times. The company officials were mortified, their belief in German punctuality, efficiency, carrying out orders, their pride and reputation were all on the line. Another car was sent to pick June up and we were taken to the executive offices of Siemens at the top of the building, to the president's private dining room. We had a delicious meal with the top executives and were presented with some fine gifts as peace offerings, including a very heavy pewter platter embossed with the Erlanger city crest and a beautiful beer stein etched with a view of the old buildings in town (Fig. 3).

5   Does Fermentation and Work Mix?

5.1   Fermentation of One Kind

Because of the increasing interest in electron beam therapy the International Commission on Radiation Units and Measurements (ICRU) formed a committee in the late 1960s, to write a report on electron beam radiation dosimetry and asked me to serve on it. For the privilege of doing so, I was requested to host a meeting of the committee in Houston, Texas. I chose a conference centre on Lake Conroe just north of Houston. The committee was comprised mainly of Europeans from various countries. To get

Fig. 3 I still have the gifts from Siemens, 50 years later but the betatrons have gone the way of the dinosaurs

the feel of Texas, I arranged for us to go to a typical Texas barbeque restaurant one evening. The best one I knew was a few miles away in the small East Texas town of Woodville situated at the entrance of the Big Thicket National Preserve, the name adequately describing the surrounding forest. The restaurant was typically Texan, and the food was excellent, but I had not realized that Woodville was in a 'dry' county and the sale of alcoholic beverages was prohibited, much to the distress of the European members. Halfway through the meal some of them decided to take matters into their own hands and approached the owner to see what could be done. After some discussion he agreed to contact the local bootlegger to provide suitable beverages. The bootlegger showed up a little later with the contraband; two ¾ litre brown ceramic bottles of Lancers' Portuguese vin rosé. There was an intense national campaign by Lancers in the USA at that time to promote this product and I'm sure the bootlegger had no problem obtaining some. However, my European friends' problems were not yet over; nowhere in the restaurant could a corkscrew be found. So followed a 15-min intensive discussion by a half-a-dozen internationally renowned physicists on how to extract the corks without a corkscrew. In the end when no viable suggestion was made, someone undertook to dig the corks out with a sharp knife trying to minimize the amount of cork that fell into the wine. Happily, the report was finished and was published by the ICRU in 1972 as: '*Radiation Dosimetry: Electrons with Initial Energies Between 1 and 50 MeV, ICRU Report 21.*'

5.2   Fermentation of Another Kind

Strangely that was not the only fermentation issue that faced an ICRU electron dosimetry report committee. ICRU Report 21 was essentially out of date by the time that it was published due to advances in linear accelerator technology, dosimetry calculations, increasing availability of computers, etc. and the ICRU constituted a new committee to prepare an update electron beam report under the chairmanship of Hans Svensson of Sweden, and many of the members from ICRU 21 were also on the committee. However, the ICRU became concerned about the length of time it was taking to finish the report. Hans Svensson is from Umeå in the north of Sweden and he arranged for the committee to meet at a conference centre inside the Arctic Circle at midsummer to finish the report, perhaps with the thought that with twenty-four hours of daylight, we would work longer hours. But at mid-night in midsummer, we did celebrate outside with laurels on our heads dancing around a maypole. The Swedish members insisted that we eat 'surströmming' which is Swedish for sour herring. The salted fish is canned and allowed to ferment in the cans to such an extent that the cans become distorted. The distinguishing aspect is the smell when the cans are opened. To my untrained ear the Swedish word 'sur' sounded like the English word 'sewer' which seemed like a good description of the smell. It is so bad that the manager of the facility would not allow the cans to be opened indoors. Fortunately, in this case a suitable can opener was available, and we all gathered in the parking lot, upwind, while one of the Swedish members gingerly opened the cans. I soon found out that surströmming is an acquired taste, proving once again that one man's fish is another man's poison or in this case surströmming. The report was finished and published by the ICRU in 1985 as: *Radiation Dosimetry: Electron Beams with Energies Between 1and 50 MeV, ICRU Report 35.*

6   How to Bring the Military–Industrial Complex to Its Knees

6.1   A New Machine

By the end of the 1960s, the clinical load at MDAH was increasing, and a new high energy treatment machine with photon and electron beam capabilities was needed. Clinical linear accelerators were becoming available, mostly with single low energy (10 MV or less) photon beams although there were some protype higher energy machines, but nothing seemed to meet our needs. Since no one at the institution knew anything about linear accelerators, Dr. Fletcher assigned Dr. Robert Lindberg, his second in command, and myself to work with consultants from Gulf Atomic in writing a set of specifications for a Request for Proposal (RFP) for a clinical linear accelerator to meet our specific needs. Gulf Atomic would prepare the specifications for the linear accelerator and would evaluate any bids. I was responsible for the radiological specifications, i.e., the beam specifications. We required it to be a dual modality machine, both x-rays and electrons, multi-energy with two x-ray and six electron beam energies. The highest x-ray energy had to match the depth dose of the Allis Chalmers' 22 MV beam. Field flatness, penumbra, field sizes, depth-dose characteristics, dose rates, monitor chamber precision, x-ray contamination in the electron beams, etc., were all specified. All state regulations regarding leakage radiation had to be met.

The RFP called for a Litton klystron to be used, the same as that in Defense Department's major radar installations; so that there would always be replacements available. (Klystrons are specialized vacuum tubes used to amplify high frequency radio waves which are needed to accelerate electrons in a high energy linac or radio waves as used in radar installations.) We required the gantry to have a rotation of just over 360^0 but not for rotation therapy purposes (Dr. Fletcher did not like rotational therapy), and the isocentre was specified at one meter from the x-ray target. The finished equipment was to be assembled in the factory in the United States and fully tested and the specifications verified before shipment to Houston.

Three proposals were received. Two, both from U.S. medical linear accelerator companies, were non-compliant with respect to many of the required specifications and were rejected. The third proposal was from a consortium between the French company Thomson CSF/CGR MeV and Raytheon in the United States. Thomson CSF had just built a high energy clinical linear accelerator (32 MeV electrons and 25 MV photons), called the Sagittaire, for a private hospital in Marseille, France, which met many of our specifications, but which used a French klystron. Their proposal called for CSF to build the accelerator sections, gantry and treatment head and Raytheon the microwave components and associated electrical components. Raytheon would also build a test facility at their factory in Waltham, Massachusetts where the machine would be assembled and tested before shipment to Houston. This proposal was accepted.

However, there were significant differences between the specifications of the Marseille's machine and our proposal which would require significant modifications, in particular to the treatment head. I was therefore dispatched to Paris, France for several weeks. I stayed in a small hotel in Paris and every morning drove down the Champs-Élysées and around the Arc de Triomphe to a chateau outside of Paris where the accelerator department of Thomson CSF was located. I met every day with Monsieur LeBoutett, the chief design engineer, who took my beam specifications and figured out what was required in the treatment head to meet them. We ended up with three carousels to rotate x-ray targets, x-ray flattening filters, light source, and the monitor ionization chamber in and out of the beam. Everything had to be precisely located each time a component was moved in or out of the beam, which was done with set screws and micro-switches. This was the age of electro-mechanical precision before the introduction of computers and micro-chips into accelerators. The treatment head also housed the primary collimators and motors, scanning magnets to flatten the electron beams, drive motor and the mechanism to rotate the collimator around the beam axis and cooling water. Assembling the treatment head required a very special order, rather like putting the various pieces of a toy wooden puzzle together to form a sphere. The last piece was the monitor ionization chamber. This was a highly specialized chamber with quadrants to check on symmetry and an annular segment to check on the penumbra for the x-ray beams. For electrons, because the beams were flattened (constant dose across the beam) by scanning, the different segments were electrically connected together to act as a simple plane parallel-plate chamber. Because it was so highly specialized, a replacement chamber would take a long time to fabricate and be expensive. We therefore ordered two with the machine, one to be installed, the other as a backup. The treatment head was assembled in France and shipped to the United States as a completed unit, except for the monitor chamber which was carefully packaged and sent separately to avoid any chance of damage in transit.

In due course, the French components were sent to Raytheon and married to the rest of the machine. I went to Waltham to conduct the acceptance tests and arrived at the same time as the monitor chamber. I was present to supervise its installation. This was not an easy operation; a very specific sequence had to be followed, the set screws set just right and the chamber angled to get it in place and then gently eased into position and screws tightened to hold it securely. I watched as the technician made several attempts to do this with my heart in my mouth scared that at any moment, he would puncture the very thin foils and ruin the chamber. I immediately knew what the problem was; he had not put the set screws at the right height, and they were preventing the chamber from going into position. Finally, my patience ran out and I picked up an Allen wrench to adjust the set screws and told him I would show him how it should be done. Suddenly the room went very quiet. I was ushered into the next room and a series of people went in and out of the test cell and gathered in small groups for heated discussions. Finally, I was told the problem. The factory was a union shop, and the technician was a union activist. I was not a member of the union and, therefore, could not pick up any tool to do any adjustments. He had called the shop steward who immediately threatened to call everyone out on strike and close the whole factory down! Once again proving the adage that the pen, or in this case an Allen wrench, is mightier than the sword. It took the whole morning to resolve the problem, and I was banned from the test cell until the machine was ready for the acceptance tests.

6.2   Use It or Lose It

The Sagittaire required three separate rooms. The microwave, klystron, and accelerator sections were in one long narrow room. The beam passed through a rotating vacuum seal into the treatment room which housed the rotating gantry, the treatment head and the treatment couch (Fig. 4). Outside of the treatment room was a small control room. Since there was no available space within the footprint of the existing basement, additional space had to be excavated next to the building. There was no freight elevator down to the basement; so, the extension was designed with a removeable top of concrete blocks through which the various parts of the machine were delivered before being sealed and back filled. I carried out the shielding design for this extension. When the original basement had been built approximately 20 years before, Grimmett had designed the shielding using heavy concrete. He had used ilmenite ore as the aggregate in the concrete to increase its density. Ilmenite is a heavy iron rich ore from Canada. For reasons now completely forgotten, far too much ore was ordered and the Institution soon found out there was no market for ilmenite in Texas; it was too expensive to abandon and so it had to be stored. But where to store it? At that time MDAH had its own surface parking lots. In the front of the building was patient parking; in

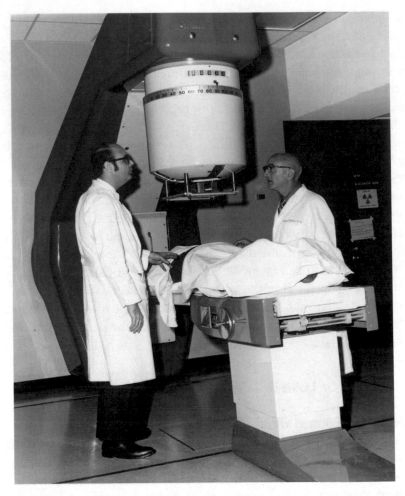

Fig. 4 Drs. Lindberg and Fletcher standing on the moveable portion of the floor. Note the treatment couch mounted in the reverse position with the top portion rotated 180 degrees and cantilevered over the moveable floor. This photo was taken from just inside the entrance

the back was staff parking. This was not covered parking, except for a dozen or so parking places next to the staff entrance over which a car port had been constructed. These parking places were reserved for the top administrators. When I arrived at MDAH in 1959, I noticed that half the car port was taken up with sacks of something stacked from floor to ceiling, but I had no idea what was in them. I was to find out. Ten years or so later as I was working on the shielding calculations for the Sagittaire extension, word came down from on high (the administration) that I had better find ways to use every last sack of ilmenite stored under the carport or my future at MDAH was in jeopardy.

I used it all and those walls are so dense that they will never come out, except with dynamite.

6.3 A-Maze-Less

I did have another, more serious shielding problem. Because of the way the excavated space was connected to the existing basement, there was no possibility of incorporating a maze as an entrance to the Sagittaire treatment room. The corridor along which the patients came went straight into the treatment room. My only option was to design a heavy sliding shielded door to close the entrance. The energy of the highest photon beam was 25 MV. The leakage and scattered photon radiation were not so much a problem, the room was large, and the entrance was at the opposite end to the gantry and the inverse square law really helped. (The inverse square law indicates that the dose reduces with the square of the distance from the source.) A lead lined door of sufficient thickness to meet State regulations was quite feasible; my concern was for the neutron leakage. At 25 MV this was not insignificant and there was a direct line-of-sight from the door to the target area where most of the neutrons were produced. Lead is not a good shielding material for neutrons and in the end, I used borated polyethylene. The hydrogen in the polyethylene was to slow the neutrons down and the boron to capture the slow neutrons. The door was a laminate of layers of the borated polyethylene followed by a layer of lead, clad in stainless steel. The door was suspended from a girder and driven by a geared down motor and chain. In case of power failure, there was a manual crank. The motor would open the door in about 30 s; the manual crank took several minutes. The gap at the bottom of the door was as small as possible, a few millimeters, since no lip could be put on the floor. I was concerned about leakage under the door but no discernable radiation, either neutrons or photons could be detected.

6.4 Drop It

The other design challenge was the height of the gantry. When everything was considered, including the height of the isocentre above the floor, the size of the treatment head, the length of the detachable electron beam trimmers, and the design of the bending magnets, the gantry could not be fully rotated without colliding with the floor. The solution to this problem, was to make a large segment of the floor under the gantry moveable. It could be lowered a couple of meters to allow the gantry to rotate.

6.5   Reverse It

One solution often creates another problem and in this case it was the treatment couch. The specifications called for the Sagittaire to be used with an Atomic Energy of Canada Limited's treatment table. With the moveable floor this could not be installed with its axis of rotation coincident with the axis of rotation of the collimators. This was not of undue concern since Dr. Fletcher did not employ isocentric (source-axis distance, SAD) but fixed source-to-surface distance (SSD) set-up treatments. We mounted the table on the fixed floor backwards, with the base aligned with the edge of the moveable floor. The tabletop was then rotated 180° so that it was cantilevered over the moveable floor (Fig. 4). It worked well and Dr. Fletcher was able to continue the four-field box technique but now with the higher dose rate and the ease of set-up, it took considerably less time per patient. Once again, necessity was the mother of invention.

6.6   Publish It

One other problem had to be solved that was due to the scanned electron beams. We needed to develop the theory to derive the correction to the ion chamber readings to account for ion recombination during calibration. When ions recombine before they are measured by the electrodes in the ion chamber, then the reading underestimates the real dose. The required correction for this was well known for continuous beams such as cobalt-60 gamma rays and for pulsed beams as produced by betatrons and linear accelerators; however, the Sagittaire electrons had a pulsed, scanned beam. I assigned the problem to my colleague, Ken Hogstrom. A few days later he showed me the solution to this problem, and we agreed he should publish it. But we were very busy and before we could write the paper, Professor Boag published a paper with the same solution that Ken had derived. He who hesitates is lost, I suppose. Fifty years later, Ken and I still talk about the paper that got away. Not that it made any difference to Ken's illustrative career.

7   When It's All Said and Done

In the Radiology Centennial, Inc.'s 1996 "*History of the Radiological Sciences: Radiation Physics*" I wrote: "It would be unwise…to try and predict what radiology, both in imaging and therapy, will be like in another one hundred years. The techniques and methods at that time are sure to be as different and as unpredictable as the procedures today would have been for the physicists and medical personnel who first used X- and gamma-rays." We are now a fourth of the way through the next century and it is clear that that prediction is being fulfilled. My hope and prayer is that the young people now entering the profession of medical physics will find it as rewarding, exciting and enjoyable as I have been privileged to experience.

Dr. Peter R. Almond obtained his undergraduate degree in honours physics from Nottingham University and his medical physics training from Bristol University in the UK. He moved to the United States in 1959 and earned his master's and doctoral degrees in nuclear physics form Rice University, Houston. From 1964 to 1985, he worked at The University of Texas MD Anderson Cancer Center (MDACC) in Houston, where he served as Head of the Radiation Physics section, Director of the Cyclotron Unit and Professor of Biophysics. From 1985 to 1998, he was Vice-Chairman of the Department of Radiation Oncology, University of Louisville, KY. In 1999, he returned to MDACC as a Distinguished Senior Lecturer. Dr. Almond participated in developing cancer treatment procedures and dose measurement techniques with various forms of radiation. He served on numerous national and international committees and councils and supervised over 25 masters and doctoral students. He authored/co-authored over 100 scientific articles and numerous chapters in radiotherapy textbooks. He served as President of the American Association of Physicists in Medicine (AAPM), as Chairman of the Board of Chancellors of the American College of Medical Physics (ACMP) and he is a Fellow of those organizations as well as the American College of Radiology and the Institute of Physics.

Awards

2017. Randall S. Caswell Award, Council of Ionizing Radiation Measurements and Standards for distinguished achievements in the field of ionizing radiation measurements and standards.

2013. Selected by the International Organization for Medical Physics (IOMP) as one out of 50 medical physicists "who have made an outstanding contribution to the advancement of medical physics over the last 50 years." This recognition was given as part of IOMP's 50th anniversary.

1997. Lecturer, VI Ramaiah Naidu Memorial Oration, The Association of Medical Physicists of India. The recipient is a person of national/international repute in the field of medical physics.

1990. Marvin D. Williams Award, American College of Medical Physics/American Association of Physicists in Medicine (AAPM) recognizing an AAPM member for an eminent career in medical physics with an emphasis on clinical medical physics.

1990. William D. Coolidge Award, American Association of Physicists in Medicine (AAPM). This is the highest award given by the AAPM and recognizes an eminent career in Medical Physics.

4

Looking Back at 50 Years in Medical Physics—The *Ugly*, the *Bad* and the *Good*

Gary T. Barnes

1 Education and Training

The cold war was part of life growing up in the 1950s. In grade school we practiced getting under our desks in the event of a nuclear attack, an exercise that younger generations born after the fall of the Berlin Wall can't envision. I finished high school in 1960. After Russia launched Sputnik 1 in October 1957, there was increased emphasis in the US on science and engineering careers. The emphasis was further intensified by John F. Kennedy's (JFK) 1961 Space Program. These national priorities helped shape my career. I went to an engineering school for college, majored in physics, went on to graduate school in physics and matriculated with a Ph.D. in December 1970.

There were mostly *Good* experiences in graduate school. *Good* was my thesis advisor and mentor, Daniel Gustafson. He unfortunately passed too young. He died in 1980 while rescuing his daughter and her friend from drowning. His book "Physics: Health and the Human Body" has been helpful in my teaching. In working for Dan, I learned patience and persistence. I remember a particularly frustrating incident with a young student. He advised don't react when angry. Wait and think of the implications of your response first. A *Good* take home point I have utilized in my career.

G. T. Barnes (✉)
University of Alabama Birmingham, Birmingham, AL, USA
e-mail: gtbarnes@radphysicsinc.com

© The Author(s), under exclusive license to Springer Nature
Switzerland AG 2022
J. Van Dyk (ed.), *True Tales of Medical Physics*,
https://doi.org/10.1007/978-3-030-91724-1_4

I had a *Bad* experience with the first quantum mechanics course. The University was on the quarter system and I found the first quarter challenging taking graduate level classical mechanics, mathematical methods in physics, and quantum mechanics. The first quarter ended before Christmas and the final course exams were given before the Christmas break. We did not see the graded exams until after Christmas and the start of the next quarter. I received a B in quantum mechanics. On my exam, a question was graded wrong and I thought I had solved it correctly. This question graded wrong was the reason I received a B rather than an A. I met with the quantum mechanics professor. *Good* was he met with me. *Good* was after looking at my solution, he agreed it was correct, although different than his approach. *Bad* was he had turned in the course grades to the registrar and didn't want to take the trouble to change my grade to an A.

Receiving a B rather than an A in quantum mechanics mattered to me. *Good* was the influence this had on my career. In my half century of teaching radiology residents and technologists, I go over graded exams the next day and use it as a teaching experience. When I have miss-graded a question, I correct the exam grade. To this day, exams given by the University of Alabama (UAB) Radiology Physics Division Faculty are reviewed with the class the next day.

Ugly was looking for a job in 1971. In July 1969, Neil Armstrong and Buzz Aldrin walked on the moon. The US Space Program had reached the goal John F Kennedy (JFK) set. In the summer and fall of 1970, NASA (National Aeronautics and Space Administration) laid off many thousands of scientists and engineers. The US aerospace industry experienced a major downturn, and the nation as a whole went into a recession. Even a number of the original members of the Peenemtinde V-2 Rocket Team that moved to Huntsville, Alabama with Dr. Wernher von Braun in 1945 were demoted or lost their jobs. The number of scientists and engineers employed in the aerospace industry dwindled from 235,000 in 1968 to 145,000 four years later. If you had an opening for a physicist, would you hire someone green, just out of school, or would you hire an individual with experience and a good track record. *Ugly* was looking for a job after four years of college and six years of graduate school when there were no jobs.

Looking for a job in 1971 was *Ugly*; however, *Good* in that it made me assess my skills and think about what I would like to do and accomplish in my career. Up to that point I had gone with the flow assuming there would be good jobs for scientists and engineers and had not given a great deal of thought to my career. It also made me realize that government priorities are fickle. A career that did not depend on government funding would be more stable than one that did. I had experience working in construction, being a janitor, working at McDonalds, and working in the steel industry on the open hearth and in the hot strip mill. Savings from the latter provided me the money to start graduate school. I could find work, but I wanted a career where I could utilize my education.

A *Good* in graduate school was that I became aware of medical physics—a fellow student, Lloyd Smith, was a resident in medical physics. Prior to 1971, Lloyd had finished his residency and was currently working in a hospital. I spent time shadowing him and realized that I could do the work. A career in medical physics would utilize my education, required a broad range of skills, and was better than building bombs. Additionally, although my thesis research involved working with ionizing radiation and nuclear instrumentation, I needed practical training. In 1971, this could be obtained by either being hired in a junior position and getting trained by a senior medical physicist or being hired in a designated training position at a larger medical centre.

I was unable to locate a junior position and contacted a number of medical centres that had medical physics groups—MD Anderson Hospital in Houston, Texas, Memorial Sloan-Kettering Cancer Center in New York, University of Toronto, and the University of Wisconsin in Madison, Wisconsin—that might provide practical training. There were few medical physics postdoctoral training positions available. I wrote and subsequently called department heads—Drs. Robert Shalek, John Laughlin, Harold Johns and John Cameron. Of note is the courtesy with which these leaders in the field responded to my letters and phone inquiries. If they had a position for 1971–72, I often made their short list, but they indicated politely that I was not at the top of the list. Their politeness made a *Good* impression—a consideration I have tried to emulate in my responses to young people interested in medical physics.

In the early spring of 1971, I received a response from the University of Wisconsin. They had received a US National Institutes of Health Grant to provide medical physics training for recent physics Ph.D.'s. The grant provided funding for three postdoctoral training positions and I was offered one. The same week, after having no positive responses to my job search for months, I received two other offers—to work in University Administration where I had been working part time, and a physics post-doctoral research position. Although the position at the University of Wisconsin paid considerably less than the other offers, and less than I received as a graduate student, I accepted the offer. Other than proposing to my wife, it was the best decision I ever made.

In 1971–1972 the medical physics postdoctoral training at the University of Wisconsin consisted of coursework and four three-month clinical rotations: one rotation in nuclear medicine, one in diagnostic radiology, and two in radiation therapy. The coursework consisted of all the medical physics graduate courses. The mentoring in radiation therapy and nuclear medicine was *Good*. There was no medical physics mentor in diagnostic radiology and the clinical rotation consisted of observing diagnostic exams, workflow, and attending conferences. Trainees also interacted with the graduate students and learned about the ongoing research through these interactions and research seminars. The coursework, although not at the mathematical level of physics graduate level quantum mechanics or electricity and magnetism courses, provided a breath of basic information. Overall, the year was *Good*, very *Good*. It provided me with the experience and confidence to look for a job in medical physics. It also etched in my mind the value of on-the-job training in the discipline.

In 1972, radiation therapy was the major focus of medical physics. The leaders in the field had laid a solid foundation, and my take was that the possibility of making a significant contribution in that specialty was small. In looking for a job, I focused on diagnostic imaging. My rationale was twofold: (1) I thought it was better to detect cancer early than treat it late; and (2) there was a greater possibility of impacting and improving health care.

2   A Job and Its Responsibilities

I interviewed for and was offered a tenure track assistant professor position in the Department of Diagnostic Radiology at the University of Alabama Birmingham (UAB). I joined the Department in September 1972 and was the second medical physicist in the State. The first preceded me by several years and his responsibilities were principally in radiation therapy. In 1970, the Radiology Department at UAB was restructured into the Departments of Radiation Therapy and Diagnostic Radiology. David Witten became the Chair of the Diagnostic Radiology Department at UAB in 1971. Prior to becoming Chair, Dr. Witten was on the staff and received his training at the Mayo Clinic, in Rochester, Minnesota, and in radiological physics by Marvin Williams. A *Good* of his experiences with Dr. Williams was the requirement he gave the UAB School of Medicine when he was recruited—a faculty position in the Department for a medical physicist.

In his letter of offer, Dr. Witten outlined my responsibilities—support the Department's patient care, educational and research missions. When I joined the Department, I had little or no feel for how to support its patient care mission. A *Good* was Dr. Witten's feel for the technical support needed. He facilitated my meeting a number of the major US screen-film system manufacturers' technical people and visiting their laboratories and production facilities. Screen-film systems consist of an intensifying screen that gives off light when irradiated by x-rays. The film is very sensitive to both x-ray photons and light photons but much more sensitive to light photons. In talking with these scientists, I developed a much higher level of appreciation of the science underlying screen-film systems, the engineering that went into their manufacture, and the quality control employed in the manufacturing process.

3   Clinical Support

Prior to my joining the Department, Dr. Witten had created a radiological technologist quality control position and initiated a quality control program. The Department also had an in-house service engineering group and did not rely on vendor service. *Good* was working with and learning from these people. They greatly helped in filling the many holes in my practical knowledge base.

For most of the twentieth century, x-ray images were detected with screen-film systems. The systems consisted of a film sandwiched between front and back of x-ray phosphor screens in a light-tight cassette. The highly x-ray absorbing screens converted the absorbed x-ray energy to light, intensifying the photographic effect on the film and resulting in an invisible latent image in the film. After an x-ray exposure, the film was removed from the cassette in a dark room and placed in wet chemistry processor which converted the latent film image to a visible film image. Important to obtaining good images was exposing the screen-film system to the correct amount of X-radiation and properly processing the exposed film.

When I joined the Department, a film processor quality control program had already been established. This was not being done at the University of Wisconsin when I was there. In attending meetings of the American Association of Physicists in Medicine (AAPM), the following and subsequent years, I heard presentations indicating daily film processor quality control was the best thing since sliced bread. It became an American College of Radiology (ACR) Mammography Accreditation Program (MAP) requirement and subsequently a US Mammography Quality Standards Act (MQSA) requirement.

Another objective of the Department's quality control program was to properly expose the film and achieve a consistent scale of contrast for the anatomy being examined, independent of the x-ray exam room and the technologist operating the x-ray unit. This depended on selecting the proper technical x-ray exposure factors for the exam and patient habitus. As the radiation output of the different rooms was not always the same, I assisted the quality control technologist and in-house service group in matching radiation output and developing technique charts for the rooms. Given the exam and patient habitus, the chart provided the technologist with the manual technique settings to expose the film properly. Later when correct x-ray screen-film exposures were achieved with automatic exposure control (AEC) systems, I assisted in matching the AEC performance of different rooms. The results of these efforts were the consistent obtainment of high-quality images and reduced exposure repeats.

In doing annual x-ray imaging unit radiation surveys and performance audits, I realized that many of the problems identified were due to the equipment not being properly installed; hence, we incorporated acceptance testing of new x-ray equipment installations in our program. Twenty percent of the purchase cost of new equipment installations was withheld until the problems identified during acceptance testing were corrected. It wasn't long before vendors knew we were going to acceptance test their installations and, not wanting to be embarrassed, did a better job installing the equipment. A result of these efforts was my being invited to give the presentation "Quality Assurance—Current Trends in the US" at the Annual Meeting of the Japanese Radiological Society in the spring of 1983 (Fig. 1) and being a co-director of the 1991 AAPM Summer School entitled "Specification, Acceptance Testing and Quality Control of Diagnostic X-Ray Imaging Equipment". Later we collaborated with the Mayo Clinic on publishing a paper in the journal *Radiology* on the importance of acceptance testing and documenting the UAB and Mayo Clinic joint experience on the problems found.

In 1976, I became a diplomate of the American Board of Radiology. To do this, I took a three-hour written exam in the morning and a three-hour written exam in the afternoon. Having passed these, I took an oral exam a few months later consisting of five 25-min exams by medical physicists and one 25-min exam by a radiologist. I had no major problems with the medical physics examiners. Such was not the case with the radiologist. The first question he asked me was about the blood–brain barrier. I was out of my element and he had to lead me to the answer, which took several minutes and had me sweating. I thought I was going down the tubes. The next question was "What could I do for a radiology department?" I spent the next twenty

Fig. 1 Certificate of appreciation for the presentation. "Quality Assurance: Current Trends in the US" given at the 39th Annual Meeting of the Japanese Society of Radiological Technology, Osaka, Japan, April 1983

minutes talking about wet chemistry processor quality control, proper screen-film exposure, techniques for different exams, and acceptance testing. I kept talking and went into details when I sensed he was about to ask me another question. Fortunately, I succeeded, and he didn't ask me another question. I went home with some degree of trepidation and was glad to learn the next week that I passed the clinical as well as the medical physics oral exams.

In 1976, Dr. Robert Fraser, a world-renowned chest radiologist, joined the Department. Associated with his recruitment was the Department's commitment to improve chest radiography. I was directed to assist in facilitating this commitment. Utilizing my knowledge and experience with screen-film systems, I helped Dr. Fraser select intensifying screens that had good x-ray absorption efficiency and sharpness.

Film manufacturers offered two options—a high contrast film and a latitude film. The high contrast film resulted in images with greater gray scale differences for given anatomical x-ray attenuation variations. The latitude film resulted in smaller gray scale differences but had the advantage of imaging a larger range of attenuation variations. In chest radiography, the attenuation differences between the lung and mediastinum (middle of the thoracic cavity) is large and a latitude (lower contrast) film was routinely employed to visualize both the lung and mediastinum. If a high contrast film was employed one could not adequately visualize both the lung and mediastinum.

To improve the visibility of the lung markings as well as the mediastinum, I constructed a filter that spatially modulated the x-ray intensity across the image with greater intensity in the centre of the image and less on the left and right. The filter allowed the use a high contrast film (Fig. 2). The filter achieved analog image processing—it suppressed the low spatial frequency contrast between the lung and mediastinum and passed the higher spatial frequency lung markings. (Spatial frequency is a mathematical description describing a characteristic of any structure that is periodic across position in space.) Of practical importance was that the spatial variation was gradual, and one filter fit all patients, both posterior-anteriorly and laterally. Gratifying to me was that a couple of years later, in 1980, one of my mentors, Murray Cleare, Head of Eastman Kodak's Radiography Research Laboratory, visited Birmingham to demonstrate a new film Kodak had formulated for chest radiography. After looking at our chest radiographs, he declined to demonstrate the new film and returned to Rochester, NY.

When I joined the Department, mammography was done with industrial film, and the film was hand processed with wet chemistry. A year later we had xerography. In 1978, we had dedicated screen-film mammography. The

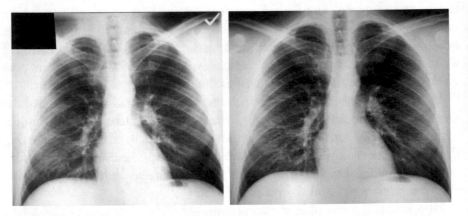

Fig. 2 Conventional 120 kVp (peak kilovoltage) posterior-anterior chest screen-film radiograph (left) and optimized 145 kVp radiograph of same patient (right). The image on the left was acquired with a latitude film. The image on the right was acquired with a high contrast film and analog image processing achieved by a beam shaping filter that increased the radiation intensity incident in the centre of the image

unit did not have automatic exposure control (AEC) and I set up manual techniques for the technologists based on measured breast thickness.

In 1982, I was asked to check out a new mammography unit at a nearby hospital. The unit was state-of-the-art with AEC. I was impressed with its elegant appearance and ergonomics. My testing indicated the unit met specifications and passed all tests with one exception—the film densities that resulted with the AEC that were good for a four cm breast phantom thickness, were too dark for a two cm phantom thickness, and too light for an eight cm thickness. I told the radiology director that he should not pay for the unit until the problem was fixed. A local firm represented the manufacturer, and the firm's salesman was not happy. He contacted the manufacturer regarding the problem, and they indicated that no one else had complained about the problem. (I have heard this from vendors many times through the years.) He then checked with the radiologists reading the mammograms. He informed me that they told him the images were good. I told him that the technologists only gave the radiologists good images and repeated the exam if the films were too light or too dark. He checked with the technologists and found that this was indeed the case—they repeated films that were initially too dark or too light. There was not an available fix for the AEC and the radiology administrator authorized payment.

The unit had a plus/minus density control. The AEC was set up so that an appropriate film darkening would result for a four cm breast and +0 density control setting. I advised the technologists to select the −1 density control setting for thinner breasts, +0 for a typical breast thickness, +1 for thicker breasts and +2 for very thick breasts. This reduced the number of repeats. I made measurements subsequently on several different mammography units and found that the AEC film density tracking with breast thickness problem was common to all the units that I checked.

X-ray imaging assumes that the x-ray beam incident on the patient originates from a point source; however, in practice they originate from a small area in the x-ray tube commonly referred to as the focal spot. Generally, the smaller the focal spot, the sharper the image; and checking x-ray tube focal spot size is an important quality control check. In 1984, we purchased a mammography unit from the same manufacturer that I tested in 1982. It was a newer model with small focal spots. Loren Niklason had recently joined the Department as a medical physics trainee, and I had him acceptance test the unit. In testing the unit, Loren found that the small focal spot failed. I contacted the salesman, the same individual that sold the earlier unit that I told the radiology director not to pay for until the AEC problem was corrected. As in our earlier experience he was not happy, but to his credit earned his commission. (These trials and tribulations resulted in a great deal of mutual respect and we became friends.) He contacted the manufacturer and we met with the x-ray tube's designer, Emile Gabay, at the Radiological Society of North America (RSNA) meeting. He educated me and pointed out we did not measure the focal spot in accordance with International Electrotechnical Commission's (IEC) recommendations. A tube engineer came and measured the focal spot; it failed, and the tube was replaced. The focal spots on the new tube passed muster. Emile's knowledge of x-ray tubes was impressive. I and the other Directors of the 1991 AAPM Summer School invited him to give a talk on diagnostic x-ray tubes. His chapter in the Summer School Proceedings is most excellent and well worth reading.

To compensate for the unit's AEC limitations and accommodate the high contrast mammography film we were using, we reduced the radiation change associated with a density step and set up a technique chart that would vary the density control ± settings based on measured breast thickness and achieved consistent film densities independent of breast thickness. We published a Technical Note in *Radiology* on our phototimer (AEC) technique chart. Robert LaFrançe and Dale Gelskey of DISC Corporation (St. Malo, Manitoba, Canada) were also aware of the problem and read our paper. Robert had developed an adjustable non-linear integration circuit that, when implemented properly, corrected the problem. He contacted me and we collaborated in identifying the physical factors causing the problem. We presented our results at the 1987 RSNA meeting and later published a paper in *Radiology*. When we gave our presentation, the room was full. After the presentation, half the people in the room left. Of clinical importance is that by 1990, the majority of mammography unit manufacturers had corrected the problem.

In 1988, I was contacted by former radiology residents who indicated their mammography image quality was subpar and wanted me to check it out. They had the same unit and were using the same screen-film combination we were using at UAB. In my report, I presented my measurements and recommendations. Their breast radiation doses were more than a factor of two greater than UAB's and I noted in my report that the cause of the high doses and poor image quality was film processing (to save money the radiology administrator had reduced the developer temperature and the chemical replenishment rate).

A few weeks later I was called by one of the radiologists and was told that their mammograms were terrible and often the AEC backup time terminated the exposure even when they increased the kilovoltage (kVp) to obtained darker films. I asked if he had read my report. He indicated that he had looked at it. I asked if he had read my recommendations. He had not and had looked only at the first few pages of my report. My measurement results comprised the first several pages of the report and my recommendations were at the end. My measurements comprising the first several pages were important to me and of little or no interest to radiologists or the radiology administrator. Putting the recommendations at the end of my report was *Bad*. I had not properly communicated my findings. *Good* was in all subsequent reports my findings and recommendations are at the beginning of the report.

Mammography was highly variable in the 1980s as was the supporting medical physics practice. In the Spring of 1990, Donald Frey and I collaborated in organizing a SEAAPM (Southeastern Chapter, AAPM) Symposium on "Screen-Film Mammography Imaging Considerations and Medical Physics Responsibilities". We edited and published the proceedings. Don and I contributed the chapter on medical physics report documentation. We stressed the importance of summarizing the findings and communicating one's recommendations in the report's cover letter. The American College of Radiology (ACR) Mammography Accreditation Program (MAP) when started in mid-1987 did not require medical physics testing or report documentation. Processor quality control (QC) documentation was also not initially required. Later ACR MAP required documentation of processor QC and of medical physics support and reports. Required at the beginning of the reports was a formalized pass/fail summary of the medical physics tests performed and the recommendations (Fig. 3).

In 1990, I assisted a former radiology resident in setting up a mammography facility in Northern Alabama. This initially involved exam room configuration and radiation shielding recommendations, and later mammography unit, screen-film, processor and chemistry selection. I reviewed and gave my advice on the wet chemistry processor and mammography unit quotes. I had good experience with the mammography unit the radiologist preferred. The pricing on the quote was reasonable and included a six-month warranty. My recommendation was that one has leverage when spending $75,000 and to indicate to the salesman, she would buy the unit at the quoted price providing the unit had a full parts and labour warranty for a year. I acceptance tested the unit, processor and chemistry, and screen-film cassettes as well as confirming that the breast doses were consistent with good radiological practice. A start up practice such as hers has substantial debt. The practice increased at a rate faster than her business plan projected necessary to pay down the debt and was doing well. However, in the 10th month, the x-ray tube failed and was replaced without cost to the practice other than

Fig. 3 1990 SEAAPM symposium proceedings cover

the small amount I charged to check out the new tube. If the replacement of the x-ray tube had not been covered by the warranty, the cost the practice would have incurred exceeded $10,000. She thanked me profusely for my advice. For the next 20 years and until she retired, I provided medical physics support to her practice.

Another interesting experience was with a focal spot failure on a mammography unit found during an annual audit. The unit was a couple of years old. I did not acceptance test this unit. It was the first time I checked the unit and didn't know if the problem existed when the unit was new. The site did not have a service contract that included glassware (x-ray tubes) and was loath to spend $15,000 on a new tube. The manufacturer's serviceman checked the focal spot and it passed factory specifications. When I find a failure, I double check my observation and did so in this case—you do not want a site spending $15,000 based on an erroneous observation you made. When informed that the manufacturer's service engineer found that the small focal spot passed, my thought was he did not know what he was doing. I met with him on a Saturday. He made his focal spot measurements and I made mine. He knew what he was doing; his results were the same as mine and the focal spot passed. I was embarrassed and confused. I had double checked my measurements earlier before failing the small focal spot. After giving the problem some thought, I realized that the only difference between my earlier measurement, and the serviceman's and my subsequent measurements was the state of the x-ray tube. I had checked the x-ray tube when it had been warm. The serviceman's and my subsequent checks were made when the tube was cold.

Subsequently, measurements were made demonstrating the small focal spot increased (degraded) as the temperature of the tube increased and failed after less than 30 min of heating associated with a busy clinical workload. The focal spot passed when the tube was cold but failed when it was hot. The site manager was not happy at paying $15,000 to replace the tube. Some months later, I was told by the manufacturer's mammography physicist that their tube factory had been aware of the problem for some time. I wondered how many of that manufacturer's mammography units were out there with focal spots that were only checked when the tube was cold and would fail if warmed up. Since then, I always warm up an x-ray tube before checking its focal spots.

An area where consistent and good image quality was difficult to achieve in the 1970s and 1980s and earlier was bedside radiography. The majority of bedside exams are done for patients in the intensive care unit (ICU). The patient habitus in this patient population is highly variable. Since employing AEC was not practical and screen-film systems had limited dynamic range, over and under exposures were commonplace resulting in poor images and high repeat rates. Fuji introduced Computed Radiography (CR) at the 1981 International Congress of Radiology. CR's digital image processing, unlike screen-film systems, delivered a constant gray scale and contrast over a wide range of exposures. The initial CR products marketed in the mid-1970s were bulky and expensive. In 1989, Fuji introduced the AC-1, a relatively compact CR system for $250,000—two CR units could be purchased for less than the price of one earlier unit. Realizing the benefits of CR, I submitted a proposal to the Department leadership and later to hospital administration to convert bedside radiography from screen-film to CR. Included in the proposal was the construction of an intensive care unit (ICU) mini-PACS (picture archiving and communication system). Hospital administration agreed to fund the project. The key aspects of the proposal included redundancy of key components (two AC-1s and dual image processing computers and file servers to store the images), sending of images to the ICU where the CR plate was exposed, and easy-to-use ICU displays. Douglas Tucker was hired to do the heavy lifting and did an outstanding job. When many medical centres spent millions on PACS systems with no net benefit, UAB Hospital spent less than a million dollars successfully converting bedside radiography from screen-film to CR and developing a functioning clinical mini-PACS for eight ICUs.

There were a number of noteworthy accomplishments: (1) Using the ICU displays required little or no learning; (2) Fuji's image processing was recreated as only raw unprocessed images could be obtained from the CR units; and (3) Fuzzy Logic and a neuro-network were used to rotated images acquired crossways to upright. (Fuzzy Logic attempts to solve problems with an open set of data that makes it possible to obtain an array of accurate conclusions, thereby making the best possible decision given the input. A neuro-network is a type of machine learning which models itself after the human brain.) The ICU displays had two monitors, a couple of control buttons, a track ball and no keyboard. When turned on or reset, the current list of ICU's patients was displayed. One would then select a patient with the track ball and his/her most recent image would be displayed on the left monitor and the previous image on the right. The window and level (these are adjustments on the images to give the best gray-scale display for the anatomy under consideration) of the images could be adjusted but this capability was rarely used. The images came up processed with good lung window and level settings and there was a button that changed the settings optimized for the lung to settings optimized for the mediastinum.

4   Teaching

Initially my teaching responsibilities were heavy and involved courses for medical residents in diagnostic radiology, and students in radiologic technology (RT) and in nuclear medicine technology. There were two medical radiology resident courses a year—a comprehensive course for the first year residents and review lectures for senior residents who would shortly be taking the American Board of Radiology (ABR) credentialing written exam. The ABR exam included a significant physics component. Initially the radiologic and nuclear medicine technology programs were hospital based. Shortly before my joining UAB, the programs were consolidated in the School of Allied Health. The RT program had two classes a year, one starting in September and the second in January. The majority of students entering the program had finished high school the previous June. I had no experience in teaching rudimentary physics as well as eighth grade math, as I was required to do, and did not do a good job the first year. To my advantage the Chair of the UAB Physics Department taught the previous year. He berated and embarrassed students in class. He was so bad that, as bad as I was, I looked *Good* to the program director and to the students who had heard the previous year's horror stories.

Teaching the radiology residents was easier than teaching students just out of high school. However, here also I had no mentoring and did not do a good job in the beginning. I was too enamored with mathematics and details. I remember in my first month that after giving a lecture on the noise propagation in an image intensifier-TV imaging chain, my wife and I attended a Department social function. A wife of one the first-year residents pointedly told my wife that her husband came home after one of my classes and threw the physics books across the room. I took the hint and spent less time on the mathematics and more time on basic concepts. A year or two later, I was discussing x-ray interactions and one of the residents asked me why we were studying this material. I realized that I had not properly tied together the different subject areas covered in the course. I drew a sketch for the resident starting with x-rays and x-ray production and ending with the x-ray image and image quality (Fig. 4). He was happier. I have used that diagram since then at the start of the course, and similar approaches in subsequent courses on ultrasound and MRI.

I listened to the RT students and residents and tried not to make the same mistakes that I experienced in my 10 years of undergraduate and graduate courses. In spite of not being able to eliminate all mathematics from radiology resident physics courses, the 1981 graduating class initiated the Gary T Barnes Annual Distinguished Teaching Award—an award given annually by the graduating residents to a member of the UAB Radiology Faculty for distinguished teaching.

The US Food and Drug Administration funded the ACR to develop a diagnostic radiology film teaching file. David Lamel headed up the group working on conveying physics concepts with images, an idea he had used in his radiology resident teaching. David invited me to join the group in 1980. Our efforts were highly successful and the Physics Section of the ACR Film Teaching File was created. My contributions were primarily images comparing the factors affecting image contrast—radiographs at different kVps, radiographs at the same kVp with good and no scatter control, etc. Although I had used images in teaching radiology residents previously, the experience working with David and the group resulted in my increased

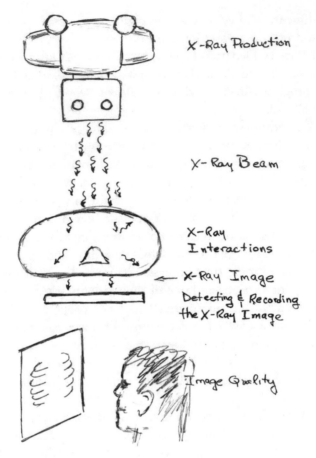

Fig. 4 Sketch to convey the relevance of the topics covered in the imaging physics course and their relationship to the image

utilization of images. For radiology residents, the saying "a picture is worth a thousand words" is an understatement.

As noted above, initially my teaching load was heavy. Teaching was only part my responsibilities and I had more contact hours than the majority of the full-time teaching faculty. Realizing this, I documented my contact and associated class preparation hours. I met with and gave the documentation to my Chair, Dr. Witten. I had learned that documentation is *Good* when there is a problem. A short time later, there was a second medical physics faculty position in the Department. Michael Yester was recruited and joined the Department in 1976. Dr. Witten asked if there should be a graduate medical physics program at UAB. Mike and I discussed the possibility. With one physicist in Radiation Therapy, who was over worked, there would only be a faculty of three, which we felt was insufficient. The UAB medical centre was large with a lot of equipment to look after (much more than I experienced at the University of Wisconsin). Yester had also been a post-doctoral trainee before joining UAB and we felt there was a greater need in the field to provide on the job training for recent Ph.D.'s. Dr. Witten agreed and a two-year postdoctoral medical physics training position was created. *Good* was the position's pay scale at the UAB Hospital's MD PG1 and PG2 resident level—much higher than the minimal remuneration I received as a postdoctoral trainee. (PG1 and PG2 refers to the first and second postgraduate year in the residency program.) Michael King joined us in 1977 and was the first UAB medical physics postdoctoral trainee. In 1990, we added a second training position and more recently the program became an accredited diagnostic medical physics residency training program.

5   Research

I was interested in the physical factors determining x-ray image quality—contrast, spatial resolution, and noise. When I joined the Department my understanding of these factors, their measurement and their interplay on image quality was limited. Although aware of various mathematical procedures such as Fourier analysis from my undergraduate electronic circuit analysis course and graduate physics courses, I did not have appreciation of its applications to medical imaging or the concept of the convolution (another mathematical concept frequently used in diagnostic imaging and radiation therapy). My appreciation and understanding came from reading the literature, attending conferences, talking with the scientists of screen-film system manufacturers, and attending presentations and talking with colleagues at AAPM, RSNA and other meetings. Murray Cleare (Eastman Kodak) and Robert Wayrynen, Russell Holland and Reed Kellogg (DuPont) were mentors. The learning experiences I remember most at national meetings were discussions with Ted Webster, Kunio Doi and Robert Wagner. These leaders in the field were extremely gracious in sharing their knowledge and insights with me.

When Dr. Witten indicated he was interested in making me an offer, I asked for a modest salary, but also asked for money to buy equipment for research and clinical support. A *Good* move on my part as it allowed flexibility in these early years to buy equipment for clinical support and research. I published a paper on the effect of x-ray tube potential on radiographic mottle in 1976 and a comprehensive theory of radiographic mottle in 1982. (In x-ray imaging "mottle" refers to the irregular gray scale variations in the image associated with random signal-to-noise fluctuations.)

In medical x-ray imaging there are two types of x-rays emerging from the patient: (1) primary x-rays and (2) scattered x-rays. Primary x-rays travel in a straight line from the x-ray tube focal spot to the image receptor and carry the image information. Scattered x-rays have interacted with atoms and electrons, are deflected from the initial path, carry no information, and form a diffuse out-of-focus contribution that degrades the image. When imaging thicker body parts the intensity of scatter is much greater than the primary x-ray intensity, and unless controlled, can significantly degrade the image. In medical x-ray imaging, scatter is controlled by an anti-scatter grid located between the patient and image receptor. An anti-scatter grid consists of an array of lead strips arranged and aligned so that the primary x-rays traveling in a straight line path from the x-ray focal spot to the image receptor see only the top edges of the strips, while the diffuse scattered x-rays see a much greater area of lead and are preferentially attenuated. Although a grid reduces the relative intensity of scatter, there is still appreciable scatter when thick body parts are imaged.

In 1974, Dr. Witten brought to my attention a Request for Proposal from the National Institutes of Health to improve scatter control in radiography. I submitted a proposal. Using published data on scatter levels for different field sizes at diagnostic x-ray tube potentials and a method I learned in my radiation therapy training for calculating radiation doses for irregularly shaped radiation beams, I demonstrated in the proposal that scanning the patient with a narrow beam (defined by a fore-slit and aft-slit) would reduce scatter compared to conventional anti-scatter grid techniques. (I later learned that this method of scatter control was first published in 1903.) X-ray tube loading (which causes tube heating) would be excessive for imaging the abdomen with a single beam but would be manageable with an array of beams. The submitted proposal had an array of beams defined by pre-patient slits and aft slits that would be able to scan the patient to reduce scatter in the resultant image. I received the grant and hired Ivan Brezovich to assist in the project. We designed and built a scanning, multiple-slit assembly (SMSA). Figure 5 shows comparison radiographs obtained in 1977 with the unit and a high quality conventional anti-scatter grid. The SMSA radiograph has greater contrast and the metastatic prostate cancer in the sacrum is better visualized.

The SMSA was efficient in controlling scatter; however, it didn't accommodate angled views, a general radiography requirement. An application where

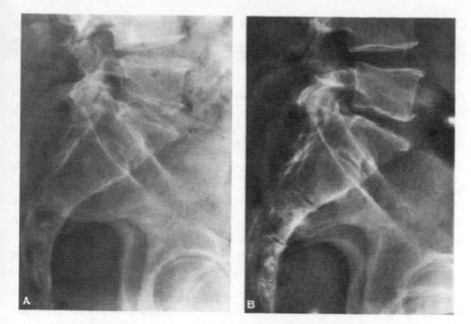

Fig. 5 Comparison of conventional 12:1 grid (**a**) and SMSA (**b**) lateral sacral spine radiographs. The patient thickness measured 33 cm. Both radiographs were taken with the same screen-film cassette and the same x-ray tube potential

the image receptor was always orthogonal to the central ray was mammography. In 1975, the general thinking was that at the low x-ray tube potentials employed and small volume of tissue irradiated, scatter was not an issue in imaging the breast. Ivan Brezovich and I measured mammography scatter levels. Our results indicated that scatter degraded contrast significantly in breast x-ray imaging. Depending primarily on breast thickness, only 50–75% of the possible primary beam contrast was imaged. In 1977, as noted above, Michael A King joined the Department as a post-doctoral trainee. With his assistance, a mammography SMSA was built. It virtually eliminated scatter. We showed our results at the 1978 RSNA meeting, the same meeting at which the company Philips introduced their mammography anti-scatter grid. Although more dose efficient and more efficient in controlling scatter than the Philips grid, the SMSA did not gain traction as it was bulky and not ergonomic.

In the 1970s, Picker International was the major supplier of dedicated chest x-ray units. In 1979, they decided to build a prototype digital chest unit. The project was headed by Mike Tesic. His approach was to scan the patient similar to a CT scout view (i.e., a 2-dimensional x-ray image acquired on the CT scanner usually used to define the scan range of the subsequent

CT images). The differences being the patient was erect rather than lying on the couch and imaged with a higher resolution detector. Mike had visited UAB to see the SMSA a couple of years earlier and I was invited to be a consultant on the project. My contributions involved the design of the fore- and aft-slits, with the recommendations that the slits be vertical and that the patient be scanned from right to left to minimize heart shadow scalloping, and the selection of the detector. Richard Sones was a physicist and a Picker employee who did the heavy lifting on the detector array and associated readout electronics. Richard and I measured the unit's scatter control, spatial resolution, and detector dose efficiency. The images were virtually scatter free; the unit's spatial resolution was consistent with expectations; the detector dose efficiency was not. It was less than half the expected result, an important observation that resulted in improvements in future detectors.

For clinical trials, the Picker prototype digital chest unit was installed at UAB in 1981 across the hall from the Department's dedicated (and highly optimized) screen-film unit. The reason for being located at UAB was that Dr. Robert Fraser was on the faculty. As noted above he was a world-renowned chest radiologist. Also, he had worked with Picker when he was at the Royal Victoria Hospital in Montreal in developing their commercially successful dedicated screen-film chest unit. The fact that I was also at UAB was a plus, but of lessor importance. The results of the clinical trial comparing the Picker prototype digital chest unit with UAB's dedicated screen-film unit were not exciting—conventional screen-film doing slightly better in the lung and the digital unit doing better in the mediastinum.

There were two limitations to the initial clinical trial. First, the digital images were compared to screen-film images obtained on an optimized screen-film system, a steep hill to overcome. Second, the digital images were viewed on interlaced 525-line cathode-ray tube (CRT) monitors—high line progressive scan, rapid refresh rate monitors were not available at that time. Somewhat later the unit was interfaced with a laser printer and the images were printed on film at full spatial resolution. The quality of laser printed film images was markedly superior to the same images viewed on the 525-line interlaced CRT monitors.

For new technology to replace established technology, it either has to do more or cost less. A commercial version of a prototype digital chest unit was unlikely to cost less. The clinical trial results indicated it did not do more. I listened to a presentation on dual energy chest radiography by Norbert Pelc at the 1982 RSNA meeting. Dual energy imaging is an imaging modality whereby x-rays are taken of the same volume using two different x-ray energies allowing the distinction of soft-tissue and bone due to their different x-ray energy dependence. The work was done at Stanford University with support from General Electric (GE). The soft tissue (with bone cancelled) and bone (with soft tissue cancelled) images that he showed were impressive. The visibility of soft tissue structures and nodules, if present, behind the ribs was obviously improved. If the Picker prototype digital chest unit could do dual-energy imaging, it would do more than a conventional unit.

The dual-energy images Norbert showed were obtained by switching the x-ray tube potential on a modified GE CT Scout View unit. There were motion artifacts associated with the time difference between the low and high x-ray tube potential image acquisitions. I proposed a simple solution to Picker for acquiring the low and high energy images simultaneously—a sandwich detector consisting of a thin, low absorbing front detector; and a thick, highly absorbing back detector. An idea I patented. The front detector would preferentially attenuate the low-energy x-rays and harden the beam (beam hardening occurs when more of the lower energy x-rays are removed from the beam); and the back detector would only see and detect the higher-energy x-rays emerging from the patient.

Richard Sones analyzed the proposed detector and his results indicated that it would work. Picker then designed a sandwich detector array, calibration methodology and upgraded the earlier prototype with the sandwich detector array in 1985. The results obtained with the Picker prototype Chest ESU (energy subtraction unit) were impressive. We worked on optimizing the image processing. The virtually scatter free images resulted in good soft tissue (bone cancelled) and bone (soft tissue cancelled) images (Fig. 6).

Previously there had been many radiographic pulmonary nodule detection comparison studies with little or no difference found—the older generation

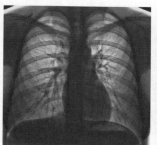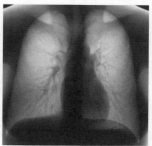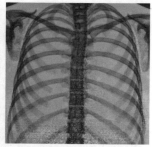

Fig. 6 Images obtained on the Picker prototype ESU circa 1985 with a single exposure and a radiation dose comparable to a screen-film dose: the digital image was obtained by adding the low and high energy images (left); the soft tissue image (centre); and bone image (right). Noteworthy is the absence of cancellation or motion artifacts in the soft tissue and bone images

of chest radiologists were as good as the current generation. Such was not the case for dual energy imaging. Compared to a highly optimized dedicated screen-film chest unit, nodule detection with the Picker prototype ESU was markedly improved.

Of clinical relevance is that calcified nodules are generally benign, whereas non-calcified nodules have a much higher probability of being malignant. With the Picker ESU, the confidence with which one could distinguish between calcified and non-calcified nodules was greatly improved (Fig. 7). My office was down the hall from the Picker ESU room. My hours often extended to 6 p.m. or later. Thoracic surgeons knew I was involved with the dual energy project and they greatly valued the unit's capability to determine if a nodule was calcified. On more than one occasion in the late afternoon or early evening when Dr. Fraser and the technologists who operated the Picker ESU had gone home, a thoracic surgeon would seek me out to display images of the patient he was going to be operating on. My experiences with thoracic surgeons through the years has been *Good*, much better than with neurosurgeons.

By the mid-1990s, dual energy imaging was in widespread use for bone densitometry (a technique that helps diagnose osteoporosis) and also for baggage inspection at airports and other security checkpoints. It wasn't until more than two decades after our work that dual energy chest imaging became commercially available. Current implementations employ x-ray tube potential switching with a short time difference between the low and high energy images. As a result, the soft tissue and bone images occasionally have motion artifacts. I anticipate the day when a system that does not result in motion artifacts is available—one in which the exposure is made at high x-ray tube potential and the low and high energy images are captured at the same time

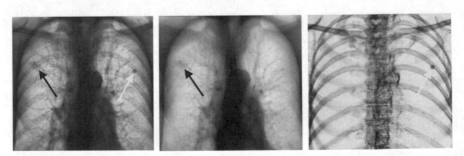

Fig. 7 Picker ESU digital radiograph of a patient with lesions in both sides of the lung (left); soft tissue (center) and bone (right) images of the patient. The nodule in lung's left side (yellow arrow) is calcified and benign. The nodule on lung's right side is not calcified and was cancer (red arrow). Note—when viewing a PA (posterior-anterior) chest projection, the patient's left is on the right side of the image

with a sandwich of large area detectors, and also one in which scatter is virtually eliminated.

In 1989, Xizeng Wu joined the Department and a year later, as noted above, Douglas Tucker. We collaborated in the early 1990s, along with Eric Gingold (postdoctoral fellow/trainee), on a number of mammography projects. The first was on developing a model for molybdenum target x-ray spectra. The model was semiempirical and the parameters were obtained by fits to experimental data. The second was on using the model along with a simulation of the x-ray interactions in the breast to estimate the average glandular dose (AGD) as a function of x-ray technical factors, breast thickness and breast composition. The simulation was bench-marked with published breast phantom scatter-to-primary ratios and measured depth dose exposures. The results demonstrated a dependence on x-ray spectrum factors that were not previously realized. The work was utilized and referenced in the 1999 ACR Mammography Quality Control Manual. The third project was on evaluating the dose efficiency of the different mammography x-ray tube target/pre-patient filter combinations that were introduced in the early 1990s.

For the target/filter dose efficiency study, Eric Gingold made contrast and exposure measurements for a number target/filter combinations, a range of x-ray tube potentials, and two to eight cm breast phantom thicknesses. Phantom average glandular doses were calculated with our methodology. The resultant paper was submitted to Radiology. It was rejected by a radiologist reviewer whose main comment was that the contrast that results with film and the recently introduced target/filter combinations was not as good as obtained with the conventional molybdenum target/filter combination. I contacted the Editor of Radiology and pointed out that the paper stated this, but also stated that the doses were significantly lower for thicker breasts; and the advantages of these target/filter combinations were not realized at present due to the contrast limitations of screen-film systems. I asked that the Editor solicit additional reviews by individuals knowledgeable on mammography physics and who would read the submission more carefully. He did and the paper was published. Of interest is that the paper's predictions have been realized. Digital mammography does not have the contrast limitations of screen-film mammography, and on digital mammography units lower contrast target/filter combinations that result in significantly lower doses are routinely employed to image thicker breasts.

UAB made millions of dollars from my patents and urged me to invent other projects to make more money. In 1998, I started a business at UAB's urging. The business model was to patent ideas with commercial potential and then license the patents. Financially, this effort was *Bad*. I do not recommend the business model. I spent most of my retirement savings on the business and shut it down because I ran out of money. I am still working to maintain a comfortable lifestyle. The *Good* is I can still work, and my medical physics skills are in demand.

The support I received from the UAB Research Foundation, UAB's technology transfer office, was minimal. No assistance other than suggesting that I develop a business plan. Medical physics colleagues at universities with more supporting infrastructure for business startups had much better and more rewarding experiences. It may have been that they had better ideas, were better businessmen or had both. You not only have to have a commercially good idea and demonstrate that it works; you have to focus on making the idea a commercial success or you have to partner with a good businessman to do this. The required financial resources I did not have and could not find. After no success in marketing business ideas and associated patents to x-ray equipment manufacturers, I worked with an engineer, who had successfully brought a number of medical imaging products to market, and we developed a business plan. Bringing a major medical imaging product to market is expensive. The business plan required four years and five million US dollars before income would equal expenses and another five years to pay back the investment. I had invested more than a million dollars, didn't have much money left, could not find five million, and, in 2014, I closed up shop.

Although the business was a financial failure, there were *Good* research and development accomplishments. In 1999, I hired David Gauntt. Our major focus was on improving image quality in bedside radiography. Bedside exams comprise a large percentage of the radiology exams performed at large medical centres, and the large majority of these are of the chest. A problem not solved by CR and digital radiography is the degradation of image quality by scatter radiation. In bedside exams, it is difficult to align the x-ray beam with an anti-scatter grid. If a grid is not properly aligned in chest imaging, artifacts associated with non-uniform grid primary transmission often result and can compromise diagnosis. Anti-scatter grids are not routinely used to image the chest at bedside. For abdominal work, grid misalignment artifacts have less impact, and grids are often employed. However, here also to minimize grid misalignment issue, the grids employed have less stringent alignment requirements than the more efficient grids used in fixed radiographic rooms where the grid and x-ray beam are aligned.

In the 1980s, Heber MacMahon at the University of Chicago demonstrated marked improvement in portable chest radiographs when an aligned anti-scatter grid was employed compared to radiographs without a grid. Dr.

MacMahon had developed a mechanical technique to align the x-ray beam and grid. Although his results were most impressive, his methodology was tedious and did not gain traction. With the exception of fluoroscopic exams or procedures, x-ray exams are performed by technologists. If an improvement in x-ray imaging was to gain traction, it had to be easy for the technologist to use or it will not be used.

David Gauntt and I developed a mobile radiographic unit that was easy for the technologist to use, and automatically aligned the grid and x-ray beam. Noteworthy, is that that accuracy obtained with our automatic grid alignment unit was comparable to or better than obtained in fixed equipment radiographic rooms. The clinical trial involved taking a conventional portable x-ray and some time later when there was a request for another portable x-ray on the same patient, an x-ray was taken with the portable with the automatic grid alignment system. The same technical factors were employed for both x-rays and the associated patient radiation dose was the same for both. I worked with technologists taking status portables in the middle of the night with the automatic grid alignment system. They found it easy to use and liked the pizzazz associated with its use and the x-ray tube moving into alignment with the grid. They also liked the markedly improved image quality. Shown in Fig. 8 are two comparison images of the same patient, acquired within 12 h of each other with the same technique and the same patient dose.

David and I developed automatic grid alignment systems for bedside radiography and demonstrated that it resulted in markedly improved bedside radiographs—radiographs that had image quality comparable to that

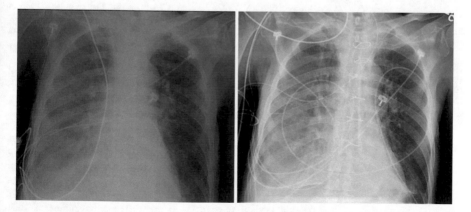

Fig. 8 Bedside radiographs of the same patient taken with the same technique and the same radiation dose. Without a grid (left), and with an anti-scatter grid (right). The grid was aligned automatically with the unit described above. As evident upon inspection, markedly better image quality was obtained with the aligned grid

obtained with fixed equipment. With such a unit there is little or no need to move sick patients to the radiology department. Radiographs can be taken at bedside with no loss of image quality. We took a *Good* step forward. I was and am still surprised that manufacturers are not interested in incorporating the technology in their product line. I am proud of our accomplishments. In the case of dual energy chest radiography, 20 years passed after its clinical advantages were demonstrated before a commercial dual energy chest unit was introduced. Now there is also dual energy CT. I believe in a few years, and hopefully in my lifetime, mobile radiographic units incorporating accurate grid alignment will be commercially available.

6   Summary

My 50 years in medical physics was mostly *Good*. *Bad* experiences helped me do *Good* in the future. Looking for a job in 1971 when there were no jobs for a recent Ph.D. physics graduate was *Ugly*, and I would not wish such an experience on anyone. However, it was *Good* in that it made me assess my skills, look for a training in medical physics, and after training find a position where I could utilize my skills.

Gary T. Barnes is Professor Emeritus of Radiology at the University of Alabama Birmingham (UAB). In 1972, he joined the University of Alabama Birmingham (UAB) Medical School Faculty. He was promoted to Associate Professor in 1976 and Professor in 1981. From 1976 to 1987, he was Chief of the Department's Physics Section and from 1987 to 2002, Director of the Physics and Engineering Division. In 2002, he retired with 30 years service. He continues to be involved at UAB part-time, teaching radiology residents, and teaching and mentoring medical physics residents.

Dr. Barnes has been active on committees of the American Association of Physicists in Medicine (AAPM), Southeastern Chapter of the AAPM (SEAAPM), American College of Radiology (ACR), Radiological Society of North America (RSNA), and American Board of Radiology (ABR).

He is past-president of the AAPM (1988) and SEAAPM (1979); a Fellow of the AAPM and ACR; and a Diplomat of the ABR (Radiological Physics). Dr. Barnes' research interests are in diagnostic x-ray imaging and include work on scatter and scatter control, screen-film systems, digital radiography, mammography and clinical medical physics. He is the author or co-author of 100+ scientific papers and 12 patents.

Awards

2011. Silver Medal Award, The Alabama Academy of Radiology's Highest Award, Given for Outstanding Deeds and Service

Throughout the Years, and Meritorious Contributions to the Practice of Medicine and the Specialty of Radiology.

2007. Jimmy O'Neal Fenn Award, Southeastern Chapter of the American Association of Physicists in Medicine's Highest Award for Outstanding Contributions to Medical Physics and the Southeastern Chapter.

2005. William D Coolidge Gold Medal Award, American Association of Physicists in Medicine's Highest Award for Outstanding Contributions to Medical Physics.

1982. Southeastern Chapter of the American Association of Physicists in Medicine's Annual Award for the Best Publication, "Radiographic Mottle: A Comprehensive Theory".

1981. Gary T Barnes Distinguished Faculty Award was established by the UAB Radiology Resident Class of 1981 and given annually to a member of the UAB Radiology Faculty for Distinguished Teaching; Recipient of the First Annual Award in 1981 and also in 1986.

Part II

Medical Physics: More than Clinical Service

"If you can't see it, you can't hit it. If you can't hit it, you can't cure it."

William E. Powers, a well-renowned Radiation Oncologist in USA and Harold E. Johns, the "grandfather" of Medical Physics both in Canada and internationally and known especially for the developments of cobalt-60 radiation therapy and the famous textbook, *The Physics of Radiology*.

Part II

5

A Day in the Life of a Medical Physicist

Arthur L. Boyer

1 "Up and Atom"

I reach out and shut off the alarm on the bedside clock. Where am I? What day is it? Who am I? Oh yeah, I am in New Braunfels, Texas and it is the fall of 1979. There is enough light outside the bedroom window to see the eager wagging of two tails. I roll out of bed and began donning jogging togs. I smile at the logo emblazoned in blue on the front of the tee-shirt I pull on. The American Association of Physicists in Medicine (AAPM) holds annual national meetings in different cities across the country each year. The meetings provide a window into the latest research and development in medical physics in the country's various teaching hospitals and academic centres. They also provide an opportunity for informal collegiality among its members. This includes an early morning 5 K fun run. One company has regularly sponsored the fun runs, replete with a tee shirt sporting a different logo each year. The logo in the shirt I had pulled out pictures an atom with a nucleus and electrons zooming around in orbits and reads "Up And Atom"—pun intended. I am met outside the back door by two joyous dogs leaping and spinning in anticipation of a loop on the asphalt road around our rural neighborhood.

A. L. Boyer (✉)
Belton, TX, USA

The fresh air is cool, but the Texas sun will soon warm it up. When I finish the loop with my little wild band, we are all ready for breakfast.

I shower and appear in the kitchen vested in the shirt and tie expected of a professional. I wolf down my wife's breakfast, peck a thank-you kiss on her cheek, and pick up my brief case along with a small paper sack containing my lunch. Soon I am sailing down the interstate toward the Cancer Therapy and Research Center (CTRC) in the San Antonio medical centre where I serve as Chief of Physics.

As I drive along, I begin to plan in my head a lecture for the radiation oncology residents that I am scheduled to deliver in the following week. Maybe not the safest way to negotiate rush hour traffic, but it is a daily habit. So far in the lectures, I have covered, among other things, the subject of dose and the instruments used to measure the dose deposited by high-energy x-rays and electrons used to treat cancer. Radiation dose is measured in units called the Gray (Gy). Typical daily doses of radiation to treat cancer patients are 2 Gy per day for five treatments per week, possibly up to 7 weeks. Dose is measured by an instrument that collects the charge created by the ionization of air by radiation in a small cavity biased with a potential of about 300 V (see Fig. 1).

I plan in this next lecture to describe the medical linear accelerators (linacs) used to create the x-ray and electron treatment beams. I will show slides of the major components of a typical linac (see Fig. 2). The beams are produced in a fiber-glass encased machine consisting of a large box firmly bolted to the

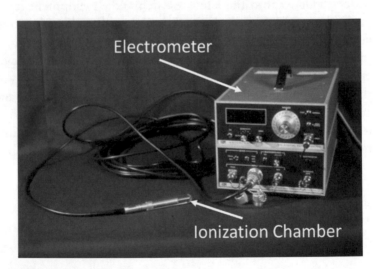

Fig. 1 An ionization chamber on the end of a cable connected to an electrometer used to measure radiation dose

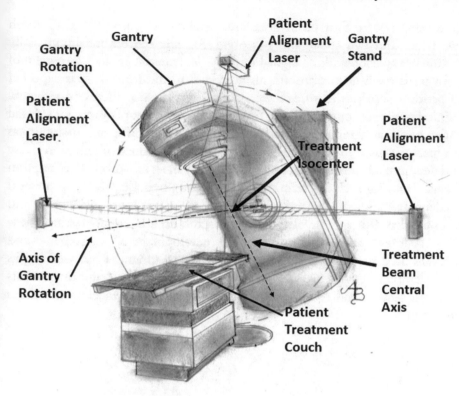

Fig. 2 Sketch of a medical linear accelerator

floor (the stand), to which is attached a component shaped like an L on its side called the gantry. The beams come out of the short leg of the L (called the collimator). The gantry can be rotated around the patient so that the beams can be directed through the collimator from any angle on an arc that encircles the patient and is centred on a point on the gantry axis (called the isocentre). High powered microwave pulses accelerate electrons down a tube inside the long horizontal arm of the gantry. Electrons exiting the accelerating tube are directed by a magnet toward the patient through the collimator. The electrons can either be used to treat the patient directly, or a metal target inside the collimator can be inserted by the push of a button into the electron beam to convert it to a high-energy x-ray beam. It is amazing technology. The challenge is to describe all this technology in terms the medically trained residents can understand.

The linacs, which were developed in the 1970s, are fully motorized, as are the older treatment machines that used cobalt-60 radioactive sources. Motorization allows treatments to be delivered efficiently. Therapy technologists can position a patient in a treatment room and deliver the typical daily 2 Gy dose using two to four beam directions in about 15 min. The CTRC has two linacs and two cobalt units (see Fig. 3). The centre's maximum daily treatment capacity working eight-hour days is therefore about 128 patients.

In most medical institutions, an academic medical physicist is expected to perform activities in three areas: clinical service, teaching, and research. Clinical service encompasses such functions as treatment planning for specific patients and checking patient treatment records, calibrating and maintaining the machines used to treat patients, as well as maintenance of radiation safety and design of radiotherapy facilities so that they meet radiation safety requirements. Teaching includes training new medical physicists, teaching medical residents in radiology and radiation oncology, and teaching dosimetrists and technologists. Research involves developing and experimentally testing scientific hypotheses as well as pursuing the development of instruments and software used for medical imaging and cancer treatment. I am working in the radiation oncology area, and my research is focused on improved treatment procedures and the software and hardware needed to carry out those procedures.

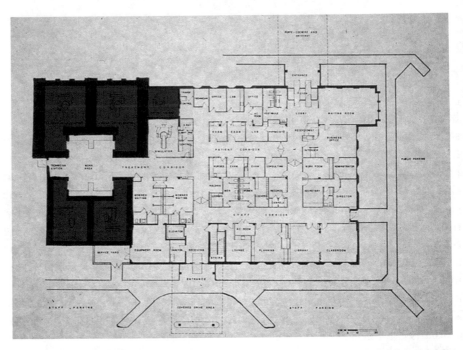

Fig. 3 Floor plan of the CTRC. Treatment rooms are in red. Note the thick shielding walls as indicated by the black lines. The thickness of these walls can be compared to the thickness of the outside walls of the rest of the centre

2   The Workday Begins

I park at the CTRC and walk into my office. I don a white lab coat with my name stitched in blue letters on its front along with my title, "Chief of Physics". My staff arrives, consisting of my dosimetrist, Ray Federico, and my assistant physicist, Ed Mok, who is also our computer system manager. The CTRC has an unusual role in the city's radiotherapy service. Along with patients of physicians holding appointments at the University of Texas Health Science Center in San Antonio, the CTRC also treats patients served by a private radiation oncologists' group whose plans are prepared by a medical physicist/dosimetrist working for the city-based group. The arrangement provides me with interesting management challenges. In addition to preparing plans for patient treatments, the physicists and dosimetrists are responsible for checking the patient treatment records once each week, looking for any errors in the sum of the total dose delivered or other compromising irregularities. I check with Ray and Ed to see how the treatment planning and chart checking are going and schedule some time to go over special problems with them.

Then I turn to another project that I have just completed. The CTRC is currently a one-story concrete and brick building with a flat roof. The CTRC Board is considering a second story expansion. But since several of its inner walls are thick and composed of reinforced concrete required to shield workers from the radiation in the treatment rooms (see Fig. 3), a second floor can be added without much structural beefing up of the first floor. However, since the high energy x-ray beams can be pointed straight upwards, a careful analysis must be made to ensure that radiation safety limits are not exceeded for the second-floor occupants. This is my task. I have been well-schooled on the calculations and regulations required by the National Council on Radiation Protection. I am certified by the American Board of Radiology (ABR) and licensed by the Texas Board of Licensure for Medical Physicists whose requirements include radiation safety expertise. The conclusion of my analysis is that a relatively thin layer of lead bricks will need to be added over the linear accelerator vaults between the existing roof and the second story floor. I have checked with the architects, and the additional weight will not compromise the ceilings of the accelerator vaults. I have written a report detailing my calculations and clearly stating my conclusions. I check over the calculations and the report one last time before putting a signed copy in an envelope to be passed on to the architects at a meeting early next week.

3   Fast Dose Calculation for Radiation Treatment Planning

In graduate school at Rice University I felt that I learned more from Dr. Sid Burras than from any other professor. Sid was in the electrical engineering department, and he was an extremely gifted teacher. I took his introduction to Fourier Transform theory as my mathematics requirement for my Ph.D. in the physics department. For the math and physics-oriented reader, the nineteenth century French mathematician Joseph Fourier discovered that a mathematical function can be reproduced by a weighted sum of geometric sine functions with ever increasing frequencies. The transformation from the original mathematical function to the infinite series of sine functions is known as the Fourier Transform. (Also see Chap. 10 by Terry Peters.) A recent innovation has been the discovery of a Fast Fourier Transform (FFT) computer algorithm. The FFT is being used in a number of practical applications. Sid was an expert on the FFT, and I had learned how to use it in FORTRAN computer programs. In electrical engineering, the FFT is used to analyze the fluctuation of a voltage over a period of time. The result of the calculation is a set of relative strengths of the frequency components of the electrical signal. More to my purposes, a mathematical property of the Fourier Transform is that it can be used to calculate a mathematical operation called *convolution*. *Convolution* is a mathematical operation that lays down a mathematical pattern (called a *kernel*—see Fig. 4) over a volume with different weights at each point at which the kernel is laid down and summed with all its neighboring weighted kernel patterns. It can be applied to one-, two-, or three-dimensional problems. I had used the FFT in graduate school and had written my Ph.D. dissertation on an ultrasound imaging experiment using the FFT to construct ultrasound images. This background has led to my research activities at the CTRC to which I now focus my attention.

I had been attempting to use the FFT to model radiation dose distributions using kernels that described the splash of energy deposited in all directions around the point at which a high-energy x-ray interacts with water. These kernel distributions of deposited energy can be computed from first principles of physics. Since cobalt-60 emits photons primarily at one energy, the calculation of the cobalt-60 kernel is more easily doable than the kernels for high-energy linac x-ray beams composed of a spectrum of photon energies. Therefore, I had been trying to model the cobalt-60 dose distribution that has been measured in a water phantom. Cobalt-60 treatment machines were the workhorses of radiotherapy in the sixth and seventh decades of the twentieth century. The dose distribution has been measured by many investigators

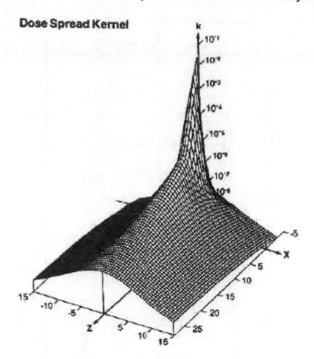

Dose Spread Kernel

Fig. 4 An isometric representation of a *dose spread kernel*. A photon with an energy of 6.8 Million Electron Volts (MeV), moving along the z-axis interacts with water at z = 0 and creates energetic electrons moving out in specific directions that deposit dose around the interaction point. The dose values are plotted here logarithmically in the k-axis direction in units of MeV-cm²/g. The energy deposition is greatest at the interaction point (x = 0, Z = 0) and falls off around that point

and is easily available in various publications. I am familiar with how cobalt-60 dose distributions look and I have precise data available in my computer files. However, my computer program to model the distribution has a bug in it. The previous day I had given up trying to find the bug and had decided to take a fresh stab at it today. Sure enough, soon after I fire up my workstation and start scanning through my FORTRAN code, there is the little culprit staring me straight in the face. I carefully correct the error and direct the operating system to compile the program again. Success—no error messages! I can now run my modeling program successfully for the first time. After a few seconds, a dose distribution flashes on the workstation monitor. I gasp! I call Ed to come into my office to have a look. We are both excited. It appears to be a reasonable representation of a cobalt-60 depth dose distribution (see Fig. 5). Ed and I begin planning how to compare the calculation with the measured data both numerically and graphically.

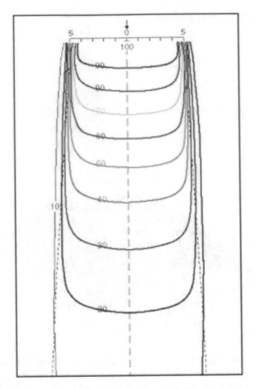

Fig. 5 Cobalt-60 relative dose distribution for a radiation beam downward directed from the top of the figure. The beam is collimated to a square 10 cm × 10 cm at the surface of a water phantom that is 80 cm from the source. The coloured lines are *isodose curves* representing points of equal dose. The relative doses are scaled to the point of maximum dose which is set to 100%

4   Back to the Clinic

However, it is now time for my weekly meeting with the Medical Director of the CTRC, Dr. Charles Coltman. Dr. Coltman is a chemotherapist who has passed through the Air Force, attaining the rank of Colonel. His military experience has left him with a somewhat imperious administrative style. But he is intelligent, politically adroit, and fair. He requires me to provide no formal written reports on a regular basis but likes to keep a finger on the pulse of the Physics Group. After brief formalities, I discuss the conclusions of my shielding report. Then I review a small administrative task I had to carry out the previous week. Ray had discovered a small discrepancy in one of the patient's charts. The technologist recording the daily doses for a patient had put a number into the record incorrectly, but the total dose had been calculated and recorded correctly. I had made an appointment with the

Chief Technologist to discuss this inconsequential irregularity. I assured her that we were generally quite satisfied with the performance of her staff but asked her to point out the irregularity to the technologist, correct it in the patient's record, and make whatever comments she felt might be helpful to her staff. I make Dr. Coltman aware of the exchange and he indicates his satisfaction with its handling. I then express my gratitude to him for his role in the acquisition of the new computer hardware and software for the Physics Group purchased from the Digital Equipment Corporation. In this exchange I hint about some of the computational research I have been carrying out using the new computer system, and how it might lead to a publication. With his usual noncommittal style, he acknowledges the compliment and encourages me to continue with my investigations as long as they do not interfere with my clinical services.

Dr. Coltman has a new task for me. I knew from attending the CTRC clinical staff meetings that the radiation oncologists were wishing to replace one of the centre's cobalt-60 units with a new linear accelerator. From their point of view, since the cobalt room was already shielded, putting a linac in the room was a simple matter of exchanging one machine for another. The radiation oncologists had approached the issue with Dr. Coltman. Their major concern was having the CTRC Board agree to providing the funding for the purchase of the new machine. They had come armed with a verbal bargain basement quote from the local Varian sales representative. Dr. Coltman had a suspicion that there were other issues to consider as well. I agreed. The shielding in place was adequate for the lower energy cobal-60 beam but was probably insufficient to meet radiation safety requirements outside the treatment room for a higher energy linac beam. Considering the location of one of the cobalt machines on an exterior corner of the building (see Fig. 3), I suggest that it is the machine to be replaced since additional shielding could be added to the outside of the exterior walls if needed. The shielding required might be minimal if the new linac produced a 4 MV x-ray beam or at most a 6 MV x-ray beam. I ask if I should begin a shielding analysis and plan for those energies so that the architects could provide an estimate of the construction costs that would need to be added to the purchase price. The change would also impact the shielding report that I have just finished for the second floor. We discuss that implication and decide that since it would probably require only a modification of the lead bricks I had already specified over the room for the linac. I could modify my report quickly and still have the report ready for the next meeting with the architects. Dr. Coltman asks how long it will take to prepare the whole shielding analysis for a new linac. I run over in my mind my workload at the moment and give him a date two weeks in the future. This will be later than his next meeting with the radiation oncologists, but he agrees to the delay.

Finally, we turn to the most delicate issue on our agenda, the annual budget for the Physics Group. Salaries are relatively straight forward since I have used hard data from the annual national survey of medical physicists' salaries conducted by the AAPM as the basis for my recommendations. The CTRC would have difficulty recruiting in the future if the salaries it offered were below the national averages.

Then we attack the capital equipment budget. This is where things usually get sticky. Among the clinical services provided by medical physicists is the measurement of radiation beams. Ionization chamber instruments are used to measure dose at a single point. The CTRC has the necessary instruments to calibrate the treatment machines at a single point (see Figs. 1 and 2). But planning a patient's treatment requires a map of the relative radiation dose across the treatment beams in three dimensions. The treatment planning process uses these data to ensure delivery of an adequate dose to a volume of tissue containing cancer cells, and to ensure that surrounding non-malignant tissues are not damaged. The measurements needed to obtain these maps are carried out in a tank of water.

Relative dose in a tank of water is an adequate representation of the behavior of radiation in the human body. Automated systems that move a small ionization chamber around programmed patterns in a water tank and deliver digital records of the ionization chamber measurement as well as the three-dimensional location of the chamber's position to an external computer are available commercially (see Fig. 6). The CTRC has a venerable automated water phantom unit that is slow and cantankerous and does not interface well with an external computer. If we are to commission a new linear accelerator in the near future, a new automated water phantom would be an invaluable asset. I mention cost estimates and vendors that supply the best units. Dr. Coltman indicates his understanding of the issue but, true to form, is noncommittal. I am satisfied that I have done what needs to be done for now.

My time is up. As I rise and turn to the door to leave, Dr. Coltman stretches his lips tightly over his teeth and with a little burst of air makes an exceptionally accurate rendition of a fart. The sound twirls me around to gaze again upon the grim noncommittal mask of an imperious Colonel that is frozen on his face. I throw him a wave and a smile as I leave his office.

I descend the stairs to a break room in the basement of the building for lunch. Jeff, our in-house linac service engineer, is there for lunch as well. I purchase a soda from the vending machine and sit down with my sack lunch.

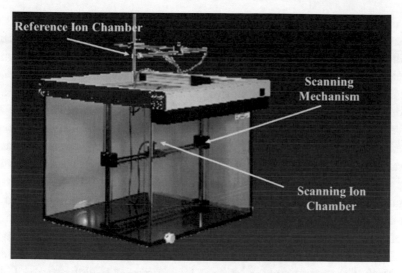

Fig. 6 Computer controlled 3-D data acquisition water phantom

Jeff was trained in microwave engineering in the Air Force. He served a stint in Vietnam until his commitment was completed. He has received additional training on linear accelerators available through our vendor. Though from humble roots, Jeff is intelligent and diligent. He is also a tad vain about his abilities. Officially he is a member of the Physics Group. However, he tends to go directly to Dr. Coltman to review the status of machine maintenance and to make requests for materials. I ignore this infraction since Jeff then immediately comes to me to brag about his conversations with Dr. Coltman. I and Dr. Coltman still get the information we need, and Jeff gets to feel he is important. He has worked his way up the hard way. I justify leaving the situation as it is by feeling he deserves a break. Over sandwiches, Jeff informs me that he plans to take a week off to go on a trip with his family during the up-coming holiday. We discuss back-up support from the linacs' vendor, Varian, should there be a break-down during his absence. I verify the necessary contact information with Jeff and wish him a nice trip with his family. I then have a further thought and ask Jeff where he keeps the detailed circuit drawings and machine descriptions in his office. He describes where the information can be found and assures me that I can use the drawings any time I need to. In fact, he welcomes me to use any of his instruments and to raid his electronic components if needed. We exchange a few ribald stories and I hurry off to my office to continue my research.

(As an aside … It was well that I had access to Jeff's instruments, diagrams, and electronic components. During Jeff's holiday absence, one of the linacs ceased functioning. Diode displays indicated there was a problem in its electronics. I called the vendor's local back-up maintenance engineer, but he was also on vacation in Hawaii. The vendor had no other service engineers available that they could fly into San Antonio. The radiation oncologists were in a flurry over the interruption in their patients' treatment schedules. Fortunately, in my early training, I had been privileged to take the same service engineering course from the vendor that Jeff attended some years later. I pulled out Jeff's oscilloscope and voltmeter along with the machine diagrams and took them to the naughty linac. From the error indicator light, I had a hint as to where the problem was located. Using the machine wiring diagrams and the voltmeter I traced the absence of a signal that should have been coming from one of the machine components. Opening the enclosure containing the offending component and using the voltmeter and wiring diagrams, I traced the problem to a printed circuit board. I called the vendor to request that a replacement be sent by express over-night to the CTRC. Unfortunately, the holiday had shut down the department that could provide the circuit board. The vendor could not guarantee getting one to San Antonio for at least a week. I returned to the linac, pulled out the circuit board, and began testing the pins on the transistors soldered into the circuit board with the voltmeter, looking for the correct electrical properties of the transistors. I found a transistor that had a short circuit between two of its pins that should have behaved like a diode. Pawing through Jeff's stock of electrical components I was lucky to find a functioning copy of the dead transistor. I unsoldered the miscreant transistor and replaced it with the good one. It tested appropriately in the circuit board. With the circuit board replaced, I fired the linac back up again and the error light did not turn on. I ran through all the treatment energies and they were all working. I cleared away my little mess of circuit diagrams and instruments, ran a quick calibration check with our ionization chamber/electrometer set, and informed the chief technologist that we were back in business.)

5   Beyond Radiation Therapy

In a corridor on the way to my office, one of the chemotherapy residents hails me down. He is animated. Earlier Dr. Coltman had listened to his proposal for a research project and had encouraged him to work on developing the concept. It involves the most common form of cancer in the world—skin cancer. He had read reports in the literature on a possible new form of treatment for skin cancer based on the porphyrin molecule. The porphyrin molecule can be activated by visible red light of a specific wavelength of 700 nm to create excited oxygen molecules that can destroy cellular structures and cause cell death. This form of treatment is known as *photodynamic therapy*. The physics of light absorption is similar to the effect of radiation passing through tissue. The resident wants to pursue the application to skin cancer cells. He wants to carry out a series of experiments in tissue cultures and in mice to verify that porphyrin in cells of a specific skin cancer line are increasingly killed by increasing "doses" of light using the red wavelength of about 700 nm. He wants me to assist him by designing and building a light irradiator, calibrating the light "doses", and assisting in conducting the experiments. He has access to funds from the University of Texas Health Science Center in San Antonio, but the funds are limited. He gives me the figure that could be devoted to the acquisition and construction of the light irradiator and asks me to determine what could be purchased with them. As it turns out, I had visited the investigator in Albany, New York who had been publishing on this research while I was still at the Massachusetts General Hospital in Boston. I would probably have tried to interest someone at the Massachusetts General Hospital in the same experiments the resident is proposing had I remained there. Of course, I agree to help him. When I return to my office, I search out the necessary components in the optical instruments catalogues that I have on hand. The funds provided by his grant can cover a light diode calibrated by the National Bureau of Standards for the wavelength to be employed. We can use one of our electrometers employed for radiation dose measurements to read out the calibrated light diode. A laser for the light source at 700 nm is much too expensive. I can make a powerful white-light projector produce the 700 nm wavelength light by using a precision filter that is relatively inexpensive. However, this will limit the light output and require rather long exposure times to achieve the light "doses" needed to reach the upper limits of cell killing. I call the resident into my office and report what I have found. He believes the response of the cell culture and the animal tumours will not be compromised by the long exposure times. I ask him when he proposes to start the project. He indicates the funding will not be available until after the first of the year. That is fine with me because between now and the end of the year I have another major commitment.

(Another brief aside … The next year I acquired the components for the light irradiator and irradiated cells doped with porphyrin. We obtained very reasonable dose response curves. The mouse experiments also showed that skin tumours growing on mice into which porphyrin was injected responded to light radiation. I helped write up a paper detailing the experiments. Unfortunately, the resident became interested in other chemotherapy drugs and did not follow up on these experiments. However, other investigators continued work on porphyrin related drugs activated by red light. Fifty years later after much research by many investigators my own dermatologist administered a widely used topical cream to pre-cancerous actinic keratosis lesion on my face. Actinic keratosis is a condition which causes scaly patches on the skin possibly from exposure to the sun over the years. After some time to allow certain biochemical changes to occur in my skin, an intense red light was used to activate the production of porphyrin. The light also caused the porphyrin to produce oxygen radicals that killed the premalignant actinic keratosis cells. The procedure had become routine in the practice of dermatology and I was one of the beneficiaries of this research.)

6   Three-Dimensional Anatomy

The roots of this effort are planted in a conversation I had years ago in the early 70s with my esteemed colleague Michael Goitein. We were standing together in the Radiology Department of the Massachusetts General Hospital. We were watching the first commercially available CT scanner collecting data to reconstruct images that sliced through the cranium of one of their first patients to undergo a CT scan. The Massachusetts General Hospital had just purchased the scanner from Electric and Music Industries (EMI), which was owned by the Abbey Road Studios in London. The Abbey Road Studios' most notable client was the Beatles. The collection of the data took about an hour, so we watched it for only a few minutes. But while we did, we speculated on the implications of this new technology, especially as to how it might affect our careers. Michael was dreaming of using the images in radiotherapy treatment planning and the prospect of being able to look at the organs in three dimensions through which a proposed treatment beam would pass. This became the "Beam's-Eye-View" concept for which he developed software and applications. It occurred to me that such remarkable visualization of anatomy would lead to a need for a higher level of communication between radiation oncologists and medical physicists. That would require medical physicists and dosimetrists to have a greater mastery of the language of human anatomy. I had taken an anatomy course previously as a post-doc at the MD Anderson Hospital that had whetted my appetite for my own development of anatomical knowledge. This urge merged with my artistic hobby and I began to make anatomical sketches of organs (see Fig. 7). Earl Adrian, M.D., Ph.D. is the head of the Anatomy Department at the University of Texas Medical School at San Antonio and is a fellow alumnus

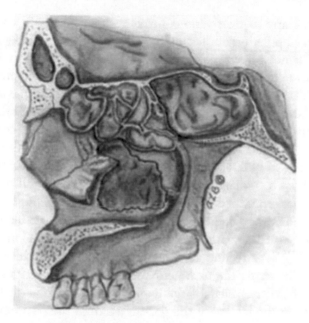

Fig. 7 One of my sketches of the paranasal sinuses which are small hollow spaces in the skull around and behind the nose. Note the teeth at the bottom. The paranasal sinuses are located in the upper half of the figure

of Rice University. When I approached him, he appreciated the need for anatomy learning for Radiation Oncology staff. Consequently, we developed a one-week anatomy short course. It is held annually in the Anatomy Department in one of its classrooms as well as the near-by cadaveric lab where medical students conduct their dissection exercises and learn their surgical skills. Lectures on anatomy geared for the non-medical student are given by Dr. Adrian and his faculty. They also conduct laboratory exercises in the cadaveric lab where they demonstrate anatomy using the prosecuted cadavers on hand. I need to complete preparations to coordinate and participate in this year's course that will be held soon, but I should be finished with the course before the end of the year. Therefore, I can agree to pursue the resident's research project.

7 Special Techniques

A patient suffering from lymphoma now requires my services. Lymphoma is a cancer of the lymphatic system of the body involving the stem cells that produce white blood cells. Total body irradiation is used in conjunction with chemotherapy to treat various forms of lymphoma. In the lymphoma patients, a subpopulation of the stem cells found in the marrow of bones has undergone a genetic aberration so that instead of the healthy white blood cells they normally produce, they produce cancerous lymphoma cells. The lymphoma cells circulate through the body interfering with the normal functions of the healthy white blood cells. When the lymphoma cell population vastly out number normal white cells and begin to invade critical organs, the patient dies. These abnormal white stem cells and the adult lymphoma cells are sensitive to radiation. Since they are found throughout the body, the entire body must be radiated to a dose sufficient to render all the lymphoma cells incapable of reproduction but less than an instantaneous fatal total body dose. Clinical data also indicate that the lung dose must be a little lower than the total body dose. (Also see the lung response discussion in Jacob Van Dyk's Chap. 2.) By directing an x-ray beam from a medical linear accelerator horizontally across the treatment room, and opening its collimator to its maximum field size, a field large enough to cover an adult human body can be formed at the far wall if the patient assumes the position shown here (Fig. 8). I wheel a cart into one of the linac rooms. The cart carries sheets of lead of various thicknesses and shapes, some miniature dosimeters, body calipers and a ruler, some data tables I need for reference, and a form I have prepared for taking measurements of the patient. The measurements of the patient are needed to select the lead sheets that will be used in the treatment. The data tables are needed to calculate the linac setting required to deliver the desired dose. I am following procedures published by the AAPM,

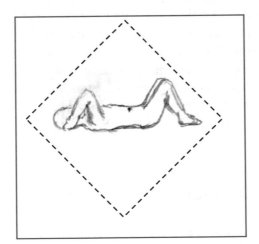

Fig. 8 Outline of treatment beam around a typical size patient to be treated with total body irradiation

a report by a Task Group chaired by J. Van Dyk, who is the editor of this book. In the room is a Lucite box I had designed that fits on a standard hospital gurney and has means to attach it firmly to the gurney. The Lucite box was fabricated at a machine shop I use in San Antonio where the German machinists sardonically named it "Schneewitchkens Sarg", translated "Snow White's Coffin". One of the side walls of the box can be removed (see Fig. 9). The patient is wheeled into the room on an ambulance gurney on which he has been transported to the CTRC from the hospital at which he is being treated. The radiation oncologist treating him introduces me. I try to make a few soothing remarks before we remove one wall from the box and transfer the patient from the ambulance gurney to the irradiation box.

Using the body calipers, I measure critical dimensions and record them on the form. One half of the exposure will be delivered with the beam directed toward one side of the patient. Since the dose needs to be uniform over the entire body, compensation for the differences in thickness of legs, arms, and head need to be provided by sheets of lead 1.6–4 mm thick that are hung on the entrance side of the box. I move the gurney up to the wall and begin to select shapes and thicknesses of the lead sheets based on my measurements. I hang the selected sheets of lead in line with the parts of the patient for which they will provide compensation for thinner thickness. The radiation exposure is determined by a setting on the linac. I have worked out how to compute the setting required to deliver the radiation oncologist's desired dose along the centre of the patient's body considering the treatment distance and

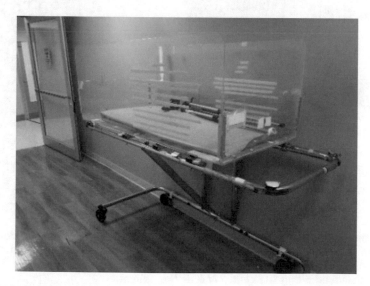

Fig. 9 Whole body treatment box

the effective uniform thickness of the patient produced by the lead sheets. I place a set of miniature dosimeters at the exit side of the patient so that I can verify the dose to the patient's centre line using standard dosimetry calculations. These measurements verify the efficacy of the lead compensators and verify that the desired dose was delivered. We tell the patient how long this part of the treatment will take, ask him to lay as still as possible, and we leave him alone in the treatment room and close the heavy shielded door. A technologist enters my calculated settings into the linac controls and turns on the beam. Given the long treatment distance, the time required to reach the desired dose is longer than standard treatments. It makes me nervous for the beam to be on so long even though I have exhaustively simulated the treatments to verify my data and calculations. Furthermore, the miniature dosimeter measurements made on previoulsy treated patients have verified the dose calculations. When the exposure is completed we reenter the room and congratulate the patient for lying so still for half his treatment. We rotate the gurney (see Fig. 10) and hang the lead sheets at locations exactly opposite their original locations. I remove the set of miniature dosimeters and place a second set of miniature dosimeters on the other side of the patient exactly opposite the locations of the first set. We repeat our request that he lie still for just a little longer and execute the treatment from the other side.

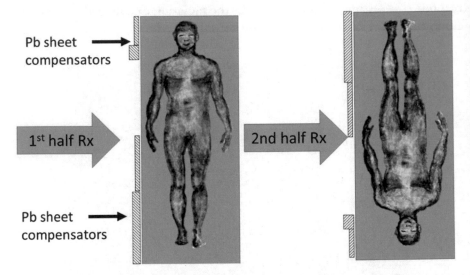

Fig. 10 Illustration of the orientations of the patient relative to the treatment beam coming in from the left. The patient is in a supine position (lying on his/her back). The one-dimensional lead compensator sheets shown shaded are used to even out the dose delivered along the centre-line of the patient

This is a sobering treatment because the patient, when he is wheeled out of the treatment room on his ambulance gurney, has no means of reproducing any blood cells, white or red, and is severely immuno-compromized. I realize that I have helped produce a walking dead. Normally, such high radiation doses to the whole body are lethal within a few days. However, before the treatment, punch biopsies of his bone marrow had harvested living blood stem cells. The cells were processed through a cell sorter to try to remove the deadly lymphoma cells from the lymphocyte stem cell population. When he is returned to the hospital, he will be placed in a special sterile procedure room and the stem cells will be reintroduced into him intravenously. The little stem cells will seek out homes in his bones and start reproducing heathy red and white blood cells. If all goes well, his red and white blood count will return to a normal range and he can manage his lymphoma. I take a deep breath, say a quick prayer, and store "Schneewitchkens Sarg" until the next treatment.

8   The Day Drags On

It is approaching 5:00 p.m. and I need to prepare for monthly calibrations on one of the linacs. I retrieve the linac's bound calibration logbook and open a closet in which a cart holds the calibration instruments and other devices and materials I need for the calibration checks. A full official calibration must be carried out in a water phantom. Monthly calibration checks can be carried out using water equivalent plastic slabs in lieu of water. A full calibration is scheduled annually and will take most of a Saturday to complete. The monthly calibration check can be completed in less than an hour if preparations for efficiency are in place. The last patient on the treatment schedule does not leave the treatment room until 5:40—par for the course. I begin setting up. I check the optical patient alignment facilities first. A light in the collimator is set outside the path of the radiation treatment beams and the optical beam it projects is reflected through a thin mirror to provide a simulation of the radiation treatment beams. The virtual location of the light source is at the x-ray target that produces the x-ray beam. A crosshair in the collimator should pass through the beam centre and accurately divide the light field into four quarters. I must verify that the squares of light produced by the variable collimators conform within limits to the radiation field and that the crosshair is aligned with the field centre. I have a plastic plate with the squares of the four quarters and the field centre marked on one side. Lead markers are embedded in the plate where the edges of the radiation fields should strike the plate. I place the plate over a sheet of x-ray film protected from exposure by a sealed paper packet, align the plate with the field light on top of the film, and expose the film. The images of the lead markers on the developed film mark the boundary of the light field, and the developed x-ray exposure shows the boundary of the treatment beam. They should align. Tonight, they do, and I am relieved because adjusting the positions of the field light and the crosshairs is a tricky and time-consuming process. But now I know I can rely on the field light and crosshairs to align my other measurement instruments.

Next, I check the laser sidelights. I have an intimate relation with laser sidelights. In my previous position working at the Massachusetts General Hospital (MGH) in Boston there were no such devices in existence as we approached the opening of a new cancer treatment facility. Having a familiarity with optical components, I designed a device that used a helium–neon laser that produced a small diameter beam. I fixed a cubic beam splitter (splits the light beam into a vertical and a horizontal beam) and a cubic beam mirror (turned the vertical laser beam 90° to be a second horizontal beam), and two cylindrical lenses (mounted to turn the two small diameter laser beams into broad, diverging sheets of red light) to project a bright red crosshair across the room. Three such units mounted on the two side walls and the ceiling of the treatment room created light sheets that were aligned to meet at the isocenter of the linac (see Fig. 2). These laser sidelights worked well and were the object of admiration of several curious visitors to the MGH. It was no surprise that within six months, at the next annual meeting of the AAPM, the Gammex Company was offering essentially identical devices for sale. These laser localization devices are now used on every linear accelerator installed in radiation therapy departments (Fig. 11).

Plastic jigs facilitate the rapid check of the alignment of the laser crosshairs with the treatment machine. I set this system up and check the alignment. One laser line from one of the wall-mounted lasers is slightly off at the machine centre of rotation. A small offset at the isocentre of the machine becomes more significant towards the lateral skin surfaces of a patient. Since the patient is aligned by marks on the lateral skin surfaces, this can lead to more serious mispositioning of the patient. The side-lights are designed so that fine adjustments of the positions of the lines are easy to accomplish. I make the adjustment and recheck the alignment. All optical guides are now satisfactory.

Fig. 11   Laser sidelight mounted on wall of treatment room (to left) and on patient phantom (to right)

A three-inch (7.6 cm) diameter circular hole has been built in the outer wall of the treatment room at an angle to prevent scatter radiation from exposing the radiation therapists working outside the treatment room. I have placed an instrument cable in this channel. I connect one end of the cable to the electrometer outside the room and the other end to the ion chamber's cable inside the room. I place the slabs of plastic used for the calibrations on the patient support assembly. A channel bored in one of the plastic slabs to fit the ionization chamber is aligned with a crosshair scribed on the top of the slab to align the chamber inside the plastic with the centre of the beams (see Fig. 12). Other water-equivalent plastic slabs are used to place the ionization chamber at a required calibration depth. I carry a thermometer and a barometer inside the room along with the logbook and measure and record the temperature and pressure for calculation of a correction factor for the ionization chamber. I use a standard format for the recording of the data and results of calculations so that the data are presented uniformly from month to month in the logbook. The calibration uses procedures and data developed by committees of the AAPM. This contributes to uniformity of calibrations across the country. The logbook will be reviewed by physicists from the Radiological Physics Center (RPC), a facility based at the MD Anderson Cancer and Research Hospital in Houston. The purpose of the RPC [now called the Imaging and Radiation Oncology Core (IROC)] is to verify uniformity of dose calibrations around the country by sending teams of medical physicists

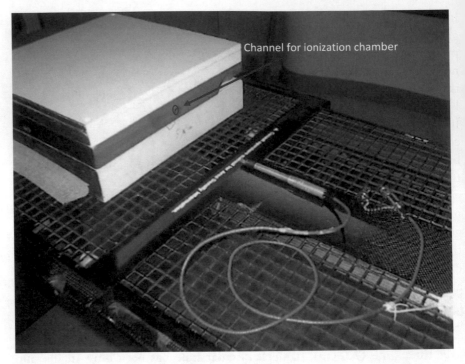

Channel for ionization chamber

Fig. 12 Water equivalent plastic calibration slabs and ionization chamber

to participating major treatment centres to perform independent calibrations of the centre's machines. The centre's calibrations are compared with the RPC measurements. The results are reported back to me and are also used to formulate an assessment of the overall uniformity of radiation dose across the country. An RPC team is scheduled to visit the CTRC in a month or so. I look forward to discussions of our procedures and constants used in the calculations with the RPC physicists whom I have known since I was a student. These discussions provide an opportunity to confirm that we are up to date with our calibration procedures.

9   The Workday Ends

Tonight, I work through the calibration checks for the two x-ray energies and the multiple electron energies of the linac. All treatment modes are found to be accurately calibrated to within 1%. If I were to find any energy mode calibration outside that limit, I would need to make an adjustment on the linac's control electronics to bring the calibration within the 1% limit. I check the logbook for any signs of systematic drift of calibrations. Such drifts would indicate a failing electronic component. I would then carry out an investigation of the electronics along with Jeff. Tonight, I am grateful to find no problems other than the small laser sidelight adjustment that has been duly noted in the logbook. I place the instruments and plastic jigs and fixtures on the cart along with the electrometer and the ionization chamber in its case. I wheel the cart back to its closet, stow the logbook and head to my office. It is quiet and cool inside the CTRC except for a gentle purring of the air conditioning system. I hang up my lab coat, drop the shielding report in my briefcase in case I feel like penciling in changes later tonight, and head for the parking lot.

It's a quiet ride home with only a few eighteen wheelers and pickup trucks to cope with along the interstate. Dinner is waiting and I feel I have done a good day's work. It has been an interesting day to say the least. The fascinating things I get to do and the technology I get to work with, along with the camaraderie of my fellow staff and the medical physicists I have come to know around the country, and indeed around the world, are why I love being a medical physicist. Surely it must be one of the most satisfying professions one can enter. Two tail-wagging mutts great me joyfully as I pull up the driveway.

Arthur (Art) Boyer is a retired medical physicist with over 40 years of experience. He received a Bachelor of Arts in Physics from the University of Dallas, and a Master of Arts in Physics, and a Doctor of Philosophy in Physics from Rice University. After a Postdoctoral Fellowship at the MD Anderson Hospital and Tumor Institute, in Houston, Texas, he worked from 1971 to 1979 in the Radiation Oncology Department at the Massachusetts General Hospital and at the Harvard Medical School. From 1979 to 1986 he was Chief of Physics at the Cancer Therapy and Research Center and was an Assistant Professor at the University of Texas Health Science Center in San Antonio. From 1986 to 1995 he was Professor of Radiation Physics at the MD Anderson Hospital and The University of Texas Graduate School of Biomedical Sciences. From 1995 to 2005 he was a Professor and Director of the Physics Division of the Radiation Oncology Department of the Stanford School of Medicine in Palo Alto, California. From 2005 to 2012 he was Director of the Radiation Physics Division of the Radiation Oncology Department, Scott & White Clinic, Temple, Texas and a Professor in the Texas A&M College of Medicine.

Awards

2007. William D. Coolidge Award—49th Annual Meeting of the American Association of Physicists in Medicine (AAPM). This award recognizes an AAPM member for an eminent career in medical physics. It is the highest award given by the AAPM.

1995. Visiting Professorship: Deutches Krebsforshungszentrum (German Cancer Research Center), Heidelberg, Germany.

1992. The Robert J. Shalek Award, AAPM Southwest Chapter in recognition of exemplary service to, or representation of, the chapter or to a member who has recently achieved landmark contributions to the field of Medical | Physics.

1979. Silver Plaque for Scientific Exhibit Achievement; The American Society for Therapeutic Radiologists 21st Annual Meeting.

6

YOU Can Change the Medicine

James A. Purdy

1   Introduction

So, you're looking for a career path are you? Well. I've got two words for you—Medical Physics. Oh, you've never heard of medical physics? Well, don't worry, I hadn't either when I started out back in the late 60s. Let me share some stories with you that may prompt you to take a look.

One of the things I liked most about being a medical physicist—particularly a radiation oncology physicist, was that you never knew what challenges you might face each day you came to work. It might be a complicated patient treatment that you were called in to help with, a new technology that was being considered for clinical use, a new research initiative, a new clinical trial that needed your input ... a never ending list.

I also found out early on that this was a profession in which there was great opportunity to make things better and safer. Yes, YOU can change this medicine. And yes, we can still do better, a whole lot better.

When I taught residents, both physicians and physicists, I made sure they heard that message, loud and clear. I wanted them to realize just how lucky they were to have found radiation oncology as a career.

J. A. Purdy (✉)
Washington University School of Medicine, Saint Louis, MO, USA
e-mail: jpurdy02@sbcglobal.net

J. Van Dyk (ed.), *True Tales of Medical Physics*,
https://doi.org/10.1007/978-3-030-91724-1_6

Now, let me tell you a little bit about myself and how I found medical physics. For the record, I grew up in Orange, Texas, a small town on the Texas-Louisiana border near the Gulf Coast. I earned my physics undergraduate degree in 1967 from nearby Lamar University; followed by a master's degree in '69 and a Ph.D. in experimental nuclear physics in '71 from the University of Texas at Austin.

With my doctorate in hand, I expected to find lots of opportunities just waiting for me. That turned out not to be the case. There'd been major cuts in research programs: NASA's (the US National Aeronautics and Space Administration) funding peaked a few years prior to '69; now its budget was half that, requiring massive layoffs. Coupled with the early 70s economic downturn, it was a NIGHTMARE for a newly-minted nuclear physicist. Application letter after letter came back with no job offer.

In December, I read in the newspaper about President Nixon's "War on Cancer" initiative—the US National Cancer act of 1971. Major funding was now available for cancer treatment research. About that same time, I found out that MD Anderson Hospital and Tumor Institute in Houston, Texas (one of the largest cancer therapy hospitals in the US) was putting on a series of courses on radiation therapy treatment planning and dosimetry. *I was desperate!* I enrolled immediately, staying at my wife's (Marilyn) parents' home just outside Houston. Marilyn and our two baby daughters (Kathy-age three and Laura-age two) remained in Austin. I spent the next two months soaking up as much radiation dosimetry expertise that I could.

While there, I found out that MD Anderson had been awarded an NCI (National Cancer Institute) Cancer Training Grant. I hurriedly sent in my application for a post-doctoral medical physics position and waited.

Finally, the letter came. I was accepted. *I jumped ten feet high—Hallelujah!* The post-doc training was intense, with lots of long hours in the evening, learning the intricacies of high-energy photon and electron beam dosimetry. (High-energy electron and photon beams are produced by linear accelerators used for cancer treatment.) The good news was this training led to multiple job opportunities.

In January '73, I joined the Washington University in St. Louis (WUSTL) School of Medicine faculty as an instructor in the Division of Radiation Oncology, Mallinckrodt Institute of Radiology (MIR). I served there until June 30, 2004 rising to the rank of tenured Professor of Radiation Physics in the Department of Radiation Oncology; for a good part of that period, I also served as MIR's Director of the Radiation Physics Division and Associate Director of MIR's Radiation Oncology Center. On July 1, 2004, I joined the University of California Davis (UC Davis) Medical Center's Department of

Radiation Oncology faculty as Professor and Vice Chair helping them revitalize their physics program as they worked to become a National Cancer Institute (NCI) Comprehensive Cancer Center.

In the following sections, I want to share with you several of my personal experiences as a radiation oncology physicist working to improve patient treatments and to make radiation oncology a safer medicine. Enjoy.

2   Change

My first week at MIR blew me away. Remember, this was my first career job. The clinic was undergoing huge changes. Installation of a major new treatment machine, the Varian Clinac 35, had begun. This caused major disruptions in staff and patient flow. Also, a new radiation beam-shaping system had been developed there and just put into clinical use—BIG PROBLEM! And, if that weren't enough, the Physics Section was in rapid flux. It was chaos.

But, that chaos taught me a valuable lesson. One that served me well over the course of my career. Heraclitus, the Greek philosopher is credited with first saying, "The only thing constant is change." And he was so right.

In 1973, the physicians were called radiation therapists; today, they are called radiation oncologists. Individuals who operated the radiation treatment machine were called radiation technicians (RTs); nowadays, radiation therapists. In fact, the medicine was called therapeutic radiology back then; now, radiation oncology. All changed—as I said, old Heraclitus really knew what he was talking about.

Dr. Bill Powers, the Division Director, was the driving force for all these MIR changes—a tall man, overweight, with a ruddy complexion and a bellowing voice that he didn't hesitate to use. He had been awarded a large NCI program project grant and had also obtained a large donation from the Maytag Family Foundation. He strived to turn the MIR radiation oncology program into a major force.

Developing a better way of shaping radiation beams was his present focus. The old method was as follows: Radiation beams were shaped by adjusting the treatment machine collimators (large lead jaws located in the treatment machine head) to a specific rectangular opening (length and width). The radiation beam passed through the collimator opening irradiating the patient's tumour volume and surrounding normal tissue while the rest of the patient's body remained shielded. In some cases, lead blocks were placed on a plastic tray attached just below the collimators to shield some normal tissues within the rectangular opening.

Dr. Powers had challenged the Physics Section to find a way to conform the shape of the beam to the irregular shape of the tumour so to minimize normal tissue damage. One of the physicists who had left (Andrezj Demidecki) had discovered a low melting-point lead alloy, Cerrobend, used in the plumbing industry. At temperatures above 158 °F (70 °C), it could be maintained as a liquid in a vat similar to a coffee urn.

Let me explain how the customized Cerrobend blocks were made. Radiograph(s) (an x-ray film) were taken with the patient in treatment position on a radiotherapy simulator. Note, a simulator mimics the allowed motions of a radiotherapy treatment machine and uses a diagnostic x-ray tube to simulate the radiation properties of the treatment beam.

The physician then drew the desired block aperture on the film(s) using a grease pencil.

Those radiographs were taken to our "block room" where a technician created a mould cut from a piece of Styrofoam™ by following the grease outline on the film with a pointer connected to a heated wire. The Styrofoam was located at the distance that the shielding block would be placed on the treatment machine and the hot wire was made such that it followed the geometry of a ray from the x-ray source (see Fig. 1).

The technician then poured liquid Cerrobend into the foam mold to harden. Once cooled, the technician bolted the block to a plastic tray that could be slid into a holder just below the treatment machine collimators. This procedure took 24–48 h from the time the radiographs were received.

This was great. Now, an irregularly shaped lead-alloy block could be created that conformed to each patient's tumour, thus reducing the amount of normal tissue irradiated (see Fig. 2).

Also, the newer isocentric treatment machines (machines that rotated around a horizontal axis, see Fig. 2 in Chap. 5) could be rotated around the supine- or prone-positioned patient (patient lying face-up or face-down), allowing conformed x-ray beams to be delivered from any gantry angle as specified by the treatment plan.

Well, it was great on paper, but not so much in practice. There were just too many steps where errors could occur. Sometimes the blocks were too heavy for the RT to lift. At other times, the block did not fit the shape drawn by the physician (due to hot-wire drag caused by moving the pointer too fast). Hence, the patient could not be treated that day—NOT GOOD! The patient got upset, the RTs got upset, the radiation therapist got upset—EVERYONE WAS UPSET! They all wanted to go back to the old way—just use simple lead blocks stacked on a tray. Remember now, I'm the new guy on the block amidst a near riot in the clinic.

Fig. 1 Technician tracing field outline on x-ray film using the hot-wire cutter. The sections cut out of the Styrofoam will be filled with Cerrobend shielding

Dr. Powers called an emergency staff meeting. He roared, "We have to try harder; we have to find a way to make this work." I swear, he was looking right at me the whole time.

I learned quickly that when things aren't working in the clinic, Physics is going to catch the heat. The procedure was complicated involving transfer of critical simulation and treatment geometry data—film distance, blocking-tray distance, and treatment distance ... all needed to be accurately set and recorded by the RT. The block-maker then needed to recreate that geometry on the hot-wire cutter and mount the finished Cerrobend block correctly on a plastic tray.

With Dr. Powers' roar echoing in my mind, I dug into the problem. It became clear to me that a much sturdier hot-wire cutter was needed—one

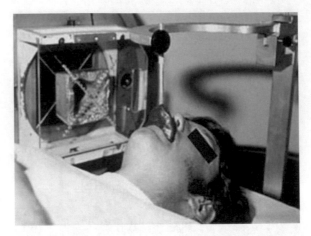

Fig. 2 Patient treated with a lateral radiation beam shaped with Cerrobend block and positioned using bite-block device

where the distance between the pivot point of the cutting arm and the radiograph could be precisely set and maintained. Scales were added to enable precise distance adjustments. The hot-wire portion of the cutting-arm was shortened to minimize lag as the block-maker traced out the grease-pencil shape. (Figure 1 shows the shortened wire version of the hot wire cutter. The original version had the wire extended from the source position to the x-ray film.) Easy to read procedures and forms were developed. All that, and lots of training sessions and feed-back mechanisms put in place.

Finally, the resisting users of the Cerrobend system became avid supporters. Result—patients were happy, RTs were happy, radiation therapists were happy. Everyone was happy, especially me, still remembering Dr. Power's intense stare and roaring voice. In fact, this effort actually led to one of my first journal publications on quality control.

Over the next several years, the MIR Cerrobend field shaping system became a mainstay in radiotherapy clinics throughout the world. It was a major step toward developing what today is called "Three-Dimensional Conformal Radiation Therapy (3DCRT)," a powerful technique that allows radiation oncologists to prescribe a much higher radiation dose to the tumour while keeping adjacent normal organs protected.

Yep, old Heraclitus was sure right, "The only thing constant is change."

3   The Board Certification Wars

When I took my first faculty position in radiation oncology physics at MIR, I thought all I needed to do in order to be successful was focus on maintaining a high level of technical expertise and pursuing innovative research to improve patient care. I hadn't really thought much about things like "board certification," "licensure," and other professional matters. Wow, was I ever naïve!

I did understand that in 1973, if you had to have heart surgery, you darn sure wanted to have a board-certified heart surgeon do it. But, if you had to undergo radiation therapy, I had no idea that you better make sure you had a board-certified medical physicist on your treatment team.

In July, I attended my first Annual Meeting of the American Association of Physicists in Medicine (AAPM). The organization was relatively young—only fifteen years old and a little over 800 members. The meeting was held in the Hotel del Coronado—a beautiful beachfront setting just across the bay from San Diego. One of the perks I quickly learned about is that medical meetings are typically held in great venues.

All the scientific and technical presentations were interesting, and I learned a lot. However, it was the heated hallway conversations that I observed that were the most eye opening. Some individuals argued strongly for the need to distinguish the differences between those physicists working in the clinic and those working strictly in the laboratory. Also, I heard passionate discussions about the current state of board certification for medical physicists. In '73, certification was being done by the American Board of Radiology (ABR)—the same board that certified physicians in diagnostic radiology and therapeutic radiology.

I listened intently as several physicists strongly supported the formation of a new clinical physics society independent of the AAPM along with a new certifying board run by clinical physicists. Others wanted to maintain the current status quo where AAPM appointed separate Councils (Science, Educational, and Professional) to address matters. They also wanted to leave board certification to the ABR.

I did my homework when I got back to St. Louis and found that this argument—science versus professional matters—had created tension in the AAPM from its very beginning. The main complaint against the ABR was that there was no real physics representation on its certifying board. In the early years, the examining physicists met with the ABR Trustees only when the results of the new physicists' examinations were being discussed. It was not until 1968 that a physicist was allowed to attend all meetings of the ABR

Trustees, but that individual participated with "voice only and no vote." *I couldn't believe it.*

But, I put professional politics aside and quickly focused on becoming board certified obtaining my ABR certificate in Therapeutic Radiological Physics in June 1976.

As the AAPM's membership grew in the 70s, it began to flex its muscle and sought to become an official "sponsoring organization" of the ABR. But, it was like pulling teeth. Finally, in 1979, the ABR approved the appointment of an official "Medical Physics Trustee" to its board. *Hallelujah!* But, unfortunately, they still wouldn't approve the AAPM as a sponsoring society.

That same year, I was elected to a three-year term on the AAPM's Board of Directors. The ABR failure to recognize the AAPM as an official sponsor did not sit well with me nor with many of the other board members. By '82, the issue reached a boiling point. That year the AAPM Board of Directors voted to support the formation of a college for clinical physicists (26 for, 1 against, 3 abstained) and in '83 the American College of Medical Physics (ACMP) was formed.

In 1984, I was elected president-elect of the AAPM and began a three-year service commitment on the AAPM's Executive Committee (EXCOM)— serving as President-elect, President, and Chairman of the Board of Directors. I and my EXCOM colleagues invested immeasurable time and effort to convince the ABR to approve AAPM as an official sponsoring organization, with three trustees like the other sponsoring organizations. However, the ABR dug in its heels and the answer was always no.

Thus, the ACMP became the vanguard for medical physics certification and licensure efforts, and in 1987, became the first sponsoring organization of the American Board of Medical Physics (ABMP). In July 1988, I am proud to say that the ABMP was incorporated in the state of Massachusetts as an independent, not-for-profit corporation with Dr. Ned Sternick as Chairman of the Board and myself as Vice-Chairman.

The ABR quickly took notice, and in that same year, invited the AAPM to become an official ABR sponsoring organization. *Yea for Medical Physics!* However, the AAPM was restricted to having only one trustee. *Boo—talk about shooting yourself in the foot!* This infuriated me and other medical physicists even more, because we believed that the one trustee restriction, compared with the three trustees the other sponsoring organizations were allowed, inferred "second-class citizenship" for the AAPM and, more importantly, for the medical physics profession.

Having two certifying boards now became the divisive issue. Many ABR certified physicists felt strongly that it was important that the medical physicist certification process be recognized by the American Board of Medical Specialties (ABMS). Others felt that this was not important because medical physicists were not physicians, and the ABMS charter is aimed specifically for boards certifying physicians. ABMP-certified physicists argued that in the future, areas of medical physics other than radiology and radiation therapy would likely present new opportunities for medical physicists. Thus, it was essential that the medical physics certification process not be limited to just radiological areas. The debate (some say war) raged on. Unfortunately, in many instances, the debate was more emotional than rational.

The good news was that board certification among practicing clinical physicists soared for both the ABMP and the ABR. And that was good for the profession and for the cancer patient. However, arguments regarding duplication of effort kept the divisiveness going. Thankfully, in '94, the ABR finally got the message and approved three Physics Trustees for the AAPM. *Yea for Medical Physics!*

That took a lot of wind out of the sails for the ABMP and eventually the two boards reached a compromise. In 2001, the ABMP leadership agreed to discontinue certification exams in the traditional disciplines of Radiation Therapy Physics and Diagnostic Imaging Physics. However, they would continue to offer certificates in Magnetic Resonance Imaging Physics and Medical Health Physics. In exchange, the ABR agreed that they would issue "Letters of Certification Equivalence" to existing ABMP diplomates if requested in Diagnostic Imaging Physics and in Radiation Therapy Physics.

On reflection, I believe that the existence of the two boards was positive for medical physics certification and should have continued. I believe that "certification" was the important issue, not the "ABR versus ABMP" situation. I was proud of our over 60 years of working with the ABR in the certification of clinical medical physicists. But, I was also proud of the establishment of the ABMP that offered medical physicists peer certification.

One thing for sure, the existence of competing boards greatly improved the certification process and helped the AAPM become a full-fledged official ABR sponsoring organization. And that was good for the profession and for the CANCER PATIENT, too!

4   Collaboration—The NCI Treatment Planning Research Contracts

The 1980s was a decade of enormous growth and excitement in the field of radiation therapy. This was due primarily to the availability of x-ray computed tomography (CT) and the increasing power of computers coupled with their decreasing cost. The CT scan provided, for the first time, a fully three-dimensional model of the cancer patient's anatomy that could be used for computerized treatment planning.

Early in that decade, the National Cancer Institute (NCI) recognized this and announced several "Request for Proposals (RFPs)" that would create "Collaborative Working Groups (CWGs)" to expedite the development of specific areas of radiation therapy treatment planning and dosimetry.

The first such proposal was focused on proton and other heavy particle radiation therapy treatment planning. The second, which is the subject of my story, was directed to external photon beam treatment planning—specifically three-dimensional radiation therapy treatment planning (3DRTP) and what came to be known as three-dimensional conformal radiation therapy (3DCRT).

At that time, 3DRTP was in its infancy. In fact, no 3DRTP computer systems were commercially available, and I'm pretty sure no one knew just what all the 3DCRT process involved.

The photon group consisted of teams from four institutions.[1] Each team was made up of medical physicists, radiation oncologists, dosimetrists, and computer scientists. The NCI Project Officer was Dr. Al Smith, a fellow Texan and a medical physicist I first met during my post-doctoral training at MD Anderson. I was the Principal Investigator (PI) for the WashU contract.

The contract listed obligatory "Deliverables" with defined deadlines. Four group meetings (three to four days each) were held each year, rotating between each institution. The meetings always went long into the evenings and were intense, with each group providing updates of their work. The face-to-face meetings were also extremely fruitful, as ideas and discussions flowed freely, and it became obvious during that first year that this effort was going to be a really "big deal."

I have to admit though, my first year was a little rocky having to deal with what I will refer to as the "East Coast Attitude." Now, don't get me wrong.

---

[1] The four funded institutions were: Harvard Medical School/Massachusetts General Hospital (MGH); Memorial Sloan-Kettering Cancer Center (MSKCC); University of Pennsylvania School of Medicine/Fox Chase Cancer Center (Penn); Mallinckrodt Institute of Radiology, Washington University School of Medicine in St. Louis (WashU).

These were highly intelligent individuals that I greatly respected, but after a while, in a supposedly "collaborative" effort, I could only take so much of "I know better than you how it should be done" attitude.

Pressure on the groups always mounted as deadlines neared. This caused tension among the groups because developmental progress was uneven. In retrospect, it was an overly ambitious undertaking and the individual contracts were grossly underfunded as they required a great amount of developmental work. In particular, each institution had to design and develop a 3D treatment planning computer system.

In addition, the group had to agree on a methodology for evaluating treatment plans quantitatively. This had never before been carried out on this large a scale. My recollection is that it was like "herding cats" before we finally got to an agreement.

To explain—3DCRT treatment planning required the physician to go through an entire set of CT slices (images) for a patient and precisely delineate (outline) the tumour/target volumes and all critical normal organs in which radiation dose was a concern. The radiation dose distribution was then calculated for the entire volume irradiated and the ensuing treatment plan evaluated. This process represented a radical change in practice for the radiation oncologist and treatment planner (and it was very time consuming).

The 2D treatment planning approach (the standard of practice at the time) required the physician to only draw the desired treatment beam shape on an x-ray film taken with the patient in treatment position. The radiation dose distribution was then calculated and superimposed on only a single slice CT image of the patient.

The contract years flew by. Egos got pushed aside and teamwork took hold. And while the collaborations were sometimes tense, I believe all participants agreed, that in the end, working together made our conclusions and recommendations much, much stronger.

With the research goals accomplished, all that remained now was to agree on "How do we disseminate the information?" The results of the first NCI cooperative group on protons had ended up merely as an NCI report—with only a few hundred copies made. It never underwent peer review nor published in the open literature. That was a real shame because important treatment planning and plan evaluation concepts and processes had been developed. I was determined not to let this happen to our group.

I still recall the rancorous meeting on deciding how to publish the findings. Unfortunately, egos resurfaced again. After lots of heated discussion, we finally reached agreement to break the work up into a series of manuscripts that would be submitted to the International Journal of Radiation Oncology,

Biology, and Physics, general known as the "Red Journal" because of the colour of its cover; it is the field's top journal. A lead author for each paper was agreed on, along with individuals from each institution that had contributed to the work. I agreed to oversee this final effort and make sure all the papers were completed on time.

Well, we did it! The results filled a full issue of the Journal, as shown in Fig. 3. There is no doubt in my mind that this effort played a key role in spurring the development and adoption of 3DCRT as the standard of practice and stimulated many other developments that now allow radiation

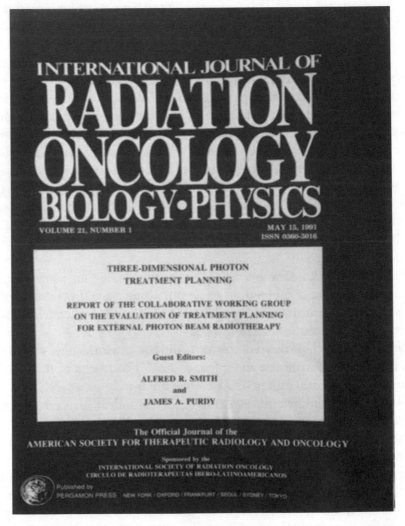

Fig. 3 International Journal of Radiation Oncology, Biology, Physics. Vol 21(1), 1991

oncologists to significantly increase dose to most tumour sites while lowering doses to critical disease-free organs.

And now, I have to give credit to all the other members of the WashU team[2]—they were great! In fact, they were so good, we were successful over the next decade in being awarded two more of the multi-institutional NCI research contracts: NCI Research Contract N01-CM-67915, "Evaluation of High Energy Electron External Beam Treatment Planning" and NCI Research Contract NCI-CM-97564-23, "Radiotherapy Treatment Planning Tools".

It seemed that once 3DCRT got going and physicians got on board defining 3D volumes, radiation oncology innovation literally exploded. Amazing conformal radiation treatment planning and delivery systems evolved: Tomotherapy at Baylor College of Medicine and the University of Wisconsin, CyberKnife at Stanford, and image-guided linear accelerators (linacs) at William Beaumont Hospital, Detroit. Systems that allowed the physician to literally paint the dose conforming to the cancer volume.

I'm happy to know that our group played a role in changing the medicine.

5   Hitting the Target—The ICRU 50 Report

"NO, NO, NO, WRONG, WRONG, WRONG," screamed Dr. Fletcher. He continued to berate the sheepish-looking young physician as he moved to the x-ray lightbox showing the radiation fields to be used. Dr. Gilbert Fletcher was the chairman of the Radiotherapy Division at M. D. Anderson (MDA) Hospital and Tumor Institute in Houston, Texas, where I was just starting my post-doc training in '72. He was a short, thin man with the thickest French accent I'd ever heard. He was also a giant in the field, pushing MDA to be one of the top cancer hospitals in the world.

Such instances like this were a common occurrence in Current Case Conference where the new patient's treatment-regimen underwent peer review. There, I learned pretty quickly how complicated radiation therapy treatments are. It's truly a medicine where flawless communication is critical. You see, the physician is not the one who actually performs the treatment. That's done by a radiation therapist who puts the patient in treatment position and turns the radiation machine on. Most importantly, the physician defines the volume to be irradiated and prescribes the radiation dose to be

---

[2] WUSTL Team: Radiation Oncologists-Bahman Emami, Joe Simpson, Mike Pilepich; Medical Physicists-Jim Purdy, John Wong, Bob Drzymala, Bill Harms: Dosimetrist-Janice Manolis: Computer Scientist-John Matthews.

delivered to that volume and also the dose limits for critical normal tissues. Another individual (the dosimetrist) calculates the dose distribution based on that prescription; another person (the medical physicist) calibrates the treatment machine so that the machine actually delivers the prescribed dose. Yeah, it's complicated. Non-ambiguous procedures for prescribing, recording, and reporting all this information are essential. And Dr. Fletcher made darn sure all of us understood that.

Consider defining the target volume. It is not easy, even with modern imaging like CT and magnetic resonance imaging (MRI). For one thing, you can't image microscopic disease, and secondly the treatment typically is not given in a single shot, thus requiring the patient to be placed in the same treatment position five days per week over a series of weeks. That's impossible without bolting the patient to the treatment table. In addition, in many anatomical sites, the tumour moves inside the patient throughout a treatment regimen due to physiological processes like bladder or rectum filling, breathing, and heart beating.

When I started my training, the radiation oncologist didn't really delineate a target volume per se. Instead, he/she specified a set of radiation fields to be used. These beam arrangements created a volume irradiated to high dose.

Over that decade, field shaping as describe previously improved as custom-made patient-specific lead-alloy blocks replaced conventional lead blocks to shield normal tissues. That put more pressure on the radiation oncologist to actually delineate a target volume. Whereas a radiologist could point to a radiograph and say, "That's the tumour," the radiation oncologist had to be precise and draw a boundary and say, "That's the volume to be irradiated to a high dose." Big difference.

In the 80s, things really changed. CT provided a fully three-dimensional model of the cancer patient's anatomy. The radiation oncologist could now delineate tumour volumes and nearby organs more accurately. In addition, relatively low-cost powerful computers became available. These technologies spurred several university groups[3] to develop powerful three-dimensional radiation therapy treatment planning (3DRTP) systems. These provided the physician with software tools for contouring the imaged tumour along with any normal organs.

It was in this early stage of the 3D treatment planning era that I received a letter from the International Commission on Radiation Units and Measurements (ICRU), an organization whose principal objective was the development of recommendations regarding the use of radiation. They invited me

---

[3] These included the University of Michigan, University of North Carolina, Memorial Sloan Kettering Cancer Center, Massachusetts General Hospital and Washington University at St. Louis (WUSTL).

to join a new ICRU Report Committee charged with updating an older ICRU report (Number 29) on radiation therapy treatment specification. They particularly wanted to address 3DRTP.

Well, I didn't know what to do. Most of these types of reports end up on a bookshelf and don't make the impact a medical journal publication would. And I knew this would entail a lot of meetings taking me away from family and work. My plate was already plenty full with my Section Head administrative responsibilities including trying to build a world-class medical physics group, my clinical physics responsibilities, teaching residents, and my research on 3DRTP. All that and my involvement in national medical physics societies. We had made real progress in the professional development of the medical physicist establishing the ACMP and the ABMP. I just didn't know if I wanted to take on this added task.

I finally convinced myself to do it. I had learned early on at MDA the importance of clear, radiation therapy prescribing, recording, and reporting procedures and 3DRTP demanded this even more so.

The committee had representatives from several countries and I and Nagalingam Suntharalingam from Thomas Jefferson University in Philadelphia would be the only U.S. physicists on the actual committee. Also, as it turned out, our WUSTL clinic was the only one of the group that had a real 3DRTP system that was in clinical use. I had gained considerable experience participating in the NCI Research Contract on external photon beam 3DRTP. I knew we had to address the tumour microscopic disease and spatial uncertainty issues.

After many meetings, we finally agreed on a methodology for volume specification that separated the old ICRU Report 29 *Target Volume* into three new volumes: the *Gross Tumor Volume (GTV)*—the part of the tumour that is visible with the use of 3D imaging; the *Clinical Target Volume (CTV)*—a clinically relevant volume that includes the GTV as well as subclinical and microscopic anatomical spread patterns (created by adding margins around the GTV); and the *Planning Target Volume (PTV)*, created by adding margins around the CTV to account for patient setup spatial uncertainties and for internal movement of the tumour inside the patient.

Finally, on September 1, 1993, ICRU Report 50 was published (Fig. 4). There was some pushback at first. Change is always hard to accept, particularly since it required a whole lot more "hands-on" work by the physician in drawing contours on each CT slice. But, over the next several years, we conducted several 3D Symposiums and American Society for Radiation Oncology (ASTRO) workshops specifically focused on the use of the ICRU Report 50 recommendations, and contouring tutorials for multiple disease

ICRU REPORT 50

Prescribing, Recording, and Reporting Photon Beam Therapy

INTERNATIONAL COMMISSION
ON RADIATION UNITS
AND MEASUREMENTS

Fig. 4 ICRU Report 50: *Prescribing, Recording, and Reporting Photon Beam Therapy.* Bethesda, MD, 1993

sites. Commercial 3DRTP systems soon got US Food and Drug Administration (FDA) approval, and momentum built quickly for making 3DRTP the standard of practice.

Once physicians got on board defining 3D volumes, it wasn't long before linac manufacturers developed computer-controlled multileaf collimators (MLC) to replace the Cerrobend blocks (see Fig. 3 in Chap. 9 by Clif Ling for a picture of MLCs). Next came planning systems that required the physician to input dose-volume constraints (i.e., how much volume of an organ could receive a specified dose) rather than just prescribing beam directions and used optimization algorithms to generate the dose distribution. This was referred to as Intensity-Modulated Radiation Therapy (IMRT) and is capable of generating concave isodose volumes that provide extremely conformal target volume coverage and avoidance of designated sensitive normal structures. The increased use of IMRT focused attention on the need to better account for spatial uncertainties, which helped spur development of treatment machines with integrated advanced imaging capabilities referred to as Image-Guided Radiation Therapy (IGRT).

The ICRU kept pace with these advancing technologies forming several more Report Committees addressing prescribing, recording, and reporting radiation therapy, and I had the privilege of serving on all of them: ICRU Report 62, ICRU Report 72, and ICRU Report 83.

Yeah, I have to admit, Dr. Fletcher really got my attention in those Current Case Review Conferences back in 1972. And I'm sure glad he did.

6   Medical Physics Residencies

"You know, it's really become a problem. We've got to do something" Faiz said.

"I know," I replied. "Linacs and treatment planning are just so much more complex today."

It was 1988 and I was talking with Dr. Faiz Khan at the AAPM Annual Meeting in San Antonio. We were discussing the current state of the training of clinical physicists—more specifically radiation oncology clinical physicists. Faiz and I had become good friends over the last decade, and often shared our views on medical physics. He was just beginning his three-year term on the AAPM Executive Committee as President-Elect, President, and Chairman of the Board.

I had completed a similar run a few years back, so I knew he was in position to really impact our profession. I also knew it would be a real challenge to make any impactful changes in training practices, particularly in a profession as diverse as medical physics. Change is really hard!

But we both knew we had to do something. The field had been rattled in the late 70s when it was discovered that hundreds of patients had been over-irradiated using a cobalt-60 teletherapy unit at a hospital in Ohio causing severe complications, including ten deaths—all due to a calibration error by a poorly trained clinical physicist. In '86–'87, the profession suffered through one of the worst radiation oncology incidents that ever occurred. Six patients were seriously injured or died after being treated using a Therac-25 medical linear accelerator. One patient described his treatment this way, "I felt a tremendous shock in my arm. I felt that my hand was leaving my body." The patient room monitor was not working, and he had to get off the treatment couch and pound on the treatment room door to get the linac operator's attention. The patient eventually lost the use of his left arm and both legs. He died from complications five months later. Horrible! This time a well-trained clinical physicist eventually solved the mystery before more patients could be harmed, pointing out a deadly "software bug" in the linac control system that came to be known as "Malfunction 54".

Yes, something had to be done! Faiz and I were both aware that entry into this field and the training practices varied greatly across the country.

"Why don't we have physicists train like doctors?" Faiz asked.

"Whattaya mean?"

"You know, real clinical rotations, just like the physician residents do," Faiz replied. "The physics resident would work side-by-side with a board-certified clinical physicist over a number of years, gradually taking on more and more responsibility in the clinical physics procedures necessary for the safe treatment of cancer patients."

"But for how long? How would you ensure that training standards are maintained?? Who would fund such residencies?" I asked.

"Well, I'm thinking about appointing a special Presidential Ad Hoc Committee to look into those issues," Faiz continued. "Would you be interested in being part of this?"

I paused and then said, "You know I would, but you also know I just agreed to serve as Vice-Chair for the newly formed American Board of Medical Physics. I'm just not sure I'd have the time to work on this."

"Yes, you do," he smilingly replied. And that's how it all started.

Dr. Ned Sternick chaired the Ad Hoc Committee. He asked me to head up the writing group for the clinical training of radiation oncology clinical physicists. Similar writing groups were established for Diagnostic Imaging and Nuclear Medicine clinical physicists. The committee met face-to-face several times and got lots of input from many experienced clinical physicists and physicians. We learned that every graduate medical education program for training physicians had to meet the standards established by the Accreditation Council for Graduate Medical Education and the Residency Review Committee. For medical physics, the situation was dramatically different, and it was worse than I had imagined. Only five of the thirty-four U.S. medical physics training programs had been accredited by the AAPM Commission on Accreditation of Educational Programs. The training programs were mostly being defined locally with no particular requirement to demonstrate to an independent accrediting organization the adequacy of facilities, curriculum, or faculty. Most were only one-year programs. *Unbelievable!*

Finally, on July 1, 1990 our report was ready. The AAPM Board approved and published it as AAPM Report 36 (Fig. 5). Bottom line—we recom-

Fig. 5 AAPM Report 36, 1992

mended mandatory completion of an accredited two-year hospital-based residency for all medical physicists before they could enter clinical practice.

The report was the easy part. Now, the hard work had to be done. An accreditation organization had to be established, financial resources had to be found, and two-year residencies had to be established.

Well, it took a while. But, in 1994, the AAPM, ACR, and ACMP came together to sponsor the Commission on Accreditation of Medical Physics Education Programs (CAMPEP) replacing the old AAPM commission.

I was able to obtain funding at WashU and changed our post-doctoral training efforts to a clinical physics residency program closely following Report 36 recommendations. And in 1997, I am proud to say that we became the first physics residency program to be accredited by CAMPEP. Today there are over one hundred CAMPEP accredited programs.

Like I said, "Change is hard." But, it can be done. You just have to roll up your sleeves and be willing to take on the challenge.

7   The ATC-The Whole Is Greater Than the Sum of Its Parts

What's the ATC? Well, you're just gonna have to wait a bit. First, I need to fill you in a little on cooperative group clinical trials back in the 90s. These cooperative groups were organizations in support of conducting clinical trials which generated data from different treatment institutions, ensuring consistency and quality of treatments and data reporting. Back then, there were about eight or nine cooperative groups. Several had their own quality assurance offices (QAO) to monitor clinical trial compliance. In addition, the NCI funded two national clinical trial compliance monitoring organizations—the Radiological Physics Center (RPC) located at the M. D. Anderson (MDA) Cancer Center and the Quality Assurance Review Center (QARC), a part of the University of Massachusetts Medical School.

Now, back to my ATC story.

Suddenly, my intercom buzzed followed by, "Dr. Purdy, Dr. Jim Cox would like to speak with you."

Dr. Cox was the chair of the Radiation Therapy Radiation Oncology Group (RTOG)—the nation's premiere radiation oncology clinical trial group.

"Jim, I need your help. What would it take to provide QA (quality assurance) for a multi-institution clinical trial using 3DRTP?" Jim Cox asked. He'd just learned that the NCI was considering making funds available to conduct a 3D dose escalation study for prostate cancer and he wanted RTOG to be in position to lead it.

Pausing for a moment, I answered, "I'm not sure, that's never been done before."

It was true that several clinics were now treating patients using university-developed 3DRTP computer systems and reporting exciting results. But, assuring protocol compliance for multiple institutions using this new technology was a whole different matter.

"I'll have to think about that and get back to you," I said.

It didn't take me long though to call Dr. Cox back. I knew the key would be having a way to verify that the 3D target volume and critical organs met protocol specifications. We'd learned from several 3D symposia and ASTRO workshops, that if you gave ten radiation oncologists a CT data set for the same prostate patient, you'd most likely get ten sets of contours for the PTV, rectum, and bladder that differed significantly. The contouring task really was the weakest link in the 3D conformal radiation therapy (3DCRT) process in those early years.

I also knew paper copies of the patient's treatment plan would not be adequate to QA the 3D volumes. We needed to figure out a way to enable the participating institution's 3DRTP computer to talk to our QA computer, i.e., we needed the patient's treatment plan in a digital data format that we could input into our computer.

Unfortunately, DICOM,[4] the digital data standard used for exchanging diagnostic digital images did not exist for radiation therapy (RT). But, NCI's 3-D Photon Treatment Planning Collaborative Working Group (which we were a part of), had worked out a way to exchange some RT digital data objects (target volumes, isodoses) using magnetic tape. *Aha, that's what we need to do.*

We just had to figure out what other RT data objects were needed for this clinical trial and write the software code to extend our existing RT magnetic tape code. Oh yeah, we'd also have to develop software for archiving and reviewing the data sets, and software for communicating back to the treating physician the results of our review. *Yeah, that's a lot of new software needed.*

Dr. Cox agreed and provided funding for me to purchase computer equipment and hire some software developers. Dr. Carlos Perez, MIR Radiation Oncology Division Chief, gave me space a few blocks away from our clinic to house the facility. We named it the RTOG 3DQA Center at WUSTL. *I was excited; history was being made.*

---

[4] DICOM: **D**igital **I**maging and **Co**mmunications in **M**edicine is the current digital data exchange standard for the communication and management of medical imaging information and related data.

Well, it didn't take long. On May 1, 1992, the NCI announced RFA: CA-92-05 (National Collaborative Radiation Therapy Trials–3D Dose Escalation Study for Prostate Cancer).

We submitted grants as both a participating institution and as a QA centre for the trial. In all, nine institutions (see Table 1) were funded. Dr. Cox chaired the group (3DOG). The ACR/RTOG Office in Philadelphia served as the Operations and Statistical Center and the WUSTL RTOG 3DQA Center provided the 3DCRT treatment planning QA.

Our first meeting focused on defining the minimum requirements for protocol participation regarding 3D treatment planning and QA guidelines. All agreed, the protocol would use the relatively new ICRU 50 nomenclature—PTV, CTV, and GTV. Most importantly, we agreed that each institution had to complete a "Dry Run" test proving their capability to develop a 3-D plan and transmit the required digital data to our 3DQA Center, prior to enrolling patients in the protocol.

Well, this caused a truckload of problems. It turns out getting eight other 3DRTP computers to talk to our 3DRTP QA computer was a whole lot harder than I imagined. Weeks turned into months, but still no success.

Physicians were clamoring to begin enrolling patients. "Without the digital data and software tools to review the target volumes, the project will fail," I argued. Those were pretty dark days, but luckily, Jim Cox and a slim majority of the institution principal investigators (PIs) supported me.

Over the next several months, we held multiple Data Exchange Workshops helping the participating institutions' software developers along with commercial 3DRTP system developers to implement what we called the "RTOG Data Exchange."

Finally, on July 14, 1994, the UNC became the first institution (outside WUSTL) to pass the Dry Run test and began enrolling patients. *Thank God!* It didn't take much longer for the others to finish implementing the RTOG Data Exchange and pass the test. Patient accrual took off.

Table 1 Prostate 3D Oncology Group (3DOG)

- Washington University at St. Louis (WUSTL)
- University of California at San Francisco
- University of Chicago
- University of Miami
- University of Michigan
- University of North Carolina (UNC)
- University of Washington
- University of Wisconsin
- Fox Chase Cancer Center

The 3DOG 94-06 protocol was eventually adopted by the RTOG and opened for all RTOG institutions provided they passed the Dry Run test. Over the next several years, we expanded the RTOG Data Exchange to provide volumetric treatment planning digital data transfer via the internet, improved the archival storage, and developed on-line QA review tools. Over a thousand patients were enrolled in the study and multiple manuscripts were published (and continue to be published) using the results of this trial. The methodology we pioneered in this first study became a major part of QA for RTOG clinical trials using advanced technology.

NCI took notice. They wanted the other clinical trial groups to get on board.

On May 11, '98, the NCI announced RFA: CA-98-006 (Advanced Technology Radiation Therapy Clinical Trials Support). This put significant grant funds in play to create a new QA resource to facilitate the conduct of 3DCRT clinical trials.

I quickly put together a grant proposal based on the experience we had gained from the RTOG studies. But this time, I knew we couldn't do it alone. My idea was to have the grant fund a consortium—the RPC, QARC, ACR/RTOG Office in Philadelphia, and the RTOG 3D QA Center—that took full advantage of existing clinical trial QA experience. I quickly talked with their PIs: Will Hanson (RPC), T. J. Fitzgerald (QARC), and Jim Cox (RTOG). They all agreed to join this effort and serve as subcontract PIs. We called it the ATC—the Advanced Technology Radiation Therapy QA Consortium.

This approach allowed the existing QA centers to work as an integrated group to develop uniform QA processes and infrastructure to address not only 3DCRT trials, but also future trials utilizing advanced techniques such as stereotactic body radiation therapy (SBRT), intensity-modulated radiation therapy (IMRT), and brachytherapy in both pediatric and adult patients. The regular meetings would promote sharing group expertise and resources—all the while providing coordination of these efforts. *This approach was new. It was big. We were all excited!*

Well, we were awarded a major U24 grant—over 1.5 million dollars yearly. But, to my surprise as well as the subcontractor PIs, a second U24 grant was also awarded to a University of Florida group—the Resource Center for Emerging Technologies (RCET) headed by Dr. Jatinder Palta. *I didn't understand.* The whole idea of our proposal was to UNIFY the development effort. Now, funds were diverted to an entirely independent group with no real experience with advanced technology clinical trials. *What the heck was NCI thinking?*

Having two independent QA centers proved to be problematic and RCET eventually lost its funding after eight years. Certainly, they tried, but their technology depended on a robust DICOM-RT being available; unfortunately, that data exchange remained a works-in-progress during those years and the RTOG Data Exchange remained the workhorse well into the twenty-first century.

Our ATC U24 was funded for over 13 years. We established technology requirements for institutions' participation in advanced technology clinical trials using 3DCRT, IMRT, SBRT, and brachytherapy that required the ability to submit the digital volumetric treatment planning and verification data to QA centres. The ATC adopted the 3DOG/RTOG 9406 protocol as a model as to how protocols should be written in terms of specifying the target volume and dose prescription, the credentialing requirements, and the QA process.

Admittedly, there were some problems at times—think two steps forward and one step back. But, we really did show that working as a group helped assure that high quality data were captured for all clinical trials utilizing advanced technology. I have to acknowledge the many individuals that made that happen: T. J. Fitzgerald, Marcia Urie, Ken Ulin at QARC; Will Hanson, Geoff Ibbot, David Followill at RPC; Jim Cox, Betty Martin, Wally Curran, Jim Galvin at RTOG; and Bill Harms, Walter Bosch, John Matthews, Jeff Michalski, Bill Straube at WUSTL RTOG 3D QA Center. Also, special recognition needs to go to Jim Deye, the Program Director in the Radiation Research Program at NCI for a good portion of the ATC years. We really worked well together.

I retired in 2011 and turned over the ATC PI responsibility to Dr. Jeff Michalski, a senior radiation oncologist at WUSTL. I had proposed Co-PIs— Dr. Michalski and Dr. Walter Bosch, a computer scientist in my group, but NCI rules required a single PI. The Co-PI arrangement would probably have been better as Dr. Bosch's expertise in data exchange including DICOM-RT development and his efforts to address interoperability issues in radiation oncology were key in much of the ATC successes.

Update following my retirement: In 2014, the NCI decided more consolidation in clinical trials was needed and ended the ATC funding—merging the cooperative groups into four adult and one pediatric group to form the National Clinical Trial Network (NCTN), with nearly 2000 participating radiotherapy facilities. The QA centres also were consolidated into the Imaging and Radiation Oncology Core (IROC), with Imaging offices in Philadelphia and Ohio, and RT offices in Houston, Philadelphia, St. Louis, and Rhode Island. The goal of IROC was similar to the ATC's—just bigger— "to provide integrated radiation oncology and diagnostic imaging quality control programs in support of the NCTN thereby assuring high quality data for clinical trials designed to improve the clinical outcomes for cancer patients worldwide."

I like to think the ATC played a major role in paving the way for NCI to establish this current QA arrangement. I'm a firm believer that the whole really is greater than the sum of its parts.

8   Summary

I hope that after reading these vignettes you have some inkling of what a career in medical physics was like over the last five decades. Improving patient treatments and making radiation oncology a safer medicine was always the focus.

Recounting these stories made me even more aware of how fortunate I was to have found medical physics and to have trained at an institution like the MD Anderson. There I saw teams of physicians, biologists, and physicists constantly working to improve cancer treatments. I also saw the importance of the close interaction of physicists and physicians that helped make the MD Anderson radiation therapy program one of the strongest in the world. I learned that an academic career in medical physics provided unlimited opportunity for meaningful clinical work, exciting research, and stimulating teaching/mentoring experiences.

I can't conclude this chapter without acknowledging and thanking all my colleagues at WUSTL and UC Davis for their hard work and dedicated efforts in making radiation therapy better and safer for the many cancer patients treated there. It was an honour to serve with you.

It is my hope that the stories in this book will inspire this next generation of young physicists to seek new opportunities in medicine. There's still plenty of work to do and I'm confident that this next generation will rise to the occasion and achieve even higher goals in the years ahead.

James (Jim) Purdy is Professor Emeritus of Radiation Oncology at Washington University in St. Louis (WUSTL) School of Medicine and the University of California, Davis (UCD). He has over 40 years of experience in radiation oncology physics. He was an early advocate of 3D treatment planning and pursued the development and refinement of technologies that advanced conformal radiation therapy (CRT). He served as principal investigator on numerous NIH and industrial grants and his research resulted in over 300 publications. He is internationally recognized as an expert in radiation oncology quality assurance (QA) and CRT. His later career focused on development of an informatics infrastructure for cooperative group clinical trials that supported QA and research use of treatment planning and verification digital data. He gave unselfishly of his time to advance the professional aspects of medical physics serving in numerous leadership positions including AAPM President and Board Chairman, ACMP Board of Chancellors Chairman, ACR Medical Physics Committee Chair, ABMP Vice-Chair, and ASTRO's first elected medical physicist member to their Board of Directors. He also served as Senior Editor (Physics) for the International Journal of Radiation Oncology, Biology, and Physics; member of NIH Radiation Study Section; and member of multiple ICRU Writing Committees.

Awards

2013. Distinguished Alumnus Award—The University of Texas MD Anderson Cancer Center Alumni and Faculty Association (Lecture given January 16, 2014).

2002. American College of Radiology Gold Medal Award for distinguished and extraordinary service to the American College of Radiology or to the discipline of radiology.

2000. American Society for Radiation Oncology Gold Medal Award, ASTRO's highest honour bestowed on revered members who have made outstanding contributions to the field of radiation oncology.

1997. American Association of Physicists in Medicine William D. Coolidge Gold Medal Award, AAPM's highest award for an eminent career in medical physics.

1996. American College of Medical Physics Marvin M.D. Williams Professional Achievement Award. This is the highest honor that ACMP bestows on one of its members in recognition of Lifetime Professional Achievement in the field of medical physics.

7

"How Good Are You?"

John W. Wong

I am passionate about Medical Physics—The application of physics and scientific principles to address the medical challenges of human conditions. More specifically, in my case, Medical Physics is synonymous with radiation therapy, the treatment of cancer with radiation. It has been an incredibly rewarding profession, one which I look forward to everyday.

I must also confess that I did not know the existence of Medical Physics as I pondered what to do in the beginning of my junior year in engineering science at the University of Toronto. As many undergraduates did then and do now, my first question was: should I find a job and what job? I count my blessings deeply as a Canadian where the lighter burden of education cost would afford a young adult to defer that scary and unclear decision. Then the next superficial, glamourous but self-deceiving question was: what about graduate school, medical school, etc.? Truth is that I was not sure that I would be good enough and had to work really hard to get there.

J. W. Wong (✉)
Johns Hopkins University School of Medicine, Baltimore, MD, USA
e-mail: jwong35@jhmi.edu

© The Author(s), under exclusive license to Springer Nature
Switzerland AG 2022
J. Van Dyk (ed.), *True Tales of Medical Physics*,
https://doi.org/10.1007/978-3-030-91724-1_7

1   Medical Biophysics (MBP) at the University of Toronto

My journey in medical physics began earnestly at the Department of Medical Biophysics (MBP) at the University of Toronto. In the summer after my junior year, I was offered a summer studentship in MBP with Professor Mike Rauth in his laboratory at the Princess Margaret Hospital (PMH). After three years of intense engineering studies, I wanted to see if life sciences might stir new interest for me. At PMH, I had no idea that I was in the mecca of Canadian cancer treatment and research at that time, walking by luminaries such as Dick Hill, John Hunt, Ian Tannock, Larry Thompson and others on the same floor. My project was to study the uptake of H^3-thymidine in CHO cells as an assay for cell culture growth. (The H^3 attached to the thymidine is radioactive tritium; thus H^3-thymidine is a radiolabeled compound.) I was blessed with the amazing patience and kindness from Steve Molnar, my graduate mentor, and Mike Rauth. I observed a great deal about wet-lab research and harvesting tumour cells from mice. Most importantly, I learned that I was not good at it, after contaminating an incubator and spilling H^3-labelled cells in a low-level "hot" room. The latter incident cast a huge impression on me on the seriousness of radiation exposure when Phil Leung from the therapy physics section came rushing up to the 6th floor in full plastic gown to survey the room. Maybe I cast a not-so-positive impression in return.

Nevertheless, and most likely with good words on my behalf, that summer experience led to my joining MBP in September of 1974 as a graduate student amongst other disciplines, although I admit that I still did not appreciate what it meant. It just seemed like an interesting place to land at that time. My journey began with a series of interviews with potential advisors. I do not remember all interviews, but I remember the one with Harold Johns (Harold). I was forewarned that he was *the man*. True to form, Harold was tough and politically incorrect by today's standard. He told me my hair was too long, to be prepared to work my butt off and social life was secondary. Most memorably, he asked me (which I doctored to be politically correct): "*How good are you?*" I did not know how to answer. I surely was not going to be *his* student.

In my not-so-ambitious state of mind, PMH felt like such a pressure cooker that I was happy to begin my graduate study at the Hospital for Sick Children (Sick Kids) on cystic fibrosis under the supervision of Norman Aspen (Norm). I soon learned the training philosophy of MBP then: If the student had physics or related background, he or she would be directed to take didactic classes on cellular biophysics and the dreaded organic chemistry. It was up to me to acquire expertise then on all matters pertinent to my research. Another important educational venue at MBP was the weekly student seminar course where the MSc candidates gave 30 min and Ph.D. candidates 60 min presentations. It was in this free-flowing intellectual environment where I learned about signal-to-noise, radiolysis, ultrasound imaging, survival curves, etc., all fundamentals for my practice of medical physics.

2   Hospital for Sick Children

As an entry level graduate student, I picked a path of study that seemed interesting, challenging and uncomplicated. The project was on the effect of gravity on the clearance of mucous by ciliary cells from the airways in cystic fibrosis patients. The subject was central to the use of postural drainage to manage cystic fibrosis. It was, however, peripheral to mainstream MBP studies on cancer therapy and diagnostic imaging. In addition to the requisite education in pulmonary physiology, I learned to make microspheres from human albumin, label them with the radioactive isotope, technetium-99m, pulse them in aerosolized saline boli into the tracheal-bronchial tree of the subjects and image the boli movement using an in-house Anger (gamma) Camera (a nuclear medicine camera used for medical imaging). The project was clearly different from the undergraduate problem sets. The tinkling, exploration and humour provided a lighter side to release the stress. Subjects would feel their heads growing after lying on a bed in a 25° tilted prone position over the detector part of the gamma camera for the 60–90 min sessions. A few of us would inhale inert SF_6 with O_2, to see if the flow of heavy inert gas would affect mucociliary clearance. No, but it got us "stoned", and more. The important revelation was that I found the direct interactions with patients and their care-givers immensely enjoyable and rewarding. I also recognized that the isolation in an off-campus environment was difficult for me. I missed and envied the comradery of the MBP students at PMH. After 2.5 years, I passed my MSc defense. I felt relief rather than excitement. I knew that my lack of passion in physiology research was a problem as I began my doctoral studies at Sick Kids.

3   Princess Margaret Hospital and Harold Johns

I consider myself blessed with good friendship and fortune. Through the weekly MBP student seminar series, I befriended many fellow students who genuinely cared for each other. Most notable for me were Aaron Fenster (Aaron) and Dick Drost (Dick). (Aaron Fenster wrote Chap. 19 of this book.) Six months into my doctoral studies at Sick Kids in 1977, Aaron called and pre-empted me that Harold needed a student for a new project. Aaron must have told Harold about me. Harold was Harold; he gave me a day to be at PMH or the project would be given to a new student. I cannot believe that I pondered a bit, but I was on my way the next day. It was a life-changing moment, although I must admit a little scary.

I thus became Harold Johns' last graduate student. It would become a badge of honour which helped me throughout my career. It was an amazing good fortune that he took me on as he neared his retirement and stricken by Parkinson's disease. Harold's rigour and passion about science and truth would cast a life-long imprint on me. My project was on the effects of a small inhomogeneity on scattered dose in a photon beam. My first meeting was in Harold's office with Jake Van Dyk (Jake) and John Andrew. It was different than the one-on-one supervisor-student meeting that I had expected. This committee-style format in Harold's office would continue, and quite frequently, for 18 months during his transition to full retirement. Besides Harold as the constant, there were others that would be "summoned" to join from time to time, including Jack Cunningham (Jack), Aaron, Martin Yaffe, and others. (Martin Yaffe wrote Chap. 18 of this book.) I often wondered why so many people were in my meetings. The story line among students was that Harold in his old age needed protection in case a student would object and react physically to his relentless demand for scientific explanation. In my case, I think that it was generational as Harold was uncomfortable with our increasing use of computer simulations and wanted reassurance.

I did not know exactly the details, but my project apparently stemmed from Harold's dissatisfaction with the empirical "weighting factors" in the Equivalent TAR (ETAR) method as devised by Marc Sontag and Jack. (See Jacob Van Dyk's Chap. 2 for some of the preliminary work.) ETAR was the first CT-based clinical algorithm for 3D scatter inhomogeneity correction. It was a novel idea to assign importance, albeit empirically, for each scatter voxel based on its geometric relationship to a dose point. Harold was insistent that the weighting factors should have measurable meaning. I am not sure, but it was probably John Andrew, Jake and Harold that came up with this ingenious experiment to measure the minute dose perturbation, of the order of 10^{-4}, due to the presence of a small cavity. Wow!!! What a gift of a project.

It was classic MBP training where discovery was based on theoretical or analytic modeling and validation of the experimental results. There were so many learning moments: the intricacies of Compton scattering; devising a 3D ring-on-ring scheme to calculate the perturbations of second order scattered photons; how to use the three separate DEC-PDP computers at PMH after hours to calculate the perturbations for 90 small void positions for three continuous months. Harold was tough and thorough. At least three mornings every week in my first six months, he would ask me: "Wong, let's see what you did yesterday?" which meant he would ask any question on the printouts of my on-going calculations. He needed to see ray-by-ray dose calculations and, with others in the meetings, reinforced that the results made sense. In hindsight, I am indebted to the attention. I learned that Harold always spent much more time with the weaker students than the accomplished ones. He was exactly what I needed then.

Later in my first year at PMH, Mark Henkelman arrived as the returning prodigy to lead the physics branch of MBP. Mark took over as my official advisor while Harold slowly recused himself from the scene. Mark was scarily brilliant; I always felt that Mark already knew the answers for all my research questions. Harold had a sly smile of acknowledgment when I remarked to others that "Mark was the tiger, Harold the lamb". I could not have been more fortunate to be mentored by these two giants. With Harold, it was always about the truth. With Mark, I learned the importance of understanding research through analytic forms. I am also forever thankful for Aaron for his friendship, wisdom and, unbeknown to him, his persistent inference of the L'Hôpital's rule which was fundamental to my numeric simulations. (L'Hôpital's rule is used in calculus to help address equations with indeterminate forms such as dividing zero by zero.)

Perhaps most influential for me was the environment of "Team Science" at PMH and MBP. Indeed, the culture was organic among the physics students on the 7th floor. Aaron, having just graduated, was the de facto gang leader that led a bunch of us, Dick, myself, and occasionally Stuart Foster, Ian Cunningham, and Brian Rutt, to frequent Hotel Isabella for beer and dinner, and to decompress before returning for the night's work. I also called these gatherings our "b… s…" sessions fondly which Aaron declared that we would be golden if each of us came up with just one good idea each year. Hotel Isabella was a dump but also my sanctuary.

It was also the fun and comradery that energized us as students. We had soccer in the warmer months where we realized that we were no match against Duncan Galbraith (radiation therapy physicist) and the machine shop boys. Touch football in the fall showed that the *ultrasonics* (Stuart and Mike Patterson) were better than the *photons* (Dick and me). Of course, the loud, after dinner table-hockey games where the 3-year tally of Drost/Foster against Fenster/Wong were like 200:0 for the bad guys. Along the way, we developed life-long friendships that would be put to good use time and time again. How can I forget the night before my doctoral defense when Ian Cunningham and Dick came to my rescue as I panicked over the workings of a volt meter?

I would receive my doctorate in 1982 after 8 years of graduate studies, longest among my peers. I did not appreciate the magnitude then that, two years prior, I was catapulted to recognition by Harold during his plenary speech at the annual meeting of the American Association of Physicists in Medicine (AAPM) when he listed me in the lineage of his students and told the audience to attend my presentation. I am sure that that was in the mind of James (Jim) Purdy, the head of medical physics at Washington University in St Louis (Wash U), when I brashly approached him about employment a year later. (James Purdy wrote Chap. 6.) Several months before my thesis defense, I had an offer from Jack Cunningham as a clinical physicist at PMH and one from Jim Purdy as an assistant professor at Wash U. I honestly did not understand the difference. I was undecided until a chance encounter in the parking lot with then retired Harold. In not so exact words, he told me to leave and find out how good I was. Two weeks after my thesis defense, I was on a flight to St. Louis with one large suitcase and two boxes of books and things.

My times at PMH were transformative. I had so many mentors that I acknowledged them all in my thesis. On top of the list was Harold Johns with his relentless quest for the truth and demand for excellence. I might have created a most virtuous persona that I tried my best to emulate to this day.

4   Washington University in St. Louis and James Purdy

When I got to St. Louis in 1982, I knew some details about the effects of inhomogeneities on cobat-60 dose in water. I had designed a scrotal shield and learned from Jake how to best use thermoluminescent dosimetry (TLD). At the urging of Harold, I attended a "nervous" cobalt-60 source exchange of the Hemitron (see Chap. 2 by Jacob Van Dyk) with the machine shop staff at PMH. But otherwise, I did not know much about the broad spectrum of clinical practice and technologies. It was probably also a good thing that I had no pre-conceived professional expectation as I arrived in a new country, the USA. I was simply there to find out what I could do (or how good I was) and to return to Canada in 3–4 years.

Joining Jim Purdy and Wash U was the next major milestone in my journey in medical physics. Just as Harold provided the rigorous mentorship, Jim Purdy instilled in me the critical foundation of responsibility, integrity and perseverance for an academic clinical career. Jim Purdy was a giant that made many significant contributions in medical physics (see his Chap. 6). Reflecting, Jim took the chance to offer me a faculty position before I graduated, a practice which I adopted later. There was no requirement for board certification at the time. I am therefore forever thankful for the tolerance and patience of Jim, Russ Gerber and Bill Harms at Wash U for my clinical training. I was most pleased after three years when they entrusted me (or at least I think they did) to be the solo physicist responsible for two off-site community clinics for a couple of years.

My arrival at Wash U coincided with the dawn of explosive advances in computer and detection technologies. Personal computers (PC) were a new thing in all offices and soon to replace main frame; CT scans were incorporated in radiation treatment planning (RTP); medical and consumer imaging were going digital; magnetic resonance imaging (MRI) was taking hold as a new volumetric imaging modality; and more. My thesis research on 3D dose calculations that incorporated explicit voxel-by-voxel scatter ray-tracing was a splendid fit to Jim Purdy's vision of 3D treatment planning. No one at Wash U seemed to be bothered that it would take more than a day to compute one meaningful 3D scattered dose distribution on the DEC-PDP11/70 at PMH.

I soon discovered that Radiation Oncology and the Biomedical Computer Laboratory (BCL) at Wash U were pioneers in RTP. St Louis was the home of the first major commercial treatment planning company, Artronix (which later became CMS), in the US. Indeed, both Roy Bentley from the UK and Jack Cunningham made substantive contributions at the BCL in the 1960s, which led to the development of the RAD8 at the Royal Marsden Hospital and the PC12 at PMH, respectively. My voxel-based method was renamed the Delta Volume (DV) method which matched up with the amazing plan at the BCL and its sister Computer System Laboratories (CSL) to produce a very large scale integration (VLSI) chip customized for voxelized 3D scatter calculations, in minutes. No wonder they did not worry about computation time.

Wash U was where I began my education on the complex relationships between academic pursuits, clinical services and commercial partnership. I learned that the university engineering and medical school departments, like the National Science Foundation (NSF) and the National Institutes of Health (NIH), would measure success differently. At the end of three years, CSL concluded that their VLSI project was successful as demonstrated through electronic communications between various gates on the chip. Through some difficult discussions, CSL decided that it was beyond their scope to integrate the DV chip into a RTP system for dose calculations. I was more bewildered than disappointed. I was, however, fortunate that BCL/CSL had implemented the DV algorithm in a high-end main frame DEC computer for experimental validation on a 25 cm × 25 cm × 25 cm volume. It would take hours for the full volume, but a few minutes or less for one point dose. Most importantly, it gave me access to an accurate full 3D dose engine, albeit limited to one single cobalt-60 beam at a time.

5   The NCI Cooperative Working Groups (CWG)

The novelty of the DV engine was instrumental in my receiving my first National Cancer Institute (NCI) grant funding. Cedric Yu, Sam Hancock and I were making progress in extending the DV framework to address the effects of inhomogeneity in the megavoltage photon range. The efforts complemented Jim Purdy leading the transition of RTP from 2 to 3D. Under his steadfast guidance, Wash U, in partnership with CMS Ltd, was awarded three major NCI contracts as a member of a Collaborative Working Group (CWG) to develop 3D treatment planning for external beam treatments using x-rays and electrons. These were heady times for me as I had a seat at the table with medical physics giants from the MD Anderson Cancer Center, Memorial Sloan Kettering Cancer Center, Massachusetts General Hospital, and the Universities of Michigan, North Carolina and Pennsylvania. Key concepts of plan evaluation, dose volume histogram, and treatment uncertainties were introduced. The impact of the CWGs efforts cannot be overstated as they ushered in 3D treatment planning as the standard of care in radiation treatment to this day.

The experience with the CWGs greatly broadened the scope of my research direction. The process and discussions took me to a deeper place. How could we be sure that we faithfully delivered an accurate treatment plan to the patient? That led Eric Slessinger to calculating a Portal Dose Image (PDI) and comparing it with the port film. (Also see Chap. 12 by Marcel van Herk.) If they did not agree, what did the patient receive? Xingren Ying then proposed novel back-projection of the PDI differences to modify the patient CT and estimate delivered dose. Then it became obvious that a PDI from a single beam projection would be insufficient in providing 3D information. Images from more projections would be needed. But the imaging dose from the megavoltage beams would be too high and the image quality poor. We then needed to address the imaging challenges. I was learning. One seemingly straightforward research question led to the next and the next, and .... These questions became multi-faceted and convoluted. I was discovering the difference between a research program and research projects.

6 Fiber Imaging Inc.

Just as 3D RTP at the time, the digital transformation was igniting a strong push for electronic portal imaging (EPI). It was clearly in our radar at Wash U. We, however, had challenges. We had preferred the 45° mirror-digital camera system (see Fig. 1) for its fluoroscopic capabilities which could not be deployed on our linear accelerators (linacs) with beam-stops. (Linacs with beam stops have a major shield on the exit side of the patient. These reduce the amount of shielding required in the walls of the treatment room.) In a chance reading of a campus-wide newsletter, I engaged the cosmic ray group in the department of physics (where Compton made his scatter discovery) where they had developed a plastic scintillating fiber technology for cosmic ray detection. The encounter soon evolved into a collaboration to produce a "fiber imager" (see Fig. 2) where an array of 256 × 256 plastic optical

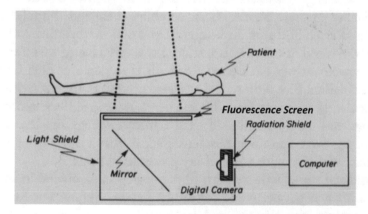

Fig. 1 Schematic presentation of the 45° mirror-digital camera system for EPI

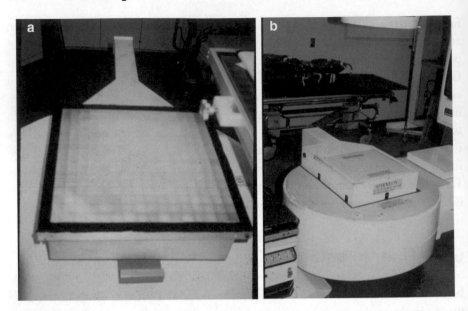

Fig. 2 Pictures of **a** a fiber-imager where of 16 × 16 fiber-reducer bundles, each consisting of 16 × 16 fibers, are used to replace a mirror for portal imaging, and **b** a functional fiber imaging EPID placed on top of a beam-stop

fibers, each 1.6 mm × 1.6 mm, spanning 40 cm × 40 cm were bundled and reduced within a vertical space of 10 cm and redirected for coupling to a 1-in. Newvicon camera for portal imaging. This novel compact design replaced the need of a mirror and offered a viable portal imaging solution for linacs with beam-stops. Great excitement ensued to garner funding from the Alafi Capital-Washington University venture capital Partnership to form Fiber Imaging Technology Inc (FITinc).

It was amazing to witness the creation of an industrial scale, manufacturing facility to draw plastic fibers, combine them into bundles of fiber reducer, and assembled into an imager. It was also eye-opening to be introduced to the world of commercialization. I was very pleased that a sponsored research agreement was established for Wash U to deploy a fiber imager for clinical validation. Five prototype units were produced. One was used in the clinic for nearly a year at Wash U. FITinc, however, did not succeed. The technology was more complicated than a mirror system and less attractive as beam-stops were becoming obsolete. I admit that I did not help when I shared this sentiment with a few of my professional peers.

I considered my experience with fiber imaging most pivotal in my career as it gave me new perspective of clinical care and workflow. For EPI, a user-friendly console with effective and efficient display and software tools was far more important than the physicist's instinct to push image quality or advanced image processing that were time-consuming. Clinical requirements of a tool would be best conceived, specified and refined in clinical operation than at an industrial laboratory. I also learned that industry was (and is) a critical partner with academia to advance clinical care. Reflecting, I was naïve to be disgruntled with FITinc committing less than 15% of the Alafi Capital-Washington University funds on research and development efforts. I learned that the push for commercial success is critical but often at odds with the "not good enough" and "discovery" mentality of academic pursuit. With industry, a "win–win" understanding must be established. I also learned, while not always possible, to be refrained from interfering with a company's roadmap.

By the time that the FITinc adventure wound down, I had spent nine years at Wash U. I also just married Pat which had the highest impact factor on me as we pondered critical balance in personal and professional decisions. With the successful three NCI CWG contracts, Jim Purdy was now moving on with great guns to develop the critical quality assurance measures to support clinical trials administered by the Radiation Therapy Oncology Group (RTOG), a group devoted to developing and executing clinical trials in radiation oncology. My academic interests were also evolving. A new research fellow, Di Yan, and I were consumed with what dose was actually delivered to a patient and how to mitigate treatment uncertainties. From the first innocent encounter when he remarked "please do not speak mandarin (Chinese) to me because your accent hurts my ears", Di became a life-long friend, collaborator and confidant. We were quick to recognize the wealth of treatment information provided by EPIDs. Di proceeded to formulate big ideas about predicting and controlling patient setup variability, and in turn dose delivery, which later grew into "adaptive radiation therapy" (ART). These ideas complemented the other "big" idea of needing more projections. We were getting restless. So, everything came to a flash point in 1992 on my 10th year at Wash U when I received an enticing offer from Alvaro Martinez to be the Director of Clinical Physics at the William Beaumont Hospital (Beaumont) in Royal Oak, a suburb of Detroit. Pat, an investment banker, knew of Beaumont's rare AA+ rating and was more than convinced that it was primed to grow. I began my next career phase at Beaumont that summer of 1992.

7   William Beaumont Hospital and a New Generation

The Beaumont's private practice environment was unique and foreign to my academic aspirations. Alvaro was a highly accomplished physician in brachytherapy from the Mayo clinic. He gave an interesting insight that a strong community health system with sound financial standing would be the best place to advance radiation treatment methodologies and technologies. I was given a clean slate to put together a new clinical physics group. Di Yan and Cedric Yu (with the Siemens company at the time) arrived within three months of me. And on schedule with the staffing plan, David Jaffray and Michael Sharpe, joined a year and two years later, respectively, both from the University of Western Ontario, London, Ontario, Canada. They were offered their positions before their doctoral matriculations. That and the fact that they could conveniently commute from Windsor in Canada to Beaumont probably played a role in their decisions.

University faculty appointment did not exist at Beaumont. That was just as well, as I was becoming cynical and perhaps a little arrogant at that stage of my career. I was amazed with the many "professors" in the medical physics community, but I was unclear about the expectations for their appointments. I did witness that for those medical physicists engaged in independent, funded research such as from the NCI, the pressure to garner continued support often left little time for them to translate their findings into clinical practice. The expectations at Beaumont were simpler. We were there to provide state-of-the-art clinical services. We were also eager and encouraged to do more.

The training at PMH probably instilled a certain amount of healthy insecurity in my scientific endeavor. I wanted critical re-assurance from others. At Beaumont, critique was welcome and came from my physics colleagues and surprisingly, from the radiation biology group headed by Peter Corry. Pete came from MD Anderson and had a long involvement in the peer-reviewed process at NCI. It used to annoy us to no end when he insisted that we must present a hypothesis on whatever we called research. Well, honestly, most of us had not formulated any real scientific hypotheses at the time. A bit like Harold Johns, Pete was relentless and at times abrasive. It took a little while for us to let down our guard to recognize that all questions, sound or baseless, were good. He was preparing us for the scrutiny of the NIH peer-review process. This was the Beaumont that set the stage for our fledgling medical physics program.

An important partnership with unexpected and far-reaching benefits was our selection of Elekta (which was Philips at that time) as our major equipment vendor. We were inexperienced and much less encumbered by the opinions and pressure from the various sales groups. I do not remember the origin of the intriguing idea to promote interactions with other advanced Elekta users beyond the cursory user-group meetings. The result was the formation of a series of working consortia modelled after the NCI-CWGs where Elekta would act mostly as the facilitator. From the initial apprehension as competitors to open engagement as collaborators, the consortia were surprisingly effective where the users drove the development of clinical solutions. The experience demonstrated again and again the synergism between the clinical community and industry. Advances were made on intensity-modulated radiation therapy (IMRT), (breathing) motion management, cone-beam CT (CBCT) and continue to this day with the MR-Linac.

I believe that the unprecedented technological advances at the time hugely complemented the eager creativity at Beaumont. First, Alvaro supported our idea of using digital kilovoltage (kV) imaging to acquire more portal images. We were also intrigued with the possibility of CBCT demonstrated by Paul Cho on a fluoroscopic simulator at the University of Washington. As an exploration in 1995, we managed to talk Cheng Pan, our research fellow, into going in-and-out of the treatment room repeatedly to acquire ninety 6 MV EPID projections of a *manually* rotated head phantom. The resultant 6 MV CBCT transformed our thinking (see Fig. 3). More amazingly, as an unofficial project, Elekta agreed to help us drill holes into the face of their linac drum for mounting a kV-charge-coupled device (CCD) digital imaging system. Dave Jaffray led a year of planning and implemented the system over two weekends four months apart. Alvaro came by the last weekend and rolled

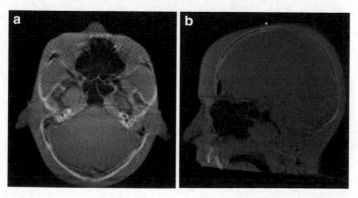

Fig. 3 A transverse and sagittal image of a head phantom acquired with 6 MV CBCT

his eyes, asking whether the linac would be available for the clinic the next Monday. Fortunately, we incurred no downtime (see Fig. 4). In 1997, we produced our first kV CBCT of three rats in a plastic container on the linac. The imaging dose with the CCD was a little high, about 25 cGy. But we all knew we were onto something. Looking back, drilling holes into a linac drum ... What were we thinking? In 1998, Jeff Siewerdsen joined us and brought with him his expertise of flat panel technology. Right on cue, Dave and Jeff replaced the CCD with a flat panel imager on the linac kV system in 2000, followed by Daniel Letourneau and Alvaro producing the first CBCTs of three patients. Thus, the era of image-guided radiation therapy (IGRT) using on-board CBCT began.

While CBCT was a poster child for Beaumont, the medical physics group was happily diverse in developing clinical solutions. In 1992, the multi-leaf collimator (MLC) was a "must" hardware. (See Fig. 3 in Clif Ling's chapter 9.) Cedric Yu created the first graphical user interface to control the MLC positioning on a simulation image. He followed up with devising various methods of intensity modulation. Most notably, Cedric called us in to observe a rotating putty glob (a prostate?) in front of a light projector. That

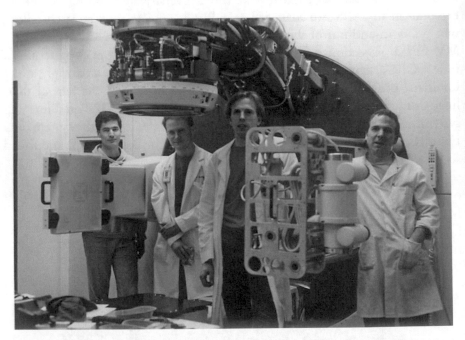

Fig. 4 Team members of the Beaumont team to install a kV EPID system for CBCT. (L-R, Michele Moreau, Doug Drake, David Jaffray, Rob Cook). Reproduced by permission of Taylor and Francis Group, LLC, a division of Informa plc.

was the first indication of his eventual seminal work on intensity modulated arc therapy.

Di Yan with his computer science background was always thinking about systems approaches to control treatment quality. In our environment where we were acquiring EPI images on every patient every day, it did not take long for Di to formulate the feedback concept of ART. Before on-line CBCT was available, Di demonstrated the off-line ART strategy where simulator-CT and EPI images acquired in the early days of a treatment course could be used to customize, and sometimes provide substantial margin reduction and dose escalation for the individual patient.

Then there was the late Michael Sharpe (Mike) who was beloved by all. He was brilliant and a funny case as he could not decide to be the centre of attention or be in the background amongst his Beaumont colleagues. Mike was well grounded in clinical reality. His method for breast IMRT was simple and elegant to ensure dose uniformity rather than turning it into an optimization research problem. To this day, his approach is prevalent in the community.

I had had a particularly fun time developing active breathing control (ABC) with Dave and Mike. From my earlier MSc training, I suspected that breath-hold at a fixed large lung volume would immobilize breathing motion reproducibly. I needed to convince Mike. So, I made him lie on a couch after which I pinched his nose with a clip and gave him a plastic tube, something akin to that used for a paper roll, to breathe through his mouth. Then I put my palm over the tube's opening and asked him to try to move his lungs. Of course, he could not breathe nor move his lungs. I could not help chuckling ever so lightly. The ensuing challenge was to show breath-hold working on images. Somehow, Mike and Dave secured a pediatric ventilator which Dave controlled with a Labview program. Learning that the dose for a CT scout-view was low at about 30 μGy, we jumped at the idea of using it to show reproducibility of controlled breath-holds using the ventilator. I was the obvious subject. All went well with many repeat breath-holds controlled by Mike and Dave. *Except for the last run*—the valve was not released, and I was held breathless for longer than I expected. I could see the devious grins on their faces through the window at the CT console. Clearly, revenge was sweet. With that, we demonstrated that an ABC device with a spirometer offered a simple solution for effective motion management before image guidance was available.

I reflect with amazement how the creative energy at Beaumont had attracted many eager talented physicists to join us. In the five years since our arrival, Qiuwen Wu, Mark Oldham, Jeff Siewerdsen, Tony Wong, Daniel Letourneau, Geoff Hugo, to name a few, came as staff physicists or research fellows. They all did great things and became established leaders in the medical physics community. From my vantage point, the most significant factor in our success was the support from Alvaro, the physicians and the department management. It also became clear to us of the critical necessity of integrating the physics service and research components in developing a harmonious environment to advance clinical solutions.

"Play" was a big part of the physics program. Whenever allowed by Michigan weather, we would have pick-up soccer games at least once a week. It seemed Dave and Jeff would always be playing against Di and me. It did not matter. Each game would end with the call to "last goal wins" which meant the last goal by my team. We loved pranks. At one time, we arranged a page to Mark during a Beaumont meeting to answer his office phone on which we had rimmed the earpiece with liquid cement. To this day, it annoyed me to no end that he never showed any emotion when he returned to the meeting. Most enjoyable for the group was when they played on my vanity urging me to seek out Steve Webb at an international meeting in Heidelberg to accept the endorsement by the European society as their candidate for AAPM president. I came dangerously close. Of course, it was false news made up most likely by Dave and Mark.

It was inevitable that we, as individuals, would outgrow Beaumont professionally. Cedric was first to leave after five years to be the head of physics at the University of Maryland. After eight years, Dave was a star and recruited to head up medical physics in Radiation Medicine at PMH, attracting with him the other Canadians, Mike and Daniel. After ten years at Beaumont, I started questioning where would technology solutions take us in advancing cancer treatment. "How much higher do we need to dose escalate?" "How much tighter can we shrink our margin when the physicians do not agree on the target?" We were basically conducting research on humans and we needed to get to the labs. It was around this time in 2001 that Dave and Jeff were characterizing the highly intense focused beams from a graphite crystal x-ray lens for a local company. The notion of small beams led to the conceptualization of a CT-guided Small Animal Radiation Research Platform (SARRP) for discovery research. As if events happened in-steps, I received a call from Ted DeWeese from Johns Hopkins University (Hopkins) to join the newly created radiation oncology department he was leading as its first chairman. The call resonated with my increasing desire to engage laboratory research, particularly given the reputation of Hopkins as a research institution. In 2004, after 12 years at the non-university Beaumont environment, I transferred the new NCI SARRP grant and returned to a most traditional academic institute at Hopkins.

8   The New Era at Johns Hopkins University

It was coming full circle for me, having left academia and returning. I was joined by new faculty, Eric Ford, Todd McNutt and Erik Tryggestad. We were enthusiastic with Ted's vision of academic excellence. We also understood that we had to temporarily forsake active research activities in IGRT and ART as we had to build the new physics division and the department from the ground up.

In the 45 years since I entered the field, the technologies in radiation oncology have advanced to a level that was unfathomable even 10 years earlier. The advents of MR-Linac and positron emission tomography (PET)-Linac are truly impressive. Yet, there is a sentiment that we can deliver far more conformal dose distributions than we can identify the target for treatment. Our new Hopkins era actually presented an opportunity for us to consider other impact directions to pursue. Before long, the group identified three emergent investigation tracks. We have to: (1) engage in the traditional technological investigations and their clinical translation; (2) invest in informatics research on big data, machine learning and decision making; and (3) expand our involvement in laboratory discovery research.

Investigations involving advanced technology are naturally very attractive for medical physicists. On the other hand, the device industries have increasingly become the necessary participants or partners in these studies. It is also not entirely clear to me whether these advances are truly transformative or incremental. Similarly, computation power has enabled many "artificial intelligence (AI)" investigations in medical physics. Many useful software tools have been created to automate or nearly automate treatment planning, contouring, decision support ... etc. Yet, with the primary focus on efficiency, it is also unclear to me whether the main driver of these efforts should lie with industry or academia. I personally believe that AI derived efficiency gain is important in improving clinical care but has not been fully utilized to generate new knowledge. Indeed, there are great opportunities with AI through data sharing and hypothesis generation to make significant generalizable discoveries that impact future practice.

We gained many insights, some unexpected, in developing the SARRP to bring medical physics to the biological research environment. Re-affirming to me the importance of industrial partnership, close to 100 commercial SARRP systems have been installed and supported world-wide by Xstrahl since 2010. It is, however, fascinating to witness the reality of its use. The biologists were simply thrilled with the capability to deliver a small beam to a target using traditional simple point prescription. Physicists, however, could not resist further developing the technology to support small animal IMRT. This gap in interest and domain knowledge is wider than that between the radiation oncologist and the physicist. The majority of the biologists are not well versed in the concept of absorbed dose (Gy), other than as a modern standard for prescription instead of exposure time. I believe the gap has led to empirical, dissatisfying concepts such as relative biological effectiveness (RBE) to describe the different treatment response to various radiation modalities. There is much work to be done to bridge the gap. In turn, huge opportunities exist to revitalize pre-clinical radiation research and treatment response that had been overshadowed by the focus on mechanistic studies on tumour biology in the last two decades.

With the tremendous and broad pace of advancements on treatment technologies and methodologies, I have become increasingly appreciative of the peer-review process. For the medical physicists, all our efforts are worthwhile. However, it is imperative for us to recognize the importance of impact in our pursuit of lasting incremental gains or to address a transformative issue. Peer review is critical in our pursuits.

9   The Interplay Between Academic and Professional Pursuits

In this my last "tour of duty" at Hopkins which ends in mid-2022, I am sensing a much more fundamental challenge concerning the career paths for future medical physicists in academic institutions. I fear that this challenge needs to be addressed thoughtfully to prevent it from becoming a significant threat.

It is now the norm in the US that most medical physicists have faculty appointments in a variety of clinical, education and research tracks, similar to our physician colleagues. For many aspiring early career medical physicists, expectations of an appointment are almost given. However, the focus seems to be on the title and not on the meaning of the appointment. Academic pursuit in medical physics is further complicated by the increasingly strong emphasis on establishing our professional status. The dominating role played by industry also impacts our practice and investigations, regardless of peer review. For many, the responsibilities of successful application of new technologies for clinical use has become synonymous with research. It appears that medical physics is in continuing conflicts between professional and discovery advancements. I fear as I have witnessed previously, that medical physicists unnecessarily divide ourselves into separate service and academic divisions with dire consequences. It is perhaps difficult for aspiring young medical physicists to recognize this self-inflicted conundrum. I believe that it lies with the senior medical physics leaders to provide guidance for our profession to pursue clinical care and research.

Medical Physics began with the recognition that we were needed to address challenges in advancing clinical care. In my view, there cannot be a distinct separation of academia and professional practice especially in the academic setting. The future of medical physics as a glorious profession lies in equal recognition to the contributions of both camps.

So, answering Harold's question on "how good are you?" It is clear, "never good enough".

John W. Wong is Professor and Director of Division of Medical Physics in the Department of Radiation Oncology and Molecular Radiation Sciences at Johns Hopkins University School of Medicine.

Dr. Wong oversees the physics and dosimetry services of the department. Dr. Wong is the primary or contributing author of over 180 peer-reviewed scientific publications and 20 book chapters. He has been a principal investigator or co-investigator on 20 research initiatives funded by public agencies and industries. He is a co-inventor of the Active Breathing Coordinator, flat panel Cone-Beam CT and the Small Animal Radiation Research Platform (SARRP) that have been commercialized as radiation therapy products for the clinical and research community. His current research focus is on the mechanism of ultra-high dose rate FLASH irradiation and molecular dosimetry to resolve biological responses to various radiation modalities. He is committed to advancing cancer treatment through education, research and collaboration.

Awards

2017. Edith Quimby Lifetime Achievement Award, American Association of Physicists in Medicine (AAPM) recognizing AAPM members whose careers have been notable based on their outstanding achievements.

2013. Science Council Session Winner, American Association of Physicists in Medicine (AAPM) recognizing outstanding scientific abstracts on a topic identified by Science Council as being at the frontier of medical physics.

2004. Fellow of the American Association of Physicists in Medicine (FAAPM) in recognition of significant contributions through service, advancing knowledge, education, or leadership.

2001. George Edelstyn Medal from the Royal College of Radiology, United Kingdom, which involves an eponymous lecture delivered every year on subjects of interest to both clinical oncologists and clinical radiologists.

Part III

Medical Physics: More than Research

"When radium was discovered, no one knew that it would prove useful in hospitals. The work was one of pure science. And this is a proof that scientific work must not be considered from the point of view of the direct usefulness of it."

Marie Sklodowska-Curie, a pioneer of research on radioactivity and winner of two Nobel prizes in Physics (1903) and Chemistry (1911). She was the first woman to win a Nobel prize and the only woman to win the Nobel prize twice.

8

Starting and Living a Medical Physics Career: Pragmatism, Pitfalls and Strategies

Paul L. Carson

1 Introduction: A Story of Setbacks or Failures as Well as Successes

For variety and, hopefully, interest in this chapter, I thought of explaining many of my most important setbacks, rather than emphasizing accomplishments and their importance. This will be a mixture of both, as we rarely get an opportunity to summarize our own work. The chapter also makes a story of how to achieve something given shortcomings and failures. Advice is spread freely here not only for your career, but also for parenting and work-life balance.

Bottom line—I was able to overcome most shortcomings and errors by having a good background in physics, hard work and persistence. Sometimes I gave priority to my work, rather than to my family and personal well-being.

P. L. Carson (✉)
Departments of Radiology and Biomedical Engineering, University of Michigan, Ann Arbor, MI 48109, USA
e-mail: pcarson@umich.edu

2   Background

A few of my early experiences provide possibly useful lessons. My schoolteacher mother gave me many advantages. With a November birthday, she had me wait a year to start school, giving me a maturity advantage over most peers in academics and athletics. The resulting boredom in school in a Northeast New Mexico (Fig. 1), town of 4500, led to great reading assignments starting in the second grade. My attention to the deep, liberal religion of Mother gave me a lasting desire to be a good person and to contribute to society. Dad was a practical man, a civil/hydraulic engineer, who believed his sons should get work experience early. From personal or sibling experience, I suggest that parents and older siblings of pre- and elementary-school-aged kids consider:

- telling the kids not to join a playground gravel fight;
- declining to supply the family grubbing hoe for a bunch of kids to use in turn to break the hard earth of a giant fox hole while kids are jumping in and out;
- telling them not to walk around the rim of a butte with your head next to the crack-filled rock face in rattle snake country;
- telling them not to make a habit of crawling in 50 gallon (230 L) drums used to burn trash in the alleys, particularly if they still seem hot.

Fig. 1 Tucumcari Mountain, 5 miles from home

After moving to the Denver area before the 7th grade, I worked Saturdays that school year, selling donuts door-to-door and then worked at odd jobs through some school years and all summers since then. That includes summer schools as work. The humbling door-to-door experience with a product few wanted, cake donuts, was most important in relieving any fear of selling to strangers, as well as in good, negative career guidance. A career-ending injury in 8th grade football finished that organized sport after two months. Lesson learned there, don't send your kids out for a sport without some preconditioning, particularly their first organized sport requiring conditioning. By 9th grade, I was leading in student government, getting into stride in athletics, music, and boy scouts, and finding classes to be not so boring.

These trends and activities continued in high school, with physical sciences and student government and politics emerging as special interests. In junior and senior years, my life was totally regimented, with government and clubs before school five days a week, athletics after, returning from home for more societies and clubs in the evening.

When elected as high school head boy, I resolved, that with all the other activities, I would give it my all with no messing around. Two-thirds through the year, I was burned out. The same thing happened in my senior year of college. Until retirement, **I never really understood the importance of planning, and sticking to, regular relaxation, entertainment and exercise time**. That lack of down-time led to limited concentration when I did get a block of time to get serious writing and studying done. The same phenomenon is probably at work with many working parents of young children and should be addressed very consciously.

3   Can You Change Ingrained Behavior?

After a National Science Foundation (NSF) Summer Science Research Program at the University of Denver, I resolved to eliminate all the extraneous activities and just spend the first year of college actually learning the material. It was wonderful, and in that small liberal arts school, Colorado College, **the professors assumed over the next three years that I had done the assignments and there was careful thought in whatever I wrote**. There was a lot going on then to distract us (Fig. 2) but improved study habits evolved. Contrary to my earlier assumptions, this was an example of the possibility of changing my behavior given adequate resolve. Eventually the possibility of real change was generally forgotten or assessed as taking too much time and effort. It wasn't until 40 years later in a conversation with an Executive

Fig. 2 Colorado College and our broader environment in the early 1960s

Coach on a flight into Detroit that I fully realized the possibilities. **Changing essentially any behavior can be accomplished, he explained, but it usually takes intensive training. Like learning to play a musical instrument, you need daily practice, and weekly review and assignments over a long time, with a coach** (sic, like him). There are solutions to many of the toughest personal challenges.

Interest in two career directions led to a major in physics and a minor in political science. The NSF and other efforts to produce many scientists and engineers to help the USA recover from our shame of Sputnik led me to choose physics graduate training. The decision to concentrate almost exclusively on academics in my freshman year was hard on my social life. To have my social expectations fulfilled required a girlfriend, irrespective of how little attention we paid to each other. Mid college was OK in this regard, but, as in high school, I was discouraged by too many diverse commitments and lack of planned social life. That discontent extended through the next year and a half despite exciting summer and graduate work in space and high energy physics in Boston, with frequent work trips to the Goddard Space Flight Center. Ways to have avoided or handled those problems productively should be obvious to the reader. **A conscious effort to lead a balanced life and confide in a close friend, family member or religious or secular counselor would go a long way to avoid such pitfalls.**

4   Pay Attention to Good Mentors and Good Advice

After a transfer to the University of Arizona, the relaxed atmosphere and great outdoor activities of more laid-back fellow graduate students made for a fulfilling and relaxing 4½ years in low energy nuclear physics. Finding a young woman and soon-to-be wife who was clearly a great match for me gave completeness and good motivation. My mentor Larry MacIntyre helped me and his other graduate research assistant and my best grad school friend, Delmar Barker, achieve Ph.D. level experimental research. A year before the anticipated completion of our dissertations in 1970, the National Aeronautics and Space Administration (NASA) had recently laid-off 35,000 physicists and engineers. New jobs for young Ph.D.s became very scarce. By now an avowed pragmatist with a wife and daughter, I joined Delmar in exploring the applied fields of medical and geophysics. While not as predictably cyclical as geology and chemical engineering, most fields like physics don't always provide jobs when and where you want them. **As you get more specialized training, you need to be flexible in the location of your work and, paradoxically, the type of work you might expect to do.**

On our holiday visit with my parents in Denver, I met the head medical physicist at the University of Colorado Medical School, William Hendee. When I had a choice between two postdoc positions elsewhere that would, at best, lead to other postdoc positions, I asked Bill whether he would give one of the faculty jobs to me if his proposals were funded. He checked my application for a few minutes and said yes, if he got both National Institutes of Health (NIH) grants. He thought that was likely, so I waited. The head of the nuclear physics group at Arizona, Doug Donahue, was very concerned about my decision to pass up the nuclear physics postdocs. He didn't know how they could support me if the faculty position did not come through, but it did. **So, it pays to take risks—sometimes.**

I had to take the job in Colorado when it became available, even though I had not finished writing my dissertation and the main paper on it. **Advice to anyone changing jobs, do not succumb, unnecessarily, to pressure to start the new job before you have cleared out necessary work from the previous one.** You will rarely have free time at the new job and the more you are removed from the old work, the harder it will be and the longer it will take to write something nearly as good as you could have while it was your primary interest.

On parting, Dr. Donahue gave excellent advice: It's not that hard to do and publish research, at least in medical physics. Just go out and do it. Also:

Buy a house that's a bit more than you can afford. Your salary and the house value will go up while the payments remain constant.

Bill Hendee was one of those very rare individuals who was noticeably outstanding in overall intelligence compared with the usual bright Ph.D. If you can work comfortably with such a person, the rewards can be very high. He was decisive, gave good advice and a lot of freedom and encouragement to those he mentored. In the medical physics M.S. training program that he established, with the help of the NIH Research Project (RO1) grant on which I was hired, we took in many students of nontraditional backgrounds and helped them acquire useful professional skills. The key, he said, was to **give students good advice on what jobs they would be capable of doing, and work for good matches through recommendation letters**. Bill was happy to take me into a faculty position without any training in medical physics, confident he could give me that experience while producing the needed work.

Dr. Hendee was an early leader in quality control of radiological instruments and techniques. While not the flashiest field for attracting top Ph.D. candidates, quality control and higher level performance evaluation of radiologic systems by medical physicists was a critical step at that time in establishing radiology as an efficient, highly reliable medical specialty.

Bill saw a need for a research symposium to bring researchers from engineering and medical physics together and started one with the services of a symposium-producing organization, the Society of Photo-optical Instrumentation Engineers (SPIE). He set me and others under him to putting together various sessions by inviting leaders in those topics to present papers, with little or no travel support or honoraria. The formula worked and the symposium for many years has been one of the best for serious researchers in medical imaging technology and basic science. The contacts and instant recognition were well worth the time for us newcomers. The secret of involvement in such volunteer activities is to not spend too much of your time at it, particularly as a young researcher. The returns probably follow a logarithmic-like (flattening) curve as a function of effort times the elapsed time.

As I became known as one of the few medical physicists working primarily in ultrasound, I was lecturing on quality control at Case Western Reserve University, Cleveland, Ohio, one of the two leading sites for development of ophthalmic ultrasound imaging. A prominent physicist there, Earl Gregg, warned how easy it is to get too caught up in the "knife and fork" club, i.e., traveling a lot giving lectures and not getting research or other work done for your employer and career. You don't want to be giving so many talks at meetings that you don't have time to listen to what other panelists are thinking or to hear other papers.

A key point I learned from the University of Colorado academic medical physics environment was to **try to make most efforts useable for multiple purposes**. Early such experience was in radiation therapy physics, and diagnostic x-ray physics, leading to certification by the American Board of Radiology, scientific publications, teaching and clinical support.

Outside of contributing to the graduate training program and learning some medical physics, I began considering how I might initiate research utilizing my experience in nuclear physics. The multiwire proportional chambers at the University of Colorado Cyclotron Lab in Boulder seemed like a good tool for energy sensitive imaging of x-rays. A graduate student in that lab worked with me to demonstrate the chamber's use for bone densitometry, one of the few medical uses of quantitative imaging other than spatial measurements. After a few months and not enough data to write a peer-reviewed paper, I applied for a large NIH grant to develop the technique. A lesson there was you do need substantial accomplishments, preliminary data and great ideas to get a large grant. Since most medical physicists came from nuclear physics, the competition was pretty stiff even then. So, the rush to do what was close to my training did not pay off.

5   Following the Opportunities—My Early Medical Ultrasound

The remaining pages mainly review my research. Most of the notes on pitfalls and opportunities will be summarized in **bold text**.

In my first year or so at Colorado, the head of one of the five programs in the world to develop medical ultrasound, Joseph Holmes, M.D., asked if I could help with the somewhat unreliable ultrasound imaging process. In earlier years, his group made the best abdominal and neck images in the world and started a spinoff company that created the first commercial ultrasound system with a flexible arm for scanning from many directions and combining the multiple views into a single image of a slice through the body (Physionics Corp., 1965). Now Dr. Holmes ran a good service and clinical research lab through most of my 10 years there.

Reading a basic book about medical ultrasound was a first step for me. Pulse echo ultrasound imaging works by having a transducer that can send sound waves at ultrasonic frequencies, i.e., above the 20,000 cycles/s or 20 kHz maximum heard by humans. We use pressure-electric (piezoelectric) transducers that vibrate when hit with an electrical pulse or oscillation and produce an electrical signal when a pressure is applied on the transducer's surface. As in Fig. 3, the transducer vibration produces pressure fluctuations in contacted tissue. The resulting microscopic motion of the tissue propagates as a wave traveling perpendicular to the wave front. A usually small fraction of the wave is scattered or reflected and the portion returning to the transducer produces an electrical signal that is amplified. In the figure, the edges of the beam formed by a flat, circular disc transducer are outlined by dashed lines. The transmitted wave is shown to the right of the gray, ellipsoidal organ, as alternating dense and sparse packing of the tissue particles, which oscillate back and forth along the direction of wave propagation. Above the beam are sketched representations of the pressure amplitude with time for the transmitted beam (right) and two weaker waves (left) reflected from the front and back of the organ.

The voltages produced across the (black) transducer element are amplified and envelope-detected to produce a positive-only signal as a function of travel time down and back in the tissue. This is called the A-mode, for Amplitude modulation, display. If you display a dot on the screen with brightness or darkness proportional to the A-mode amplitude (top left display), this is called B-mode, for Brightness modulation, display. If you move the transducer up or down the body, or wobble it, and display the B-mode signals in their proper relative positions, that is a B-mode, pulse echo image. If you move the transducer to image the same points from multiple directions, that is called compound imaging.

Dr. Holmes' request for help fit in well with the quality control interests of our medical physics division even though I had never had a course in

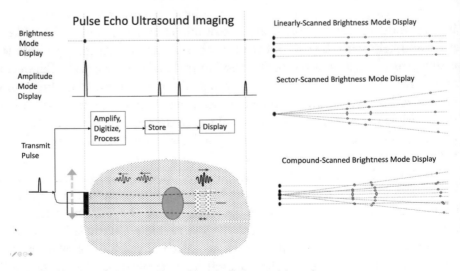

Fig. 3 Principles of the initial modes of ultrasound imaging

acoustics and essentially no coverage of it beyond freshman or sophomore physics. Did I also mention that I have never had a biology course beyond 9th grade general science? **Continue to learn what you need in order to get things done**.

To test the resolution of the ultrasound systems throughout the imaging plane and the accuracy of size measurements, it had been proposed to use a test object (phantom) composed of rigid steel wires lying parallel to each other between two plastic plates. It was difficult to get the angle of the ultrasound image plane defined by the scanning arm to lie perpendicular to the steel rods in the test object. The alignment difficulty, plus keeping a bubble-free alcohol and water solution with the correct speed of sound for correct measurements, led me to make an enclosed tank. There was a thin plastic window on three sides and you only had to adjust the image plane so that the ultrasound transducer face sat flat on the window. You could see how well the images of each wire overlapped as the transducer moved over the three sides (Fig. 4).

On a trip to Japan for an ultrasound conference, I took one of the phantoms to show to Ken Erickson, the head of the Standards Committee of the American Institute of Ultrasound in Medicine (AIUM), at his work at Lockheed Aircraft, Corp. south of Los Angeles. That went well, although on my way back from Japan, I learned **it is good to remember to have any required visas before leaving on a long international flight**. Ken and I did include the enclosed as well as the open versions in the AIUM Standard 100 mm Test Object. **More advice: Work very hard to get official reports and standards published in academic journals, not a slick bulletin of a small society that might not be archived. Usually, the principal authors are not listed on standards. If peer reviewed publication is not realized, get the data supporting the document or an executive summary, published in an archival journal**. To avoid any interference with the enclosed test object being accepted widely, I chose not to apply for a patent. A few million dollar's worth of them were sold and a typical 5% royalty for that sort of thing would have been handy in the dollars of those days.

The biggest rewards from a new clinical quality control effort come when the procedure is applied widely enough to get information on how equipment in the field is performing. Using the enclosed AIUM Standard 100 mm Test Object, we checked ultrasound scanners at numerous hospitals in the Denver area and elsewhere, as part of a regional technologist training program (multipurpose again). The terrible performance observed in some units initially and over time (Fig. 4b) was a shock to at least one major manufacturer and they determined to implement better calibration of the units in the field.

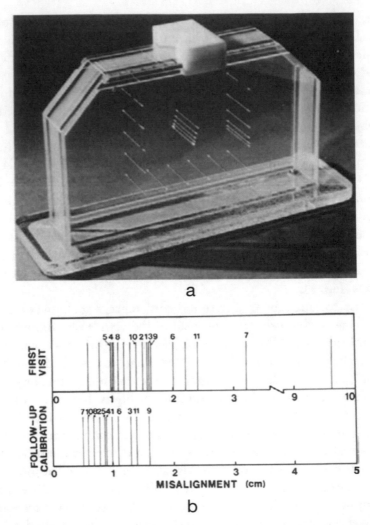

Fig. 4 **a** One version of the AIUM 100 mm test object, with a transducer guide and Teflon tape for smooth sliding of the transducer. **b** Maximum misalignment of echoes from the same wire when scanned from different directions. Reprinted with permission from Christensen and Carson (1977)[1]

Numerous systems were corrected with service calls (Fig. 4b). Even distance measurements from a single direction of viewing, so critical to fetal maturity assessments, were found in three cases to suddenly jump out of calibration by up to 10%.

---

[1] S. L. Christensen and P. L. Carson, "Performance survey of ultrasound instrumentation and feasibility of routine monitoring," Radiology **122**, 449–454 (1977).

This quality control work satisfied my multiple obligations enormously well, **but it is a delicate matter to convince most hospital administrators that doing any publishable development of techniques is worth their paying part of your salary**.

In my clinical training in radiation therapy physics, I was able to use ultrasound to measure the contours of a breast to be treated for cancer. This improved my treatment planning using beams at grazing incidence to the chest wall and minimizing exposure to the sensitive lung. I also started a study on multiple patients with prostate cancer using ultrasound localization of the prostate and adjacent sensitive structures including the rectal wall, bladder and a feelable landmark, the pubic bone (Fig. 5). All the ultrasound images here are historical. Take a look on the web at modern ultrasound images and realize how much better they are now. There is still room for substantial improvement. (Also see Aaron Fenster's Chap. 19.)

It was too hard to train the radiation therapy or ultrasound techs to use care to not distort the skin surface and find the best contours of the relevant structures using both the front approach and one between the scrotum and the rectum. Thus, I had to do the scanning myself. Choosing my next application area, I went back to the breast. Having become an "expert" in a short

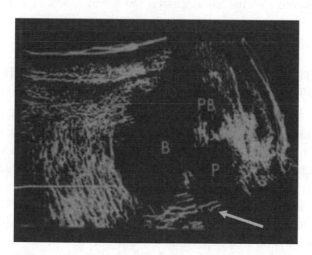

Fig. 5 Front to back image parallel to the spine, showing the urinary bladder (B), prostate (P), pubic bone (PB) and rectal wall (arrow). This early binary, and even early gray scale, ultrasound image showed primarily reflecting surfaces, rather than the fine structures in the tissues producing low echoes. Reprinted with permission from Carson et al. (1975)[2]

[2] Carson, P. L., W. W. Wenzel, P. Avery and W. R. Hendee (1975). "Ultrasound imaging as an aid to cancer therapy, Part I." Int. J. Radiation Oncology. Biol. Physics. **1**: 119–132.

time on ultrasound for radiation treatment planning, I collaborated with a truly expert ultrasound radiologist at the leading radiation therapy facility in Denver. We wrote a two part review of our work on that application of ultrasound.

Safety of ultrasound was discussed deeply in AIUM committees and various journals. After seeing considerable performance defects in early ultrasound systems and seeing the role of medical physicists in calibrating the output of x-rays and radiation delivered with nuclear medicine radioactive tracer administrations, it was natural to consider the safety of the acoustic outputs in commercial systems in use and with what caution they should be used.

We studied and developed portable measurement systems and completed a fairly large survey that helped define possible output levels for the first Food and Drug Administration (FDA) guidelines and a series of user (AIUM) and industry (National Electrical Manufacturers' Association (NEMA)) standards on measurement and reporting of acoustic outputs. We obtained a contract from NEMA to define the detailed methods for the first such standard. That effort is still continuing now in numerous international standards, often under USA leadership. Similar evolution of the evaluation of system image and measurement quality has occurred throughout my career as discussed later.

6   Ultrasonic Transmission Tomography (UTT) of the Breast

To study making ultrasound imaging more quantitative and displaying a much wider range of signal levels, I scanned a tissue specimen with the raw transducer signal going into a nuclear physics digitizer with 4192 gray levels, similar to the 12-bit digitizers used in many modern ultrasound scanners. This led to the idea of studying what would limit pulse echo ultrasound in the human body if a much wider range of echo levels were displayed. We found that even echo-free regions of the body would be filled in with weak signals from sound scattering back and forth between tissues overlying the region of interest. This multiple scattering limitation is only recently beginning to be suppressed by software beamforming and artificial intelligence in the image reconstructions.

Considering the big discovery of x-ray computed tomography (CT), Bill Hendee, a medical student Gary Thieme and I decided to make a CT system at a very high gamma ray energy, the same energy as that in cobalt-60 radiation therapy machines. (Tomography means imaging a slice.) Thus, the mapping of the body tissue properties was that of determining the dose in the therapy beam. I found a then-obscure, but later dominant reconstruction algorithm for that project. As that work was proceeding, I asked, why not try ultrasonic CT? We used the scanning mechanism from one of the four versions of one of the earliest medical ultrasound imaging systems. Developed by D. Howry, J. H. Holmes and colleagues, the system had the motions necessary for an ultrasonic transmission tomography (UTT) scanner. The transducer translated back and forth over 20 cm and rotated around the object.

We merely attached a frame to the transducer mount, for a receiving transducer on the other side of the scanned object and added our digital signal acquisition. It was disappointing when I learned at a meeting that we had been scooped in producing the first UTT system by a paper presented in 1973 and published in the meeting proceedings. **Special hint to the dense: when working on new research ideas, spend some time checking meeting abstracts as well as the published scientific literature to learn what has already been done. But in the rush, remember a quote from the founding head of medical physics at the University of Wisconsin, John Cameron: "Anything worth doing is worth doing poorly." It was understood that this referred to initial research to show some new principle or approach.**

For some new contribution, we scanned not just soft tissues, but a phantom approximating that of soft tissues and bone. The potential of scanning through bone is just beginning to be realized with high quality UTT systems. The transmission method offered much to ultrasound quantitative imaging of tissue properties, as the amplitude of the pulse echo ultrasound signal depends on the strength of the back-scattered wave as well as the two-way attenuation by tissues between the tissue and the transducer. Extrication of these two phenomena from only backscattered data is still a very challenging process. In addition to imaging the attenuation of the sound per cm of tissue, it was soon shown that a highly accurate image of the speed of sound can be reconstructed by detecting the transit time for an ultrasound pulse between the two transducers being scanned back and forth and around the tissues. These were done with small transducers to receive only waves following a straight line.

We proceeded to obtain the first attenuation images in patients with our larger, focused transducers, which also gave pulse echo images of surprising detail in some women such as in Fig. 6. Here, the first published pulse echo, attenuation and speed of sound images of a living human are shown. Clinical studies showed a good, but not perfect ability with these semiquantitative UTT images to discriminate cancer from noncancerous abnormalities and normal tissues.

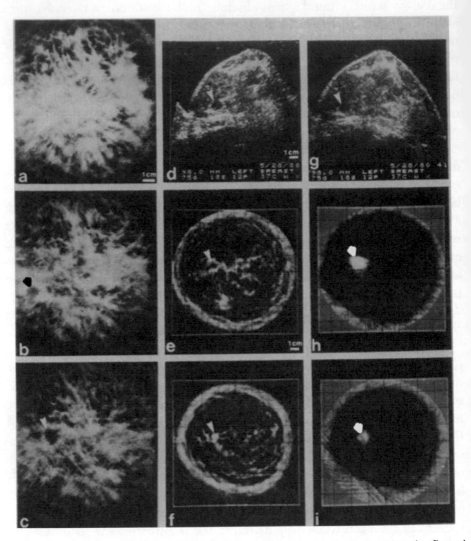

Fig. 6 The first combination of many-view pulse echo (**a–c**), attenuation (**e, f**) and speed-of-sound (**h, i**) UTT images of a breast. A cancer is shown at the arrow heads. **d, g** Single-sector and dual-sector images of the patient from a then-new commercial scanner, in a plane perpendicular to the UTT images. Reproduced with permission from Carson et al. (1981)[3]

---

[3] P. L. Carson, Meyer, C. R., Scherzinger, A. L., Oughton, T. V., "Breast imaging in coronal planes with simultaneous pulse echo and transmission ultrasound," Science **214**(4525), 1141–1143 (1981). Reprinted with permission from AAAS.

7 Fetal Lung and "Tissue Characterization"

During these exciting times in Colorado, Dr. Michael Johnson replaced Dr. Holmes as director of clinical ultrasound. He was a year behind me in our Colorado College days. He immediately became a leader in cardiac ultrasound and later, department chair. He suggested an important clinical problem that might be solved with quantitative ultrasound—determination of when the lungs of the fetus were mature enough to survive in a necessary early delivery. With Chuck Meyer, Ph.D. in Biomedical Engineering working on data transfer and signal processing and Gary Thieme, subsequently M.D., performing surgery, we demonstrated in sheep fetuses, extracted, but still connected by the umbilical cord, that the rate of change of ultrasound attenuation with frequency changed with lung maturity as expected. We had funding from a leading ultrasound company, Rohe Scientific and obtained a series of NIH grants to study sheep fetal and human fetal lungs and livers. The results in only eight additional sheep were also positive in seeing the expected change on the average, but the results were not good enough to make solid predictions in individuals. The size of tissues in the lungs producing backscatter were also reported. The liver results showed the increased ultrasound attenuation from molecules in the liver providing energy storage near birth.

8 The Big Move

The first decade, of the now five in my career, was passing. My leadership of the equipment evaluation, transmission breast imaging and quantitative fetal lung/liver research and attraction of research support for the last two, made me a prime candidate to be lured away by departments looking for researchers that might support their own operations almost immediately with external funds. I interviewed twice each at Harvard (cold, far from home in Colorado, getting tenure was very unlikely) and Vanderbilt (didn't like having to put kids in private schools for decent education). In an interview with imaging research at the FDA, it was noted that I should get out from under the mighty oak of my mentor at Colorado. **It is often worthwhile to move between the site of your Ph.D., postdoc and faculty positions**. I had skipped the postdoc, but so many opportunities had come to me through Bill Hendee that I thought that issue was awash. The main issues were: (1) my concerns that renewals of the NIH grants did not seem likely with available collaborations and resources at the poorly funded University of Colorado

Medical Center; (2) my parents and my brother's family lived there, my wife preferred not to move and our seventh-grade daughter hated the idea. When decision time came to start and lead a medical physics team at the very large and stable University of Michigan Health System, at considerably increased salary, I eventually took the opportunity. **My main advice on this is that the best advancement in a specialized field usually does require and result from some mobility. It can be hard on some in the family**.

The chairman of radiology had only been there a year and was anxious to get more basic research going. We negotiated a major package of startup funds, new faculty positions and essentially all the minimally used space of the department. A recent Ph.D. graduate from our Colorado group, Jim Mulvaney, was to head the medical physics clinical support team. Dr. Charles Meyer, the heart of our ultrasound hardware and software development, decided to join as faculty. Tom Chenevert was finishing his Ph.D. in physics on our breast transmission tomography before joining us. A talented research and clinical physicist from Cleveland, Dr. Michael Flynn also joined our faculty. After my position at Colorado had been offered to one of the young faculty in my ultrasound group, I learned from the Michigan chairman that he was being ousted by a rebellion in the Michigan department. If I still wanted the job, I should come to Michigan and renegotiate the agreement with the person who was likely to be the interim and probably next chair of the department. That negotiation with the interim chair didn't work; however, I got the agreement signed by the dean and the hospital director. The medical school and the hospital kept all their promises. Life with a boss who had me forced upon him was tenuous and not particularly fun the first couple of years. Things eventually worked out and he gave us pretty good support over the years. **This type of problem can be common and disruptive in industry, academia or government**.

At Michigan, we got fetal maturity and breast transmission tomography NIH grants transferred or refunded. **Company support for the fetal maturity work was dropped, as it was the combined clinical and medical physics leadership at Colorado which they supported. It is too bad we did not try to work out a joint approach to continuing the project**. We also looked at heart-beat-induced motions in the fetal lung, seeing how much the tissues deformed, quantified as strain. This again did not result in the diagnostic test we were looking for but became two of the first papers on elasticity or strain imaging of tissues. Strain imaging with magnetic resonance imaging (MRI) was introduced, as was the now-clinical shear wave elasticity imaging with ultrasound or MRI. Measuring these mechanical properties has now become one of the frequently used diagnostic tests by ultrasound and magnetic resonance imaging, with particularly significant ultrasound applications in measuring the deformation of heart tissues as they beat.

As we joined Michigan radiology in 1981, there was a battle between internal medicine's nuclear medicine division and radiology for control of future MRI services. I started contributing by performing MRI system evaluations, site planning, safety and quantitative imaging research and began serving on several MRI committees of the American College of Radiology and American Association of Physicists in Medicine (AAPM). Keeping up with the research in two modalities, directing the Radiologic Physics and Engineering Division, including the already established in-house service group, and teaching the radiology residents was challenging. Passing the resident teaching to our clinical x-ray physicist was good for all. After replacement of our first MRI, with one of the first superconducting clinical systems in the country, we hired our first full time MRI physicist. I was able to reconcentrate most of my research and clinical support on ultrasound. **Do what you must, but try to choose, so you can do your work well. When primarily a researcher, whatever teaching you do, try to do it to build your own skills as well**.

In the mid 1980s, national radiology leaders decided that too much effort had been spent on automated breast ultrasound and quantitative ultrasound. Both had not lived up to the high threshold set for clinical use. It was decided there should be no more big, government-funded clinical trials of the water path automated breast ultrasound systems nor grants for quantitative pulse echo ultrasound. **It is hard for a new imaging modality to replace an established one, just because it is less expensive or marginally better in some circumstances**. It took decades to develop diagnostic capabilities accumulated by the breast imaging radiologist community. There was some evidence that one of the automated pulse echo ultrasound systems would save more life-years than mammography in the large percentage of women with dense breasts, but that was not enough. Recently this has been borne out strongly with modern ultrasound.

9   Colour Flow and Power Mode Imaging

Both of my main research topics got caught in that funding drop; so, I was primed for new topics when colour flow ultrasound imaging became commercially available. When Jonathan Rubin, M.D., Ph.D., the director of the ultrasound service, was invited to Seattle to evaluate the first USA system with colour flow imaging, we were impressed by this evolution of tracking blood motion by the round-trip shift in ultrasound pulse travel time to groups of moving red blood cells. Such measurements of the speed of blood or other scattering particles toward and away from the transducer was traditionally measured using the Doppler effect, the change in frequency that is proportional to such motion. We injected into a rabbit hip, some cells from a cancer type known to have increased blood flow. With the rabbit in a cat carrier, we boarded a plane. His presence in the passenger compartment was to protect the rabbit, which might easily have died in the cold luggage hold. Dr. Ruben was very concerned our little deception might be discovered when I chatted with agents and passengers about my cat. Hopefully the statute of limitations has been reached on that little infraction.

That quick study is an example of doing something just to be first and get a lot of citations of the resulting paper (not police tickets, but references in the literature). That and resulting work led to an NIH grant for semiquantitative imaging of perfusion (blood volume flow through the small vessels per gram of the imaged tissue region). To get the perfusion from ultrasound imaging, you would need to record the average speed of the blood flow at each pixel in the area of interest multiplied by the signal level at each pixel from moving blood, and be sensitive to flow from approximately 20 cm/s to less than 1 mm/s. (It is hard to avoid counting the small artery and major feeder artery flow and the major vein outlet, thus double or triple counting.) The ultrasound systems could not cover that necessary range of velocities nor output the signal power, so we often imaged at two velocity settings and came up with measures like speed-weighted colour pixel density and nagged the ultrasound system manufacturer for the signal power. They suddenly provided it. The postdoc working on the project, Xie Li, was a recent physics Ph.D. graduate and didn't like all the crude approximations, pushing me along to help him make it as quantitative as possible. **He left before writing up his research, so after a year or so, I eventually put out a quick proceedings article, taking first author on this start of quantitative color flow imaging. People said the color flow systems were simply not quantitative. We know, however, that such non-quantitative imaging measurements work in many clinical situations. Imagine what we could do if the systems were engineered to maximize their quantitative abilities. That is just starting to happen in a wide range of measurements**.

Dr. Rubin worked with the ultrasound company, Diasonics, to make images of the power signal pleasing and diagnostically useful (Fig. 7). Power Doppler imaging (technically the Doppler effect is not used with short pulsed measurements) was born and was popular almost immediately and was fruitful as a visual and semiquantitative research tool. We followed with

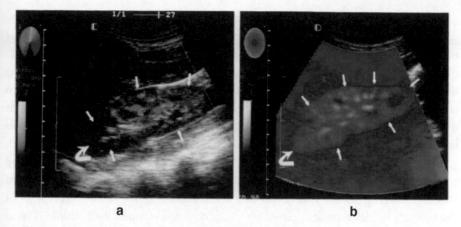

Fig. 7 **a** Colour flow and **b** power Doppler long-axis images of a normal kidney. Straight arrows denote the kidney margins in both images. Note the coloured oval in the far left upper corner of **b**, which represents the signal level map. Note the decreased signal in parts of the kidney that are mainly tubes for urine filtering and storage. The power Doppler is a bit more sensitive than the colour Doppler that keeps track of the direction of flow. Reprinted with permission from Rubin et al. (1994)[4]

other use of power Doppler measure blood flow in large vessels and the capillaries. This led to methods of splicing information to make images of much larger segments of the body in 2D or 3D.

10   Bubbles

Let's skip rapidly through the next 20 years. Nearly all x-ray CT scanning, most MRI and all nuclear medicine is done after introduction of a liquid or other "contrast agent" into the vascular, digestive, lymphatic, urinary, cerebral-spinal or respiratory system to increase the contrast of the injected systems and tissues. Ultrasound has more contrast naturally (intrinsic contrast) between and within most organs, and many would like to keep the exams simple. But ultrasound discrimination of the blood vessels or digestive tract can also be enhanced by introduced agents. The most important of the ultrasound vascular agents contain microbubbles with a fatty or protein or polymer coating to keep them stable in the blood. They are probably safer than x-ray and MRI agents because of the volume of fast injection required in the former and the dangerous heavy metal, bound in

---

[4] J. M. Rubin, R. O. Bude, P. L. Carson, R. L. Bree, R. S. Adler, "Power Doppler US: a potentially useful alternative to mean frequency-based color Doppler US," Radiology **190**(3), 853–856 (1994).

a protective compound, for most of the latter. Miniscule quantities of microscopic bubbles are required for ultrasound contrast injections because of their extremely strong, resonant ultrasound scattering. Making the exam results quantitatively reproducible has been a major challenge and the burgeoning ultrasound contrast agent research field in the US was decimated by a cautious warning from the FDA. We continued with research on quantitative contrast imaging and the FDA has removed the onerous warning. It will begin to take its place, but slowly.

A more lonesome path our group followed was creating microbubbles only where we wanted them in the body, using high amplitude ultrasound. This was for imaging where you could not conveniently get contrast agents, as in the urinary bladder, or for delivering therapy. Such bubbles also could serve as well-separated point targets for refocusing the ultrasound through the skull or measuring the speed of sound in the tissues. With large numbers of bubbles, we could even occlude the vessels to stop minor bleeding or trap drugs in a tumour. This work followed from Brian Fowlkes' and Oliver Kripfgans' background in the physics of acoustics (physical acoustics), which included a lot of study of bubbles in ultrasound. Out of this work, to some extent, came a direct form of doing deep internal surgery without infection and major bleeding. Named histotripsy, this expected breakthrough in surgery and interventional radiology was invented and developed by Prof. Charles Cain and his group in biomedical engineering at the University of Michigan, including Dr. Fowlkes.

The internally-generated gas bubbles we were creating were composed of tissue vapour and air, without much oxygen. They diffused back into the tissues or fluids fairly rapidly. For more lasting bubbles, we injected into a vein, microdroplets of a pure, chemically inert liquid, that would usually be a gas at just above body temperature. These droplets were somewhat easier to vaporize than the tissues and lasted much longer in body fluids. With bubbles from these droplets, we used the increased absorption of sound around the bubbles to show how to enhance ultrasound thermal therapies; we painted planes and U-shaped tunnels of bubble walls to reflect and contain the ultrasound, increasing the treatment rate where desired and protecting sensitive tissues in ultrasonic therapies.

We took the droplet vaporization further with the expertise of Mario Fabiilli, chemical and biomedical engineering student, and former drug company scientist, then faculty. Storage in droplets of various concentrated chemotherapy or other agents for precisely targeted drug delivery were demonstrated.

11   Photoacoustics

The work with drug delivery by ultrasound vaporization of loaded micro-droplets included using a new imaging modality, photoacoustic tomography to monitor the drug delivery. Photoacoustics works by sending an intense, short pulse of infrared light into the tissues where it is absorbed, causing instant heating. The rapid expansion of the darker tissues (usually blood) produces a pressure pulse (ultrasound). An ultrasound system in receive-only mode can image that weak ultrasound wave from the borders of a very slightly heated region, or from throughout that region if the transducer can detect low enough frequencies. With contrast mechanisms different from those of other imaging modalities, photoacoustics is very promising, though even the most penetrating, red, or infrared, light does not penetrate far in the tissues. I recruited a postdoc, Xueding Wang, Ph.D., who stayed on as faculty. He recruited others and involved me in exciting research with colleagues here and in China. Many techniques were developed, and applications explored.

Applications included assessment of prostate cancer and arthritis in joints. Work was performed with optical contrast agents and development of flow measurement. Photoacoustic microscopies with resolution defined by light for shallow targets and by ultrasound for deeper targets were implemented. A plot of the wave amplitude or power as a function of frequency is referred to as a frequency spectrum. As with frequency spectra from ultrasound imaging, processing the frequency spectra of the absorbed light was shown to provide additional tissue properties including sizes of interacting structures that are smaller than the image resolution. We and others have worked to help commercialize photoacoustic systems, though achieving wide clinical acceptance has not come rapidly.

12   Automated Ultrasound with 3D Mammograms

Alignment of 3D image sets taken at different times is an important part of detecting changes with time, comparing results from different modalities, using one set to correct artifacts in another, and live monitoring or guiding of treatments. We did this by relatively simple methods initially and then by image registration software such as that first developed by Dr. Chuck Meyer and his lab that allowed warping of one image set to match another.

When discussing colour flow and power mode imaging, I mentioned some of our work on breast cancer discrimination and tracking of response to treatment. This was performed mainly in the nineties, with the help of NIH and Department of Defense grants. Around the turn of the millennium, we agreed to help with specifications, testing/improvement and obtaining federal funding of a combined ultrasound and mammography system. It would provide images of the breast in a single compression for easy comparison of results from the two modalities. After agreeing in that meeting, Dr. Kai Thomenius asked whether I would like to include in the proposal interfacing the automated ultrasound to a prototype second generation x-ray tomosynthesis system they were working on. This 3D mammography machine would work much better for definitively identifying the same objects in the two modalities but would put stress on the maximum budget available from the relevant Biomedical Research Partnership grants of the NIH. I said yes and that formed the direction of most of my research and much of that of my radiologist colleague, Dr. Marilyn Roubidoux, for the next 16 years or so. The combined system (Fig. 8) required a lot of practical improvements like using hair spray to eliminate air between the breast and plastic compression plates without their slipping out when using slick ultrasound coupling gels. **We learned that with large instrumentation development and human studies, nontrivial papers don't always come rapidly**.

We were in competition primarily with groups at Siemens and Philips and their human studies colleagues. It became clear that imaging through the entire compressed breast was difficult using high ultrasound frequencies that could come close to competing with conventional scanning by hand from above, with the breast flattened by gravity. The other two groups have never gone beyond scanning from one side because of the difficulty of scanning from above and below the breast. GE Global Research got a partnership grant with us and Butrus Khuri-Yakub's lab at Stanford to develop arrays of hundreds of thousands of little transducer elements, called CMUTS, to replace the two compression plates. Those arrays never achieved acceptable reliability, but during their development, we at U of M built, with GE's grant and a new dual sided scanning system for ultrasound, including photoacoustic tomography. This was to demonstrate how ultrasound scans through both compression paddles could work with x-ray tomosynthesis performed on the same or a separate machine, but in the same views on each study participant. Coverage of the breast with high resolution was much improved, but reliable acoustic coupling still required more engineering. Grant reviewers were unwilling to support taking it to the next level, which we thought would show full coverage with adequate convenience. Larger companies are usually

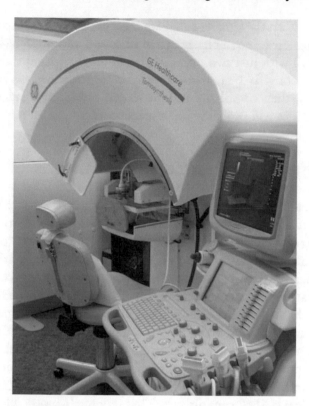

Fig. 8 Research, automated, 3D breast ultrasound performed under slightly reduced mammographic compression in the Gen II Prototype 3D mammogram system called digital breast tomosynthesis (DBT). A breast mimicking soft polymer phantom is the yellow object being scanned. The x-ray tube moves in an arc in the enclosed housing above and GE's first prototype of its sensitive digital x-ray detector is below the phantom. A conventional ultrasound unit performs the ultrasound imaging

unwilling to spend a lot of their own money to develop a system to replace ones that are selling perfectly well. GE did develop a new prototype, single sided scanning ultrasound system for their commercial x-ray tomosynthesis unit, based on their standalone automated breast ultrasound system (ABUS). With its long, curved transducer, we did get near-side ultrasound images as good in most of the breast as any obtained in mammographic compression, but it still suffered single-sided and other coverage difficulties.

For photoacoustic tomography of the breast, we worked with a laser company to double the previous record for energy in a short pulse for this type of human imaging and had a special transducer array and electronics constructed to receive the signals. The latter was not as sensitive as expected, but we were still able to image thin tubes of blood in a breast specimen to five cm deep. With the double-sided scanner, we worked to show that ultrasonic transmission tomography of speed of sound and attenuation could be performed well with just two opposing linear arrays, without views from all directions in a ring. If x-ray mammographic tomosynthesis is to survive, as I think it will, then combination with several pulse echo and transmission ultrasound modes will probably be implemented.

13   Active Retirement and Conclusions

As I got close to retirement at age 75, I wanted to keep doing what I had been doing, at a reduced pace, yet maintaining a sense of contributing to society and exploring interesting ideas that I appreciate. That is easiest to do in a field you already know. I began working with an active group in radiation oncology and biomedical engineering as the ultrasound person on radiation acoustic imaging. That is photoacoustic imaging, using radiation therapy beams as the photons that heat the tissue. Being at extremely high energies, not many photons are involved and most pass through the tissues, so the heating is exceedingly small. We specified and obtained a transducer with 1000 active elements and preamps at a frequency 6–30 times lower than typical diagnostic transducers, to match the signals coming from the radiation beam. We can image two edges of the beam directly. That is enough, along with knowledge of the beam shape and real time pulse echo imaging to quantify dose rates as well as to track the radiation beam relative to the tissues to provide further reliability in the radiation treatments. Another current collaboration is with colleagues at Wayne State and Nanjing Universities on designing UTT ring array systems and algorithms to provide spatially-selective and self-verified warming of cancer tissues during chemo- or radiation-therapy, allowing at least a halving of the treatment dose.

You might note in this chapter that there was a lot of pragmatic response to needs and opportunities, taking what life allowed, while trying to do good for society and individuals. I've included only some of my embarrassing missteps that could serve as useful lessons for others. I will note that some of my most productive times were in research with new, diverse collaborators. Some greatest missed opportunities were in not accepting or continuing collaborations. Finally, in today's times, be particularly aware of others' political and religious beliefs, and how different belief systems might lead to disagreements or conflict. In times of major changes or conflict with colleagues, try to see the issue from their side and use your knowledge to be helpful to those around you and contribute to society in a productive way.

Paul L. Carson received the B.S. degree from the Colorado College, Colorado Springs, CO, and the M.S. and Ph.D. degrees from the University of Arizona, Tucson, AZ, in 1969 and 1971, all in physics. From 1971 to 1981, he served in the Department of Radiology at the University of Colorado Medical Center, Denver. Since 1981, he has served as Associate Professor, Professor, BRS Collegiate Professor and Active Emeritus Professor in the Department of Radiology, and as Professor and Emeritus Professor of Biomedical Engineering and member of the Applied Physics faculty, University of Michigan, Ann Arbor, MI. He was founding Director of Basic Radiological Sciences in Radiology from 1981 to 2008. He is Concurrent Professor, Nanjing University. His responsibilities have been in research, clinical support, and teaching of radiological sciences. Research interests include medical ultrasound quantitative imaging, functional imaging, equipment performance, safety, new or improved diagnostic and therapeutic instrumentation and applications including microbubble creation in body fluids in vivo, various combinations of breast imaging, including x-ray tomosynthesis, ultrasound, microwave, and photoacoustic tomography.

Awards

2013. FJ Fry Memorial Lecture Award, American Institute of Ultrasound in Medicine (AIUM) recognizing a current or retired AIUM member who has significantly contributed to the scientific progress of medical diagnostic ultrasound.

2008. AAPM Coolidge Award, American Association of Physicists in Medicine's highest award for outstanding contributions to medical physics.

2001. Basic Radiological Sciences (BRS) Collegiate Professor, University of Michigan.

1997. J. H. Holmes Basic Science Pioneer Award, American Institute of Ultrasound in Medicine (AIUM) honouring an individual who significantly contributed to the growth and development of diagnostic ultrasound.

1987, President, American Association of Physicists in Medicine (AAPM).

9

Rainstorm Over Kyoto: Technology Development and Interdisciplinary Collaboration at Memorial Sloan Kettering Cancer Center

C. Clifton Ling

Memorial Sloan Kettering Cancer Center (MSKCC) is considered to be one of the top cancer hospitals in the United States. The institution began as the New York Cancer Hospital in 1884. In 1912, it became known as the Memorial Hospital for the Study of Cancer and Allied Diseases. With a donation of land from John D Rockefeller, the hospital moved to its present location in 1939. In 1945, the industrialists Alfred Sloan and Charles Kettering of General Motors provided funds to establish the Sloan Kettering Institute, a research arm for Memorial Hospital. The two entities were finally combined into MSKCC in the 1980s. The following three stories describe major technological developments and the evolution of interdisciplinary collaboration as they occurred at MSKCC from 1989 to 2012.

C. Clifton Ling (✉)
Varian Medical Systems, Palo Alto, CA, USA
e-mail: Clif.Ling@varian.com

© The Author(s), under exclusive license to Springer Nature
Switzerland AG 2022
J. Van Dyk (ed.), *True Tales of Medical Physics*,
https://doi.org/10.1007/978-3-030-91724-1_9

1   Rainstorm Over Kyoto: The Story of Intensity Modulated Radiation Therapy (IMRT) at MSKCC

1.1   Cancer and Radiotherapy

Cancer has long struck fear in the minds of patients and their families. Through much of human history, the scourge of cancer most often led to death. But thanks to many factors, a large number of cancer patients today can expect to recover from this increasingly treatable illness.

The National Cancer Institute of the United States estimates that the relative five-year survival rate for people with cancer has increased to about 67%. This achievement is due, at least in part, to significant advances over the last 50 years in radiotherapy, particularly the technology for treating cancer with radiation.

1.2   Early Forms of Radiotherapy

Radiation therapy was first introduced as a palliative treatment for cancer in the early 1900s. Devices, using primitive x-ray tubes, generated low energy radiation that was not sufficient to provide a cure or to even to penetrate the body very deeply. The best hope for a patient was for the radiation to shrink the size of the tumour and provide some measure of pain relief.

1.3   The Medical Linear Accelerator (Linac) and the Start of Modern Radiotherapy

Modern radiation therapy traces its origins to the invention of the "klystron" tube by the Varian brothers, Russell and Sigurd, in 1937. First used in radar systems, the klystron could accelerate electrons through a vacuum tunnel to nearly the speed of light.

In the early 1950s, Dr. Henry Kaplan, head of Stanford University's Department of Radiology, proposed that a linear accelerator, or "linac," be specifically designed to generate high energy x-rays to bombard a cancerous tumour. The Varian brothers and several colleagues, who had formed a company called Varian Associates, developed a machine based on the klystron tube. It accelerated electrons and crashed them into a metal (usually tungsten) target to generate photons, or x-rays, with very high energies. The first Varian

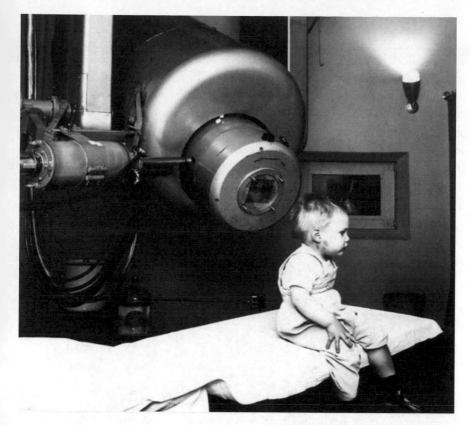

Fig. 1 The first patient treated with the first medical linac at the Stanford University Medical Center in 1956

medical linac produced an x-ray beam of 6 MV,[1] and was used to treat the first patient, a 2-year-old boy, at Stanford University (see Fig. 1).

While radiation therapy technology has progressed considerably in the last half-century, the basic goal of such treatment is unchanged: to target and kill cancer cells while exposing the surrounding healthy tissue to as little radiation as possible. Virtually all the advances in radiation therapy have worked to achieve a more successful clinical outcome by increasing the intensity, precision, and accuracy of a radiation beam.

---

[1] MV is a short form for megavolt, which is the unit of energy for an x-ray beam. A 6 MV x-ray beam is produced by electrons with an energy of 6 million electron volts (MeV) striking a target.

1.4   The Development of Modern Radiotherapy

The medical linac subsequently underwent a variety of transformations, in terms of size, cost, and complexity. In the 1970s, the linac could produce x-ray beams of up to 18 MV, and by the early 1980s, it was possible to operate the linac at either of two energy levels, e.g., 6 and 18 MV. The x-ray source was mounted on a gantry which can rotate 360° around the patient so that radiation can be delivered from different angles aiming at the cancer (see Fig. 2 in Chapter 5). In addition, computer-control was introduced, and radiation intensity or dose rate was increased.

Fig. 2 Low melting point metal alloy blocks designed specifically for each patient and each radiation treatment field

1.5   Beam Shaping Technology

During the evolution of the linear accelerator, another line of development focused on strategies for improving accuracy. Early linacs generated x-ray beams that were rectangular or square in shape and were directed at the tumour from two to four different angles. Naturally, tumours themselves are not square or rectangular, and so, these beams invariably encompassed some of the surrounding healthy tissues, resulting in unwanted side effects. This made it necessary for doctors to minimize the damage by using less-than-optimal therapeutic doses.

Improvements were attained in the 1970s, when custom-molded, lead-alloy blocks were placed in the path of the beam to shape them so that they more closely matched the two-dimensional profile of a targeted tumour. This spared some healthy tissue, but the process was highly labor-intensive and time-consuming. The blocks were very heavy, had to be individually manufactured for each and every patient, and then, during treatment, loaded in and out of the machine by the radiation therapist. Figure 2 shows three lead-alloy blocks. (Also see section 1.2 in James Purdy's Chapter 6.)

In the 1990s, a significant advance in beam-shaping was the development of the multi-leaf collimator (MLC), in which computer-controlled slats or "leaves" can be individually adjusted to shape the aperture through which the radiation beam passes. By changing the beam shape while delivering radiation from different directions, clinicians achieve very fine control over how, and where, the radiation is administered, and precisely conform the beams to the shape of the targeted tumour. The enhanced precision enabled radiation oncologists to boost the dosage of the x-ray beam to more effective treatment levels, improving outcomes and limiting side effects. A picture of an MLC is shown in Fig. 3.

1.6   Three-Dimensional Imaging and 3-Dimensional Conformal Radiotherapy

Important advances in medical imaging and radiology, particularly computer-tomography (CT) and subsequently magnetic resonance imaging (MRI), permitted clinicians to visualize the internal anatomy of cancer patients in 3 Dimensions. This capability facilitated the development of 3-Dimensional Conformal Radiotherapy (3D-CRT) in the 1980s. 3D-CRT, in enabling accurate delivery of radiation to the tumour while avoiding the critical normal tissues as much as possible, brought an incremental advance to the efficacy of radiotherapy in treating cancer. The MLC, described above, was an important

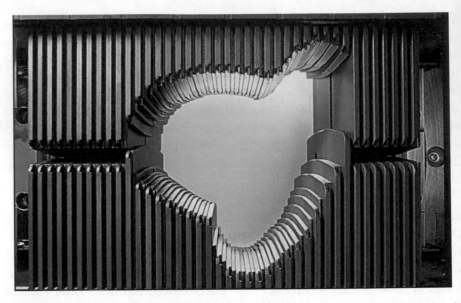

Fig. 3 A multi-leaf collimation system. The individual leaves can be independently positioned to shape the radiation treatment field

device in the implementation of 3D-CRT. A 3D-CRT treatment planning computer system, developed by a team led by Radhe Mohan, Ph.D. (see Chapter 22), was used at MSKCC in New York for patient treatment in the 1990s.

Throughout the developments described above, the radiation fields were uniform. That is to say, the intensity of the radiation was approximately the same across the exposed area. Then, with another advance described below, decades of striving for homogeneous distribution with radiation fields of uniform intensity was upended by the seemingly preposterous idea of intentionally delivering intensity-modulated (non-uniform) beams to the tumour.

1.7   The IMRT Revolution and How It Happened at MSKCC

In the mid-1980s an important paper by Anders Brahme, Ph.D., from the Karolinska Institute in Sweden, described the basic idea underlying what is now called intensity-modulated radiation therapy (IMRT). Dr. Brahme proposed to plan the radiotherapy treatment using the so-called "inverse" technique. Specifically, a computer algorithm considers the relative geometrical relationship between the tumour and the surrounding normal tissues and comes up with a dose distribution that optimally irradiates the tumour while sparing the abutting normal organs. The optimal dose distribution is achieved using intensity-modulated radiation fields. This paper stimulated many groups, including MSKCC, to study various aspects of this method, with the goal of actually implementing it for cancer radiotherapy.

In the spring of 1992, Professor Jerry Kutcher of MSKCC was invited to give a lecture at the MD Anderson Cancer Center (MDACC), in Houston, Texas. The MDACC is also considered to be one of the top cancer hospitals in the United States. While there, he met Thomas Bortfeld, a post-doctoral trainee from the University of Heidelberg, Germany, who had developed an inverse planning algorithm for his Ph.D. thesis. At that time, Dr. Bortfeld and Professor Arthur Boyer of MDACC had just demonstrated a successful delivery of inversely planned IMRT to a phantom using multi-leaf collimators (see Arthur Boyer's Chapter 5). Bortfeld was subsequently invited to spend a week or two at MSKCC, on his return to Germany, to see whether his algorithm could be linked to the MSKCC's 3D-CRT treatment planning system.

During Bortfeld's visit, Radhe Mohan and his team worked with Bortfeld to integrate his inverse planning algorithm to MSKCC's 3D-CRT planning system. Subsequently, upon further modification and improvement on the algorithm, their joint effort created a viable inverse planning module ready for clinical use. What was needed was a method to deliver the intensity-modulated radiation fields.

Concurrently in the early 1990s, a graduate student at Columbia University, Spiridon Spirou, applied to Professor Chen Chui of MSKCC to become Spirou's mentor for his Ph.D. thesis project. Their collaborative research resulted in the so-called sliding window technique for delivering an IMRT radiation beam using multi-leaf collimators. Unlike the previous use of MLC, with the leaves staying in static positions to shape the radiation field, in the sliding-window method the leaves actually move during radiation delivery. Using the MLC in the "dynamic" or sliding-window mode (DMLC for dynamic MLC), the individual leaves sweep across the radiation field, with the opposing banks of leaves forming the sliding-window. By properly adjusting the position of each leaf and the speed of its movement across the radiation field, the adjustable leaves of the MLC are used to control not only the shape of the beam, but also the exposure duration for small segments of the tumour, effectively "modulating" the dose within the treatment area.

The dual abilities to perform inverse planning and to deliver intensity-modulated beams set the stage for implementing IMRT at MSKCC. Unfortunately, the existing linacs were unable to deliver radiation while the multi-leaf collimator (MLC) leaves were moving.

1.8   Rainstorm Over Kyoto

A serendipitous rainstorm saved the day. I was on my way to attend the June 1993 International Conference on Radiation Oncology in Kyoto. It was raining heavily as I got out of the train station to look for a cab, but Tim Guertin and Marty Kandes of Varian Medical Systems, the company based in Palo Alto, California, which manufactures linacs, came to my rescue with a limousine. During the trip to our respective hotels, I told them about MSKCC's abilities to perform inverse planning and the use of DMLC to deliver intensity-modulated beams, but that the Varian linacs were unable to perform DMLC to implement IMRT. I was aware of Varian's so-called Kyoto software which was specially designed for the University of Kyoto to deliver conformal arc therapy. In conformal arc therapy, the MLCs were controlled by the gantry angle, and the leaf positions changed as the gantry rotated to conform to the shape of the cancer. I suggested to Tim and Marty that the MLC leaf positions could be controlled by the monitor units[2] (MU, which is a surrogate of dose) instead of the gantry angle, thereby performing DMLC for IMRT.

Several months after the meeting in Kyoto, Varian provided MSKCC with the control system for a Varian linac, the Clinac 2100C, to perform DMLC for IMRT. In October 1995, more than two years after the Kyoto meeting, the first IMRT patient was treated at Memorial Hospital. During those two years, the MSKCC team performed important preliminary studies to verify that each step in the IMRT process was accurate and reliable. These included the work of Radhe Mohan's group on inverse treatment planning, of Chen Chui's team on DMLC operation and dose calculation, of Thomas LoSasso's and Chandra Burman's contribution to physical and dosimetric quality assurance methodology, and Jerry Kutcher's overall guidance in the clinical physics program.

At the same time, Zvi Fuks (Chairman of Radiation Oncology) and Steve Leibel (Clinical Director of Radiation Oncology), were heavily involved

---

[2] In the head of the linear accelerator, near the source of radiation, there is a radiation detector known as a monitor ionization chamber. The readout from this detector is known as monitor units (MU) and it determines the length of the radiation exposure for the patient.

and spearheaded the clinical aspects of the IMRT program. Finally, and importantly, Varian provided the control systems for using DMLC for IMRT.

The superiority of IMRT, relative to 3D-CRT and 2D treatment can be visualized by the illustration in Fig. 4, showing the dose distribution of a patient's head treated for a cancer.

Fuks recollected that because there was no method of physically visualizing the treatment fields delivered by DMLC, physicians and physicists found it difficult to check and sign off each treatment field prior to delivery. He indicated that "the transformation into trusting the computer record and verifying system for DMLC function constituted a psychological barrier that had to be overcome". Furthermore, in his words, the "first delivery of DMLC-medicated IMRT constituted an act of crossing the Rubicon that psychologically initiated a process that facilitated image-guided radiation therapy (IGRT)". "Since then," he added, "we have adopted the routine of on-line dynamic treatment procedures totally controlled by computers".

IMRT ushered in a new era in radiotherapy, because of its ability to deliver a high dose to the tumour while minimize dose deposition to organs-at-risk abutting the tumour. This ability permitted dose-escalation to the tumour, which was hypothesized to increase local tumour control. This hypothesis was tested in clinical trials conducted at MSKCC and elsewhere, initially in prostate cancer and then extended to other disease sites. At MSKCC, the

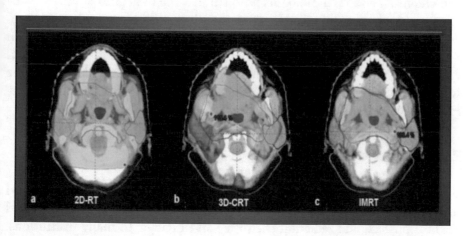

Fig. 4 A comparison of the dose distributions achieved by the three methods of treatment planning and delivery for a patient with a cancer in the head. Two-dimensional radiotherapy (2D-RT) delivers the same dose to the tumour and the normal tissues. Three-dimensional conformal radiotherapy (3D-CRT) delivers a high dose to the tumour while avoiding the normal tissues. Intensity-modulated radiotherapy (IMRT) further improves the dose distribution, to better conform to the geometry of the tumour and the normal tissues

dose delivered to the prostate cancer was escalated from 71, to 76, 81 and eventually to 86 Gy[3] using IMRT. Fuks and Leibel subsequently reported improved outcome for patients receiving the higher doses, thus confirming the IMRT hypothesis.

The IMRT technique using inverse planning and dynamic MLC on a linac, pioneered at MSKCC, was widely adopted by other centres within a few years as commercial planning software became available. Subsequently the technique evolved into VMAT (volumetric modulated arc therapy), which combines gantry rotation, variable dose rate, and dynamic MLC, to increase the speed of IMRT delivery. IMRT and VMAT are the standards of radiotherapy today. This story indeed confirms the title of this book, that medical physics is a life-saving specialty.

2   Growing Apart and Coming Together

2.1   The Multi-Disciplinary Science of Radiation

From the very beginning, following the discovery of x-rays by Wilhelm Roentgen, and radium by Marie Curie, the science of radiation had always been multi-disciplinary. The research of radiation initially involved physicists and chemists, and then biologists who studied the biological effects of radiation. And, almost immediately upon the discoveries of x-rays and radium, astute physicians began using x-rays for the diagnosis of diseases, and both x-rays and radium for the treatment of cancer. At the Memorial Hospital in New York City, the radiation scientific team during the early decades of the twentieth century included Gioacchino Failla, PhD, Edith Quimby, MSc, Henry Janeway, MD, and others.

2.2   Growing Apart

As the study and clinical application of radiation grew, specialties evolved into disease diagnosis, treatment of cancer and allied diseases, radiation and medical physics, and radiation chemistry and biology. In many institutions, at least in the United States, these scientists and physicians were homed in the same department, the department of radiology. Initially, physicians were trained in general radiology which included both diagnosis and treatment.

---

[3] Gy stands for the unit of absorbed dose, "gray", representing the radiation dose delivered to the patient.

With time, however, as each of the disciplines became more specialized and complex, it was more and more difficult for a physician to become a 'general' radiologist and master the skills of diagnosis and treatment at the same time. Over the decades of 1960s and 1970s, the trend was for the formation of independent departments of radiation oncology, separate from the departments of radiology. With this split, the medical physicists also became divided into two groups, with one in the department of radiation oncology, and the other in the department of radiology.

However, the infrastructure at MSKCC evolved into three independent clinical departments: radiation oncology, diagnostic radiology, and medical physics. Within the medical physics department there were scientists studying medical imaging (e.g., magnetic resonance imaging or MRI), improving cancer radiotherapy (e.g., 3D conformal radiation treatment), advancing computation methods (e.g., Monte Carlo probabilistic calculations), and investigating the biological effects of radiation. That these multiple and diverse scientific experts were in the same department was a tremendous benefit, with cross-fertilization and up-to-date communication of advances in each sub-specialty. In their interaction with the departments of radiology and radiation oncology, medical physicists avail innovation for clinical implementation to benefit cancer patients.

2.3 New Role for Medical Imaging in Cancer Radiotherapy Treatment Planning

In the decades from 1980 to 1990s, important advances in cancer imaging brought about somewhat of a re-integration of imaging and radiotherapy sciences. It was in that context that in September, 1997, the USA National Cancer Institute (NCI) organized an "Oncologic Imaging Workshop," convened by Carl Mansfield, MD (NCI Radiation Oncology), David Bragg, MD (NCI Diagnostic Imaging) and Robert Wittes, MD (NCI Medical Oncology). The workshop's purpose was "to identify new opportunities for imaging in treatment planning ... new research approaches and specifically find ways to enhance the rather limited practical clinical interactions between the imaging and radiation oncology communities when rich opportunities exist to tap the new advances in technology in both specialties".

Among the workshop's participants were medical physicists, CT/MRI diagnostic radiologists, radiation oncologists, and nuclear medicine physicians. I was invited to give a summation presentation and lead the general discussion to develop recommendations and action items for NCI to consider. Initially I declined the invitation, explaining that I was inadequately informed in the diverse spectrum of discussion. It was with some reluctance, and after much arm-twisting by the NCI staff, that I accepted the invitation. A good deal of time and significant effort was devoted to becoming better prepared for the task, with valuable input from many members of the department. Here, fortunately, was an opportunity to explore the horizon of combining radiological imaging, radiotherapy and medical physics, and beyond.

Upon my return from the NCI meeting and having reported what I learned to the faculty members, the enthusiastic reaction led to many research workshops to consider the development of a research program focused on the utilization of advanced imaging methods to guide radiotherapy. These culminated in a two-day workshop, April 9–10, 1998, in which David Bragg, MD, and Robert Sutherland, Ph.D., were invited as Laughlin Visiting Professors, and respectively spoke on "Imaging in the 21st Century: A New Paradigm" and "Tumor Hypoxia and Gene Expression: Implications for Malignant Progression and Therapy". Several MSKCC faculty members also made presentations during that workshop.

2.4   Finally, Together Again: Imaging and Radiation Sciences

From that point on, momentum gathered for a multi-disciplinary effort focusing on biological imaging, with team members from imaging physics, radiation biology, and clinical physics. These endeavors led to funding from the United States Army Medical Research to study "Non-invasive PET Imaging of Human Breast Xenografts: Influence of Tumor Hypoxia". With that fund, M Urano, MD, a research scientist from Massachusetts General Hospital was recruited with the goal of developing animal models for biological imaging. The ensuing research efforts resulted in a five-year NCI Bioengineering Research Partnership grant (2001–2006) on "Multi-modality Biological Imaging of Cancer and Tumor Hypoxia" with funding of US\$2.6 M for laboratory research. Finally, in 2006 an NCI Program Project Grant on "Tumor Hypoxia Imaging—Laboratory and Clinical Studies" was awarded with a direct cost of US\$5.75 M over 5 years.

The participants of the afore-mentioned NCI Program Project Grant were truly multi-disciplinary, in a sense reflecting the composition of the scientific teams in the early days of radiation sciences. Our team members included radiation biologists, radio-chemists, scientists in both clinical and animal imaging, medical and radiation physicists, and two physicians, one a radiation oncologist and the other a surgeon. As suggested by the title of the project, the research focused on imaging of tumour hypoxia with projects conducted both in the laboratory and in the clinic.

Perhaps a brief explanation of the tumour hypoxia and its importance is needed at this juncture. Hypoxia, in tissues and tumours, is the phenomenon of low or inadequate oxygen supply. Hypoxia exists in many human tumours, because the cancers grow so fast that the development of needed blood vessels cannot catch up. Important for radiotherapy, hypoxic cells are more resistant to radiation than oxygenated cells and may be a cause of failure for radiotherapy to eradicate the tumour. Thus, it is important to find methods to determine whether a patient's tumour is hypoxic, and also discover strategies to treat hypoxic tumours.

The goals of our NCI Program Project Grant were to study methods to measure hypoxia in tumours and to explore techniques to target hypoxic cells. Of course, our efforts represent a very small step in the global endeavor to conquer cancer, but its espousal of the multidisciplinary spirit reflects a return to the origins of the important discoveries of radiation, and of its seminal research. Our fledgling beginning eventually led MSKCC to establish the IMRAS program of Imaging and Radiation Sciences.

3   Beginning of the End and a New Beginning—From Physical to Biological Conformality with Dose-Painting in Radiotherapy

In a sense, IMRT represented the beginning of the end in terms of improved conformality of the physical dose of x-rays to the tumour and the surrounding normal tissues. In the two and a-half decades after the implementation of IMRT, further advances have yielded only marginal improvements in terms of physical dose distribution conformality. However, its ability to shape the dose distribution, when combined with the emerging, evolving and improving methods to derive tumour biology information with non-invasive imaging, position us to explore another horizon.

Over the last several decades there have been significant advances in medical imaging. The modalities include ultrasound, computed tomography (CT), single photon emission computed tomography (SPECT), positron emission tomography (PET), and magnetic resonance imaging (MRI). Other chapters in this book tell stories related to some of these imaging modalities (see Table 2 in the *Prologue*).

Until the development of PET and MRI, the radiological images are largely anatomical. In other words, they are images that show the physical anatomy of the patient, which are in fact very important for radiation treatment planning. The International Commission on Radiation Units and Measurements, in their Report No. 50, defined the concepts of gross tumour volume, clinical and planning target volumes for radiotherapy. The gross tumour volume represents what is visible on anatomical images. But because there are likely microscopic extensions beyond what is visible on these radiologic images, the doctors expand the gross tumour volume somewhat to generate a clinical target volume, based on their clinical experience. Another expansion is added to the clinical target volume to yield the planning target volume to account for other estimated physical uncertainties which are inherent in the process of treatment planning and patient irradiation, accounting for things like patient set-up variation from day-to-day.

With PET and MRI, additional information became available about the tumour, and of normal organs. For example, the increased metabolism of the tumour relative to normal tissues can be deciphered using PET with the radioactive reagent fluorodeoxyglucose (FDG). Other methods, some already developed and others under investigations, can provide information about the genetic make-up and functional characteristics of the tumour. Given the wide spectrum of information that the "new" imaging techniques can provide, we suggested the descriptor "biological" for this class of images (in contrast to anatomical). Biological images broadly include those in the metabolic, biochemical, physiological, and functional categories, and they should also encompass molecular, genotypic, and phenotypic images. For radiation therapy, images that give information about factors (e.g., tumour hypoxia) that influence radio-sensitivity and treatment outcome can be regarded as radiobiological images. These ideas led to a publication, in the year 2000, with the concept of biologic target volume (BTV), as shown in the Fig. 5.

The schematics in Fig. 5, in advancing the concept of the BTV, projected a "blue-sky" scenario. It was hypothesized that the biological target volume, as derived from biological images, may guide customized dose delivery to the various parts of the tumour volume. Specifically, there have been suggestions that non-uniformity within the planning target volume, specifically regions of increased dose, may actually increase the control of tumours. The ability of IMRT to deliver detailed non-uniform dose patterns by design, brings to fore the question of how to "dose paint" and "dose sculpt." In this regard, we suggested that biological images may be of value. As depicted in the figure, regions of low oxygen level from the PET study, high tumour burden from

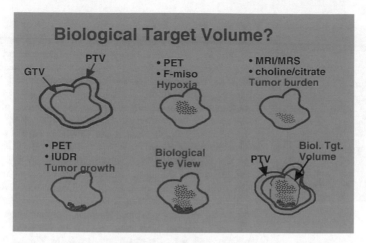

Fig. 5 The concept of the biologic target volume, with biological information provided by the various imaging methods using PET and MRI. GTV = gross tumour volume; PTV = planning target volume; F-miso = fluoromisonidazole, a PET radiotracer for imaging hypoxia; MRS = magnetic resonance spectroscopy; IUDR = iododeoxyuridine, a PET radiotracer for measuring tumour repopulation

MRI data, etc., may be incorporated into the biological target volume, based on which "dose painting" or "sculpting" strategy may be applied.

Our paper concluded with the statement that "Whereas IMRT may have initiated the beginning of the end relative to the physical conformality of dose distribution in radiotherapy, biological imaging may launch the beginning of a new era of biological conformality". During the two decades since the publication of that paper, there have been significant research efforts, and some clinical studies, on this topic. The term biology-guided radiotherapy is now commonly used for this approach. These developments were possible as a result of the cooperative efforts with medical physicists, radiation oncologists, imaging scientists and radiobiologists—clearly a multidisciplinary collaboration.

C Clifton Ling was born in China and received his Ph.D. in nuclear physics from the University of Washington, 1971. He held positions at Massachusetts General Hospital and Harvard, George Washington University, University of California San Francisco, and EA Haupt Professor/Chairman of Medical Physics at Memorial Sloan Kettering Cancer Center. Dr. Ling was active in American Association of Physicists in Medicine (Board of Directors 1982–1987, Scientific Program Chair 1983–1987, Science Council Chair 1991–1993). He chaired the Radiation Physics Committee of American Society for Radiation Oncology (ASTRO), and was a physics councilor in the Radiation Research Society. He was a grant reviewer of the USA and Canadian NCI and was on the Nuclear/Radiation Studies Board of the USA National Academies.

In addition to his numerous awards, he was Ray Bush Visiting Professor of Princess Margaret Hospital, Suntharalingam Lecturer of Thomas Jefferson University, Speaker of the Royal College of Physicians and Surgeons of Canada, Ira Spiro Visiting Professor of Harvard Medical School, Franz Buschke Lecturer of University of California, San Francisco (UCSF), and James Purdy Lecturer of the Washington University.

He has authored nearly 300 peer-reviewed papers. He received numerous grants from US National Institutes of Health (NIH), Department of Energy (DOE), Department of Defense (DOD) and the American Cancer Society.

Awards

2014: Distinguished Alumni Award—Memorial Sloan Kettering Cancer Center.

2013: Selected by the International Organization for Medical Physics (IOMP) as one out of 50 medical physicists "who have made an outstanding contribution to the advancement of medical physics over the last 50 years". This recognition was given as part of IOMP's 50th anniversary.

2006: Gold Medal, America Society for Radiation Oncology—the highest honour bestowed on an ASTRO member.

2004: Coolidge Gold Medal Award, American Association of Physicists in Medicine (AAPM)—the highest honour bestowed on an AAPM member.

1998: Honorary Member, the European Society of Therapeutic Radiology and Oncology (ESTRO)—the highest honour bestowed on a non-ESTRO member.

10

From Fourier Transforms to Surgical Imaging

Terry M. Peters

1 Foundations

As a child I was fascinated by maths and physics. Electronics became a hobby and building a sensitive radio receiver (that could pick up signals from Australia, 4000 km from my home in New Zealand) using two or three transistors was an exciting achievement. In my final year of high school, I recall being introduced to complex numbers and marvelling at Euler's identity (i.e., $e^{\pi i} + 1 = 0$). This, of course, branded me as a geek, but since I was gravitating towards a science (or maybe engineering) career, then perhaps this was not such a bad thing!

The structure of the New Zealand (NZ) secondary education system at the time was somewhat binary. By the age of 16, one was expected to choose between a liberal arts curriculum and science. So, during my fourth year of high school, I abandoned History, Geography, Biology and French (English was compulsory) in favour of Mathematics, Physics, "Advanced Mathematics", and Chemistry, in preparation for first year university studies in science and engineering. Fortunately, today's high-school curriculum allows a much broader educational experience!

T. M. Peters (✉)
Robarts Research Institute, Western University, London, Ontario, Canada
e-mail: tpeters@robarts.ca

© The Author(s), under exclusive license to Springer Nature
Switzerland AG 2022
J. Van Dyk (ed.), *True Tales of Medical Physics*,
https://doi.org/10.1007/978-3-030-91724-1_10

217

2   Undergraduate Years

I had received a scholarship from the NZ Post Office, which at the time was also responsible for all telecommunications, to study Engineering. Associated with the scholarship was the opportunity to work as an engineering intern during the summer, and one project to which I was assigned involved writing a computer program that would model the population growth in a small town in Southern NZ, in an effort to predict the telecom resources necessary to support an increasing population. This involved learning computer programming in FORTRAN from scratch and gaining access to the computer lab at the University of Otago. So, my first exposure to computer programming was using their brand new IBM 360/20 with 16 KB of memory and a 1 MB hard drive.

While the closest full engineering school was at the University of Canterbury in Christchurch, some 400 km away, it was nevertheless possible to take the first year engineering courses in mathematics, applied mathematics, physics and chemistry, in my hometown of Dunedin at the University of Otago. Following completion of this year, I moved to the University of Canterbury to commence my undergraduate degree in electrical engineering. In the second year of this program, I was introduced to my academic mentor, Professor Richard Bates, whose profound influence on my undergraduate and graduate studies at Canterbury set the direction of my future career.

Richard had a background in electromagnetic wave propagation and had recently arrived from the UK after working in the US and Canada. Although at the time he had only a bachelor's degree (he was subsequently awarded a D.Sc. from the University of London, England), he had nevertheless already established a prolific publication record and had made a name for himself. While his interests were primarily in electromagnetic theory, much of this work was based on the Fourier transform. He recognized that many systems that could be described in terms of Fourier theory, could perhaps be applied to problems in medicine. He had begun discussions with clinicians at the local hospital, who were modeling systems dealing with human physiology, and these conversations sparked his long-time interest in using the Fourier transform to describe these models in terms of linear system theory. This interest later extended to the description and analysis of medical imaging technologies.

2.1   The Fourier Transform

Since the Fourier transform (FT) played such a role in my career, it is probably worth taking the time to explain what this miraculous mathematical tool is all about. Simply put, any signal (e.g., an electrical waveform describing a patient's brain activity such as an electroencephalogram (EEG), or an image from an x-ray or computerized tomography (CT) or magnetic resonance (MR) can be constructed from a large number of waveforms added together. In the case of a one-dimensional (1D) signal, these waves are sines and cosines varying with time, each with different frequencies and amplitudes. In the case of images, the waves are also sines and cosines, but this time are intensities varying in space, such that they look like black and white stripes of different widths and laid down in different directions (Fig. 1). The Fourier transform in each of these cases is simply an array of numbers that describes the frequencies, and amplitudes of these sine (or cosine) waves, which must be added together to reconstruct the original signal or image. One-dimensional signals have a 1D Fourier transform, while 2D signals (or images), have 2D transforms. The Fourier transform is not simply a mathematical curiosity, however,

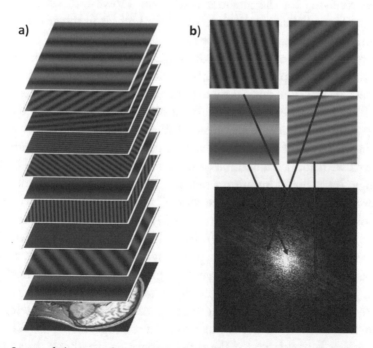

Fig. 1 **a** Some of the waveforms, that when added together, result in the image at the bottom of the stack (an MR image of the brain). **b** The 2D Fourier transform map of the image, and four of its components with the arrows indicating the position in the transform array that represent the frequencies and amplitudes of these waves

since it exists as a measurable quantity in an optical system. For example, if an image transparency is placed at one focal point of a convex lens and illuminated with coherent (laser) light, the image at the other focal point of the lens is the Fourier transform of the original image. If a second lens is placed a further focal length from the first, a 180° rotated version of the original image is recovered. The FT shows up again as the array (K-space) containing the raw data collected in an MRI scanner. Figure 1 shows how a number of "spatial" waves can be added together to create an image, and the Fourier transform "map" that describes the frequencies, amplitudes and phases of these waves.

3   Graduate School

It was during my final year as an engineering undergrad that Richard suggested I become one of his new intakes of graduate students the following year. I liked the idea and agreed to join his group. I should mention at this point that this decision entailed reneging on my commitment to the New Zealand Post Office—which meant paying back the scholarship funding I had been receiving for the previous four years. However, they did allow me to keep the fancy Faber Castell slide rule they had given me at the beginning of my university studies. (I still have it!).

Richard was of the opinion that any Ph.D. student worth their salt should be able to become expert in more than one discipline. So, in keeping with his interest in electromagnetic theory, he asked me to investigate the scattering of electromagnetic waves from infinite dielectric wedges, and also to come up with a solution, using time-of-flight ultrasound, to measure bone density in the bones of the arm. This latter project had been motivated by his discussions with local endocrinologists about the role of calcium in human physiology. As a final admonition before I began, he reminded me "… and don't forget to use the Fourier transform!" The Fourier transform domain would subsequently play a major role in Medical Imaging—not least in the magnetic resonance imaging (MRI) field where it would become well known as "K-space". In retrospect, this was probably the best advice he ever gave me.

I have to admit that the first year of graduate studies was not all that pleasant. I found it difficult to get my head around the complex mathematics required for the electromagnetic inverse scattering problem, not to mention the necessity of developing computer programs in FORTRAN to solve complex integrals and differential equations. In those days, software libraries were few and far between, and, since the research was entirely mathematical, there was no way of validating any results. In parallel, I began looking at the ultrasound/bone problem, but because commercial ultrasound scanners did not exist at the time, I was forced to build a rudimentary time-of-flight ultrasound device from scratch. This involved mounting 2 cm diameter piezo electric crystal disks facing each other at the ends of a U-shaped structure made from plexiglass. The device was mounted above a water bath containing a shin bone of a sheep. By exciting one of the transducers with a high voltage impulse to generate an ultrasound wave and observing the response from the other transducer after the sound wave had passed through the bone, one could measure the ultrasound transit time from transmitter to receiver. While the received pulse was fairly broad, the prevailing wisdom was that the first instance of the received pulse would represent the wave traveling the shortest distance through the bone. This time could ultimately be interpreted in terms of the total amount of bone in the path of the beam. However, ultrasound doesn't travel in straight lines, and it was difficult to obtain any meaningful measurements using this rudimentary device.

Throughout this process, a thought had been nagging me. Even though this work was taking place in the late 1960s, x-rays had already been a diagnostic imaging modality for over 70 years. X-rays do travel in straight lines, and they clearly record the presence of calcium in bones as opacities on a film-based x-ray image. However, these images were 2D projections of the 3D structure being imaged. Could this be used to measure bone density? If we took many views from different angles, could we figure out the 3D distribution of the material in the bones and maybe even calculate their density on a point-by-point basis in a cross-section? But how could we relate the 2D projections to the 3D object?

4   Fourier to the Rescue

This is where Richard's knowledge of Fourier transforms and radio astronomy came into play. A decade or so earlier, Ron Bracewell, an Australian engineer who was subsequently on faculty at Stanford University in California, had published a paper that inferred the 2D distributions of radio sources in the sky from 1D interferograms of their radio emissions. These interferograms were recorded from two widely separated radio telescopes whose signals were then cross-correlated. He had shown that if the 2D Fourier transform (i.e., K-space map) of the of the radio distribution was computed, then a single interferogram was equivalent to the 1D Fourier transform that would be obtained if this 2D distribution was sampled along a diameter of the 2D Fourier space. As the earth rotated with respect to the radio source distribution, similar signals were generated, but they corresponded to different diameters through Fourier space. In this manner, the individual, 1D signals (interferograms) could be interpolated into a 2D grid to reconstruct the full 2D Fourier transform of the radio source. The actual source distribution could be recovered simply by computing the inverse 2D Fourier transform. This operation seems trivial today with the availability of multi-dimensional fast Fourier transform (FFT) algorithms, but in the early 1960s Bracewell had to compute these transforms using a mechanical hand-operated calculator!

This problem bore a remarkable similarity to the one with which we were faced. Instead of an interferogram, we had the 1D projections of a 2D object. We knew from Bracewell's work that the Fourier transform of the projection of a 2D object was the line, at the same angle as the projection was acquired, through the 2D Fourier transform of the 2D object. So, all we had to do was to record our projections, compute their Fourier transforms, interpolate these signals at appropriate angles within a 2D complex plane, and compute the inverse 2D Fourier transform of the result (Fig. 2).

This process turned out to work extremely well! When formulating it mathematically, we realized that the equations could be factored in various ways. In addition to interpolating the Fourier transforms in Fourier space, we could obtain the same result by high-pass filtering the projections with what became known as a rho-filter (since the filter's magnitude was proportional to ρ, the radial coordinate in 2D Fourier space), and back projecting the filtered signals onto the image plane. We could also back-project the initial

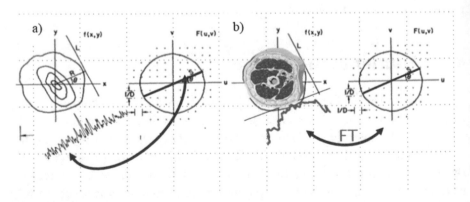

Fig. 2 a Interferogram of radio sources interpolated into 2D Fourier domain. **b** One-dimensional Fourier transform of limb projection interpolated into 2D Fourier plane. In both cases, the original cross-section of the radio source or limb is obtained through the 2D inverse Fourier transform of the 2D array

data directly onto the image plane, which resulted in a rather blurry image, which we called a "layergram", because it was the result of different views being layered upon one another, and then perform a 2D high pass "rho filter" on this image. Although these three approaches were mathematically equivalent, each offered a different approach that had advantages and disadvantages from a practical computational standpoint.

To illustrate the role of the Fourier transform properties of lenses in tomographic image reconstruction, we built a prototype apparatus that recorded the original unfiltered back-projection on a piece of x-ray film directly, resulting in the layergram. Then, following the theory that if we high-pass filtered (sharpened) the layergram image in exactly the right manner, we could reconstruct the desired image using only optics. With a laser, an optical bench and some lenses to perform the mathematics of Fourier transforms and filtering, we were able to shine a laser through a miniaturized version of the original layergram, and using a couple of lenses with an optical filter placed between them, reconstruct the original cross-section image instantly without computers.

Perhaps the most useful outcome of this work, however, was the back projection of the filtered projections onto the final image plane. This technique, which we called filtered back projection, became the mainstay of CT image reconstruction for the next few decades. We had little appreciation of the potential impact of this work, but thought it sufficiently interesting to warrant a publication, which appeared as "Towards Improvements in Tomography", in the New Zealand Journal of Science in January 1971.

Now that we had shown this approach would work in theory, we turned our attention to the reconstruction of real objects. And what better item to use as a test object in a country of 80 million sheep? A bone from a leg of lamb! To obtain radiographic projections of this bone, we mounted it in a jig enabling rotation by 9° at a time, while acquiring an x-ray image of the bone on film at each pose. This resulted in 20 projections of the bone equally spread over 180°. The task then became how to input these data into a computer. Somewhere within the University, we unearthed a scanning densitometer that could scan one line of each projection on the film and convert the optical density to a curve on graph paper. These twenty plots then had to be converted to numerical values that could be input to the university mainframe computer (IBM 360/44) via punch cards. For the reconstruction, we implemented the 2D Fourier interpolation scheme adapted from Bracewell's paper mentioned above.

The result of this exercise exceeded our wildest expectation, with the reconstructed cross section (Fig. 3) looking remarkably like the radiograph we

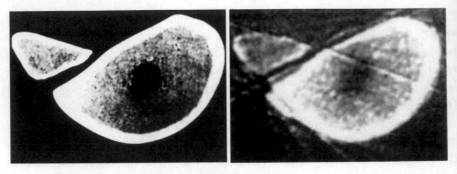

Fig. 3 Left: photograph of original bone cross-section. Right: reconstruction from 20 projections using Fourier domain interpolation

had taken of the same slice of the bone that had been subsequently excised. With great excitement I rushed to the Radiology Department of Christchurch Hospital and waved the image of our reconstructed bone in front of the chief radiologist. Expecting great excitement and congratulations, I was greeted by a distinct lack of enthusiasm and his comment that existing x-ray films were much higher resolution than this and that he really couldn't see the need for cross sectional imaging. Not a very encouraging response!

Shortly afterwards, somewhere into my third year of my Ph.D., I came across a news release, that showed Godfrey Hounsfield, standing beside his EMI CT scanner-the first in the world. This news of course was devastating, as there had been no other indication in the literature that such work was so close to clinical application. So how was he computing his reconstructions? After much digging, we discovered that he had been using an iterative reconstruction approach. Nevertheless, I felt that I had been scooped. If someone had already built a CT scanner, what was left for me to do? The interesting thing was that Hounsfield had never published on this topic before building this machine that was hailed as a revolutionary contribution to medicine. Indeed, along with Allan Cormack, Hounsfield later received the Nobel Prize for his work.

5   First Job!

In spite of this perceived setback, my thesis was eventually completed (and the work on electromagnetic scattering abandoned—I think Richard had become so excited about CT that he hadn't noticed)—apparently the first to have tackled the reconstruction problem for CT! The next task was to find a job!

While I had made some half-hearted attempts to find a postdoctoral position somewhere throughout my Ph.D. studies (a response from Harold Johns in Toronto was "sure we will have you if you bring your own funding!"), I meanwhile had been making a nuisance of myself at the local hospital. I had sought encouragement from the radiologists(!) and help from the medical physicists, one of whom, Chief Physicist Jack Tait, had taken an interest in my work, and made a place for me within his small but rapidly growing team. While at this point the Medical Physics department had been mostly concerned with planning for radiation oncology procedures, after the advent of Hounsfield's CT scanner and some early ultrasound devices, they were becoming increasingly involved in diagnostic imaging as well. I was simply in the right place at the right time, and even though I came with a newly minted Ph.D. in Engineering, I was accepted warmly by the group of four card-carrying physicists. However, a postdoc opportunity also arrived around this time. A colleague, who had been a postdoc in the Physics Department at the University of Canterbury while I was completing my Ph.D., and with whom I had collaborated on several papers, had since moved to the Biozentrum of the University of Basel in Switzerland. This institution had an affiliation to the Basel University Hospital, which had recently acquired one of the first commercial versions of the EMI scanner to be installed in Europe. Would I like to join him for three months and work with their new toy? This was not an opportunity to pass up, so with my wife and 6-month-old daughter, we left in the height of a NZ summer to the depths of a European winter. And while the entire stay lasted only three months, I had the experience of working first-hand with the EMI scanner and the clinicians using it.

Following my return from Europe, and after a year at Christchurch Hospital being involved in the development of radiation treatment planning software using the latest in mini-computers (a Digital Equipment Corporation PDP-11 GT-44), I floated the idea of building on my Ph.D. work and constructing a CT scanner in-house.

By this time, the chief radiologist had changed his tune about CT and was lobbying management to purchase an EMI scanner for the hospital. Such efforts were clearly being made at other hospitals in the country as well, and the NZ government decided to take the lead in evaluating and funding CT machines for the major centres across the country.

6   A Pivotal Moment

Being the only individual in the country with firsthand experience of CT at the time, it was suggested that I make a trip to the first international CT conference in 1975 being held on the campus of Stanford University in California. This event transpired to be life-changing for me! Arriving at a pre-conference event, fresh from a 15 h flight from NZ, I was met by a sea of unfamiliar faces. However, from the other side of the room, I detected and gravitated towards a familiar accent. I introduced myself to Chris Thompson, a New Zealander by birth, who had moved to Canada some years earlier and had been working as an engineer at the Montreal Neurological Institute (MNI or the "Neuro"). We quickly discovered we were from the same home-town of Dunedin and had several mutual friends. As a result of this meeting, Chris invited me to visit the MNI to evaluate the scanners they had recently installed—an opportunity that would come to fruition a little more than a year later.

The other significant event at the Stanford meeting was listening to Paul Lauterbur speak about his fledgling imaging method—"zeugmatography"—that relied on magnets and magnetic field gradients to elicit radio signals from objects containing water molecules. The images were very crude—no more than blobs—and there was a lot of head scratching going on amongst the audience. Magnets? Gradients? This was an x-ray CT meeting! Was he at the wrong conference? However, I did recognise the Fourier-based modified back projection approach he was using to reconstruct his image, and I realized the Fourier transform was important to magnetic resonance imaging (MRI) reconstructions as well. It was still some time before Fourier space became universally known as "K-space" in MRI circles. However, neither I, nor any of the audience, had any concept of where this seminal work would lead. The rest, as they say, is history, with Lauterbur also being awarded the Nobel Prize along with Peter Mansfield in 2003 for their contributions to MRI.

The following year, I was invited by the NZ Government to join a senior radiologist to undertake a month-long whirlwind tour of the US and Europe to evaluate CT for New Zealand. First stop was the Radiological Society of North America (RSNA) conference in Chicago, which even then in 1975 had tens of thousands of attendees, and was the largest meeting, by several orders of magnitude, that I had ever attended. The aim here was to evaluate the many CT scanners that had suddenly materialized. There were over 20 companies exhibiting and at least half of them represented the drug industry. Armed with an extensive check list, we attempted to compare the character-istics of all the available systems. While the EMI system and several others had used a "translate-rotate" design, there were a number of systems that had embraced "rotate-only" configurations, with either the x-ray tube and detector mounted on the same rotating gantry (3rd generation—similar to the design we would eventually adopt for a home-grown machine) or a 4th generation design that had the x-ray tube rotating inside a ring of fixed detectors (Fig. 4).

Following RSNA, we visited several manufacturing facilities, as well as a number of hospitals in the US, Canada and Europe. As part of this tour, I visited the Neuro in Montreal as the guest of Chris Thompson, armed with a large plexiglass phantom that I had lugged all the way from NZ, to test the newly installed EMI "body scanner". These visits were the beginning of a life-long friendship with Chris and prompted an invitation from him to spend a couple of years at the Neuro.

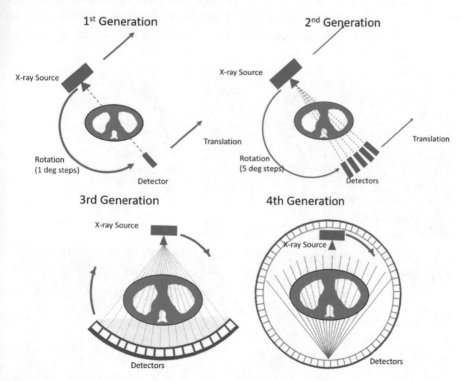

Fig. 4 Four generations of CT scanner available in the mid 70s

7   A Home-Grown CT

Back home in NZ, I continued my work developing radiation planning systems and continued to push the Hospital administration for funding to build our own CT scanner that would be used for radiation treatment planning. Finally, a sum of $20,000 NZ (equivalent to ~ $140,000 US today) was made available for this project. This was quite a tidy sum for a public hospital to part with at the time. The only way of working within this budget was to repurpose equipment that had become redundant within the hospital. Following the construction of a prototype device (Fig. 5a) that we used to test the system and troubleshoot the many bugs, we re-built the gantry of a de-commissioned cobalt radiation therapy treatment unit, retaining the motor drive, but replacing the cast iron c-arm with one of a much lighter sheet steel construction (Fig. 5c). However, the detector posed a problem. Through my trip to Stanford the previous year, I had become aware of work there to build CT detectors, and we commissioned them to construct an arc-shaped

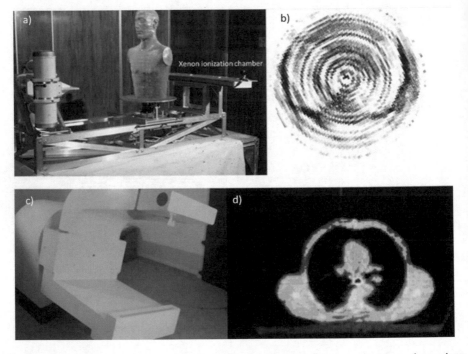

Fig. 5 a Prototype scanner with first thoracic image from a phantom (complete with "ring artifacts" as seen in (**b**). **c** Final CT gantry powered by motor-drive from obsolete cobalt-60 machine and resulting patient image of the thorax **d** used for planning radiotherapy treatments

aluminum chamber with 120 radial wire electrodes. When filled with pressurized xenon gas, this device became a multi-wire ionization chamber with 120 detectors. Each channel required its own current-to-voltage convertor, pre-amplifier, analogue to digital converter and digital storage buffer, but on tallying the costs of these electronic components, we realized we were significantly over budget, having only the resources to purchase half the required components. The rings seen in the first reconstruction of our chest phantom (Fig. 5b) were caused by poor calibration of the detector, prompting the development of an extensive calibration protocol to address the problem.

Necessity, as they say, is the mother of invention. While we were building a fan-beam data acquisition system, we decided the reconstruction problem would be much simpler if the data were simply interpolated from the fan-beam data array onto a rectangular grid that would have represented the dataset acquired from a translate-rotate system. Normally, to acquire a set of data similar to that produced by a translate-rotate system, which rotated its source over 180° around the patient, the fan beam system required a rotation of 180° plus the fan angle of the beam. However, if we rotated the fan beam by 360°, while at the same time off-setting the fan symmetry by one-quarter of the angle between individual detector views, we could double the spatial sampling rate of the data. Thus, by instrumenting only every second detector element, we were effectively able to meet the performance of 120 detectors as long as we implemented the "1/4 detector offset" technique and rotated the beam by 360°. A sample image is shown in Fig. 5d. Shortly after publishing this method, we were impressed to learn that General Electric (GE) had adopted this approach and retro-fitted it to their own fan-beam CT scanners. It had never occurred to us at the time that there was potentially valuable intellectual property wrapped up in this simple idea!

8   To Canada

In 1978, I accepted Chris Thompson's invitation to join him at the Neuro for two years, taking the rather bold step of selling our home, and moving to Montreal with three children under five in tow! As it would be only for two years, we decided we should savour the experience that had been offered. This did of course also involve moving to one of the largest French-speaking cities in the world! Surprisingly however, the three years of French taken in high-school 15 years earlier, had served me well and I was able to pick up the language again without too much difficulty.

On arrival in Montreal, I was immersed in a new world of leading-edge research and clinical imaging equipment. The MNI had been one of the first institutions in North America to receive a CT scanner—the EMI Mark-1 head scanner, but now they also possessed an EMI-5005 body scanner. In parallel to the clinical CT work, MNI's Director, Dr. William Feindel also had encouraged the development of radio-nuclide scanning—particularly the new modality of Positron Emission Tomography or PET. Dr. Feindel, although a neurosurgeon, had been fascinated by the use of radionuclide scanners to detect intra-cerebral tumours, and as a surgical resident, had worked in Saskatoon with Canadian Medical Physics giants Harold Johns, Bill Reid and Sylvia Fedoruk to develop such an instrument. Now Chris Thompson, along with nuclear medicine physician Lucas Yamamoto, had begun the journey that would establish many of the foundations of PET scanning as we know it today, and would produce four increasingly sophisticated "Positome" brain scanners for use at the Neuro. I quickly became involved in projects relating to image-analysis and reconstruction for CT and PET images.

Besides his work with PET, Chris had another string to his bow. He had worked with Dr. Feindel's neurosurgical colleague, Dr. Gilles Bertrand to develop image-guided deep brain implantation of stimulation and recording electrodes, based upon anatomical atlases registered to the patient. This work profoundly influenced my later work with stereotactic imaging and surgery, and as Chris' energy became increasingly consumed by PET, I assumed the role of integrating these concepts with CT, and later MRI.

A highlight during my early years at the Neuro was an invitation to a Retreat on Image Reconstruction, hosted at the Oberwolfach Research Institute for Mathematics, nestled in the mountains south-west of Stuttgart in Germany. Where else to share a dinner table and scribble mathematics on table napkins with recent Nobel Laureate Allan Cormack, who had earlier that year shared the honour with Godfrey Hounsfield for their independent contributions to CT developments—work that had pre-dated our own by several years!

The two years in Montreal passed quickly and as we were preparing to return to NZ, the opportunity arose to apply for a position as Scientist at the MNI and faculty position, in Medical Physics and Biomedical Engineering, at McGill. My future, it seemed, was to remain in Canada.

Now that CT had become an established clinical imaging modality, the next new player on the block was, of course, Nuclear Magnetic Resonance or NMR. It was still called "Nuclear" back then, before public pressure forced the industry to drop the "N-word" and the technique became MRI. It had only been a few years since the first Lauterbur paper, but now industry, led by the companies that had been successful in the CT market, was on a roll marketing these new imaging devices. Just as my NZ radiology colleague had spurned CT, the Chief of Neuroradiology at the MNI was equally dismissive about MRI—"Who needs magnetic fields when we already have CT"? Nevertheless, MNI Director Feindel was convinced that MRI was the next big thing and he set me on the quest to select the most appropriate machine for the Neuro. Although Canada already had three MRIs, they all had a magnetic field strength of 0.15 T. Those in Toronto and London, Ontario used resistive magnets, while a unit in Vancouver used a superconductive magnet. Feindel's aim was to install Canada's first "High-field" (0.5T) superconductive magnet.

After an extensive search that mirrored the NZ CT evaluation experience almost a decade earlier, we eventually settled on a 0.5T superconducting magnet from Philips. Since one of the primary applications for MRI at the Neuro was for pre-operative planning for stereotactic neurosurgery, I volunteered as a human phantom to demonstrate that a stereotactic frame (loosely) fastened to the head and equipped with plates containing MR compatible "Z-bar" markers, could be successfully imaged with this system. So, on a visit to Eindhoven in the Netherlands, stereotactic frame under my arm, I became the first human subject to be imaged by the magnet that would not be delivered to Montreal for a further six months (Fig. 6).

The magnet finally arrived and was to be installed on the ground floor (actually designated "3rd basement") of the MNI. Its installation, however, was not without tense moments. An opening had been designed in the side of the building to accommodate the magnet and its top-mounted turret, with 10 cm to spare. Unfortunately, between ordering and delivery, the specs of the turret had changed, and the entire assembly was now roughly 10 cm taller than anticipated. There was much anxiety as we finally saw the magnet clearing the space by less than 5 mm!

A catastrophe early on in the operation of this device was the experience of a magnet "quench", which occurs when the super-cooled cryostat spontaneously vents its liquid helium and nitrogen, the super-cooled windings lose their zero-electrical resistance characteristics, the magnet loses its field, and the super-cooled cryogens (helium and nitrogen) are vented into the atmosphere. This quench occurred under very strange circumstances. One morning I observed that the images on the cathode-ray tube monitors in the control room were all rotated clockwise by around 30°. This would be normal behaviour for such displays in the presence of a magnetic field, and under normal operating conditions the electromagnetic electron-beam deflection yolks surrounding the neck of the cathode ray tube (CRT) monitors

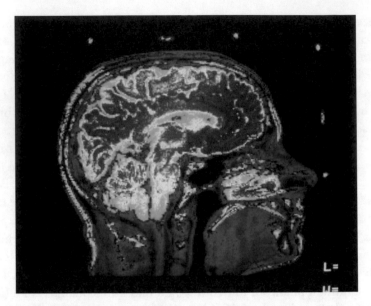

Fig. 6 Sagittal view of first MR image (my own brain!) on the soon to be delivered 0.5 T "high-field" superconducting magnet. Note the bright spots at the top and right side. These are fiducial markers on the stereotactic frame that would provide reference points for the registration of the patient wearing the frame

were rotated slightly such that the image was displayed correctly, even in the presence of the magnet's fringe field. However, if the field was turned off, the images would rotate back in the opposite direction. Observing this phenomenon, I was afraid that the magnet had quenched. However, to be sure, I approached the magnet with my key-ring dangling from my finger, with the knowledge that without a field, the keys would not be attracted towards the magnet. Surprisingly however, I did feel the tug on my finger as the keys aligned themselves with the magnetic field. This was totally unexpected! I subsequently activated the "liquid-helium level sensor" on the magnet control panel to check the status of the cryogen. This process indicated that the cryogen level was indeed low. However, the very process of making this measurement, which entailed injecting a tiny external current into a loop of conductor submerged in the liquid helium, was sufficient to induce a second quench. Now the magnetic field was indeed zero—the MR room suddenly filled with freezing fog as a consequence of the expelled super-cooled helium and nitrogen, and the images on the CRTs were rotated several more degrees anticlockwise! This meant that the magnet had actually quenched in two stages, a phenomenon that I have never seen reported, and one that violates the accepted physical principles relating to magnet quenches.

During my early years at McGill, I became involved in the activities of the newly formed Canadian College of Physicists in Medicine (CCPM), which had begun the process of credentialling Medical Physicists in Canada. While I had Engineering, rather than Physics, training, I felt that my background had prepared me well for the roles I was playing in Diagnostic Radiology at the Neuro. Nevertheless, I felt it would do no harm to formally study radiation physics and its practical implementation in Medicine, so in 1983 I embarked on a mission to devour "The Physics of Radiology", by Johns and Cunningham, in preparation for the CCPM Fellowship exams in 1984. The examination process had evolved into an oral session, and a closed book examination—where three questions were selected from a previously published catalogue of questions covering all aspects of Medical Physics. Every Friday afternoon for almost a year I studied "Johns and Cunningham" with Gino Fallone, then a physicist at the Montreal General Hospital, who had also decided to take the certification examination. A gruelling process, but finally successful—we both became CCPM fellows that year. Part of my role within the Neuroradiology Department at the Neuro was to teach the radiographic technologists about the new imaging modality of MRI in preparation for what was to soon arrive. Part of this task was to talk about K-space (No—not "Case Base!") or Fourier transforms. I recall two of the responses to the examination question "Describe in your own words what you understand by the Fourier transform". The first was—"I frankly have no idea, but it must be important because Dr. Peters gets awfully excited when he talks about it!", However the second—"the Fourier transform splits an image into waves, just as a prism breaks up white light into its component colours!", let me know that at least someone had been listening!

9   A New Paradigm

The advent of MRI at the Neuro immediately opened a whole new world of image-guided neurosurgery. Now we could easily visualize the boundaries between white and grey matter and observe the many intricate structures within the brain that previously had to be inferred by the registration of the image to an atlas. The images actually reflected the anatomy of the patient rather than images depicted by atlases that were derived from multiple post-mortem sliced brains, which then had to be squashed and stretched to match the patient. The availability of true 3D imaging meant that procedures could be planned precisely, specifically for each patient.

Simultaneous with the acquisition of MRI, we were also exploring stereoscopic digital subtraction angiography. For context, it should be pointed out that all angiographic imaging at the Neuro had historically been performed using bi-plane stereoscopic film changers. These devices employed two sets of dual x-ray tubes, set orthogonally to each other and also with a pair of mechanical film changers that were capable of changing film cassettes several times a second. A typical angiographic sequence would therefore yield around 60 frames comprising 15 stereoscopic pairs from the lateral view, and a further 15 from the anterior–posterior perspective. These images, when laid out on viewing screens in the viewing room, could sometimes fill the entire wall. While there were available hand-held Wheatstone stereoscopic viewing devices to assist the radiologist to appreciate the 3D structure present in the angiograms, most were able to view the stereograms simply by viewing image pairs with crossed eyes. Legend has it that stereo vision was a pre-requisite for being hired as a radiologist at the Neuro!

Digital subtraction angiography had been around for several years at this point. Initially, stereo angiograms were performed by delivering repeated contrast injections and re-imaging the sequence from a slightly different angle to simulate the inter-ocular distance required for visualizing the image pairs in stereo. This of course introduced increased risk to the patient by the excess radiation exposure as well as the double dose of contrast. Motivated by the desire to overcome this problem, we worked with Siemens Medical to develop a dual focus x-ray tube, whose beam could be delivered alternately from each focal point. These images were captured sequentially by a single image intensifier tube, and after some analogue image-preprocessing that removed the still fading previous image from the current one, stereoscopic image pairs resulting from a single injection of contrast were collected and stored on a hard drive. Display of these images was via a monitor equipped with an electronically switchable polarizing screen and viewed using spectacles that polarised incoming light to the left and right eyes with mutually orthogonal polarization respectively, so that each eye only saw the appropriate image. It was mildly amusing to note that the director of the Siemens project to develop this technology was blind in one eye and could not appreciate the incredible 3D images that this system was able to acquire.

With the advent of this technology, pre-operative MRI and CT images could be registered to the stereoscopic angiograms to allow the identification of safe zones when planning the implantation of deep brain electrodes for epilepsy diagnosis or the treatment of Parkinson's Disease. Motivated by the enthusiasm of both the radiologists and neurosurgeons with whom I was working, we developed several platforms that allowed simultaneous visualization of 3D MRI images and stereoscopic digital subtraction angiography (Fig. 7).

The advent of Stereoscopic Digital angiography, alongside MRI and CT, opened up many new vistas of image-guided neuro intervention. These ranged from guiding catheters into brain aneurysms and arteriovenous malformations (AVMs) to introduce glue or platinum coils into the regions to prevent blood flow to the regions, to stereotactic radiosurgery to treat arterial-venous malformations with a focused beam of radiation using a standard radiotherapy linear accelerator. This latter project, undertaken in collaboration with Dr. Ervin Podgorsak at the Montreal General Hospital, provided a realistic alternative approach to the "Gamma-knife" using cobalt-60 sources for the ablation of deep brain lesions.

With the collaboration among our neurosurgical colleagues, this work evolved into a comprehensive 3D surgical planning system for the treatment of Parkinson's Disease. For reason's buried in history, this became known as

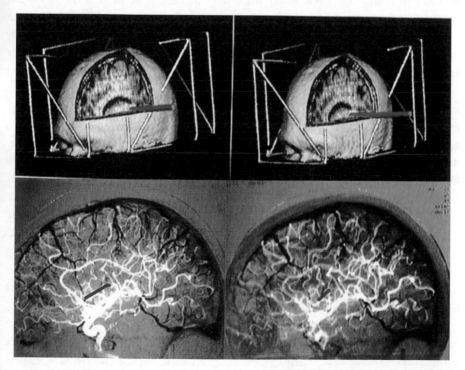

Fig. 7 Registered 3D MRI (top) and digital subtraction angiography images (bottom), depicting an electrode being implanted in a deep brain target. The red line in the lower image pair represents the electrode, white vessels are arteries, black the venous system. The images are set up for cross-eyed viewing in stereo

the Viper Platform. Students of the day referred to the lab as the Viper Pit, name it still retains at the Neuro 25 years later!

10   The Robarts Years

Sometime in the mid 1990s, I attended the annual Canadian Organization of Medical Physicists (COMP) conference in Vancouver, where I happened to take a stroll along the beach with Aaron Fenster, a long-time colleague, who had several years earlier decamped from the Princess Margaret Hospital in Toronto to start up a Medical Imaging group at the new Robarts Research Institute in London, Ontario (See Aaron Fenster's Chapter 19). At the time, this was the only private Medical Research Institute in Canada, although it was located within the campus of Western University. He indicated that they planned to establish a theme in Image-guided Interventions at Robarts, and would I be interested in moving to London to help set it up? After almost 20 years in Montreal, I was ready for a new challenge, so after some thought, agreed to join the group, which already numbered amongst its ranks two of my former students from McGill (and was soon to be joined by a third). So, in 1997, my wife and I made the trip 750 km west to Western University and Robarts.

In London, my first task was to quickly establish a lab, re-create the tools from Montreal and begin interacting with the neurosurgical community. By this time the Viper platform had become somewhat bloated and unwieldy, and so my new batch of students (one of whom was a Neurosurgery Resident with a strong computing background), decided to re-build the software from scratch, basing it on the recently announced open source Visualization Toolkit software library or VTK. This approach resulted on a much more reliable and extensible package, that, following its reptilian genus, was aptly named "ASP". When asked at a scientific meeting shortly thereafter what ASP stood for, its architect, student David Gobbi, replied without missing a beat—"Acronym Still Pending"! Over the years, ASP evolved into a quite sophisticated platform for the planning and guidance of stimulator implants for the treatment of Parkinson's Disease.

This was also the time when the adjacent London Heath Sciences Centre had begun investing in Surgical Robotics. This was initially to support minimally-invasive heart surgery and it wasn't long before the lab was visited by one of Robotic Cardiac Surgery's pioneers, who recognised the potential of our image-guided surgery software to assist in minimally-invasive cardiac surgery. After all, the problem was basically the same as in the brain—finding an unseen target within the body. Except for the minor complication that the heart was moving! So, I established a second research theme in minimally-invasive cardiac surgery, and the lab became known as the laboratory for Virtual Augmentation and Simulation for Surgery and Therapy—or the VASST lab. With a suite of core technologies for tracking and visualization, over the years the lab has become involved in projects in image-guided neuro-surgery; minimally invasive cardiac surgery; vascular interventions; kidney and liver therapies, prostate surgery, and spinal procedures. One of the most satisfying aspects of this work has been the incredible level of interaction the lab has enjoyed with our clinical colleagues. Facilitated by the proximity of the Robarts with the University Hospital of the London Heath Sciences Centre, we have been able to take advantage of clinical interaction that is difficult or impossible to achieve for similar research groups that are located far away from their clinical colleagues.

One of the underlying themes of the work in the lab has been the improvement of the surgeon's interface with the patient, and Virtual and Augmented Reality has played a major role in this endeavour. Most of the time, virtual or augmented reality is employed to partially or fully align the surgeon's visual field with their proprioceptive field. In many cases of laparoscopic or ultrasound-guided surgery for example, the view of the target from a video camera or ultrasound transducer inserted into the patient's body is displayed on a monitor that is placed far from the surgeon's operating field of view. Under these conditions, instrument maneuvers depicted on the display may bear no relationship in terms of orientation or direction to the surgeon's view of a patient, and cause constant distraction to the surgeon as they continually switch their focus from the monitor to the surgical field. An example of how serious this situation can become, and how simple visualization technology can mitigate this problem is demonstrated below.

The problem in this example is the repair of the mitral valve in a beating heart by introducing an instrument into the left ventricle via a small incision between the ribs. Intra-operative imaging is provided by an ultrasound probe, which provides two orthogonal ultrasound planes covering the mitral valve, introduced into the patient's esophagus. The standard of care is to perform the procedure while visualizing the probe's motion within the two ultrasound images (Fig. 8a). Under these conditions, we tracked a sensor at the end of the probe to record the trajectories five cardiac surgeons used to reach the target region. These coloured trajectories are presented in Fig. 8b.

However, by incorporating tracked sensors into both the tool and ultra-sound probe, virtual models depicting the relative positions of these devices could be constructed and visualized. We could re-format the display to

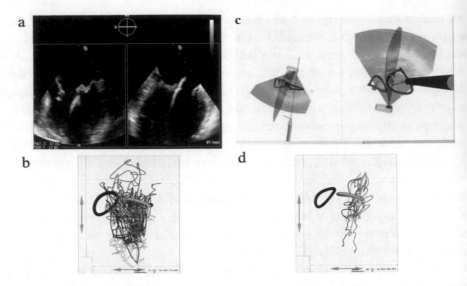

Fig. 8 a Bi-plane transesophageal ultrasound; b probe trajectories from five surgeons using (a) as guidance; c reformatted display showing virtual representational targets, US probe and surgical instrument; d trajectories from five surgeons using (c) for guidance

demonstrate what the ultrasound beam is actually seeing, and display virtual representations of the probe and ultrasound transducer to the surgeon in a much more intuitive manner (Fig. 8c). Performing the same task as before, but now with the enhanced display, produced a new plot of tracked tool trajectories as shown in Fig. 8d. This study demonstrated that the use of simple tracking and visualization technology could not only make the procedure significantly safer (Fig. 8b shows the probe intruding on danger zones 40 times more often than in Fig. 8d), but also that the procedure was executed four times faster!

This example is but one of the many the lab has worked on in various organ systems, including the brain, heart, spine, kidney, liver, prostate and the vascular system, but it serves to demonstrate the improvements that can be gained by the application of simple visualization and tracking technologies.

11   Concluding Remarks

Looking back on a career spanning almost 50 years, I am humbled to have witnessed at such close range, not only the history of computerized imaging in medicine in general, but also its implementation for surgical guidance. This career path opened opportunities to work and travel internationally, to meet colleagues all around the world, and to work with an amazing group of graduate students and colleagues, both in Montreal and London. Perhaps the most rewarding aspect has been the opportunity to work closely with clinical colleagues, and to share their excitement when technology developed in the lab had a direct impact on patients. Many of my professional colleagues and former students have become life-long friends, and these include several of the authors of chapters in this book. A particular shout-out to Jake Van Dyk who encouraged us to tell our stories! One of the highlights of my career has been my involvement with the Society for Medical Image Computing and Computer-assisted Intervention (the MICCAI Society), which first saw the light of day as an annual meeting in Cambridge MA in 1999, and as an international Society in 2003. It has since become the most prestigious annual international meeting for applied Medical Imaging research. I had the privilege and fulfilling experience of acting as General Chair or Program Chair for three of these meetings in the past two decades.

At both Western and McGill, I was fortunate to have been able to recruit an amazing group of students, many of whom have gone on to establish their own successful careers in academia and industry, and some using their time in the lab to springboard into clinical careers. It has been a privilege to have been part of their own career journeys.

So, what have I learned over the past 50 years?

- Not to become discouraged if you feel you have been scooped! If you had an idea similar to someone else's, it was probably a good one, but your perspective was surely different. And there is always room for refinement.
- Not to always believe the "experts"! After all, wasn't IBM's Thomas Watson alleged to have said in 1943 that he thought the world could maybe use five computers? Or Bill Gates to have opined that 640 K was surely enough memory for anyone. Even EMI believed that there would be a global market for only half a dozen of their EMI-Scanners!
- To provide your students abundant freedom to explore. They will reward you handsomely.
- To grasp opportunities offered to you; follow them passionately; don't look back.
- And finally—you never know where a stroll in Fourier space may lead you!

Terry Peters is a Scientist at the Robarts Research Institute and Professor Emeritus of Medical Imaging, Biomedical Engineering and Medical Biophysics at Western University, London, Canada. He has more than 45 years of experience in Medical Imaging and Image-guided Interventions, beginning with 5 years at the Christchurch Hospital in New Zealand following his Ph.D. at the University of Canterbury, where he made early contributions to the development of algorithms for CT reconstruction. He then spent 19 years at the Montreal Neurological Institute, where he was involved with the development of CT and MR-guided planning systems for the surgical treatment of neurological disease. For the past 24 years he has continued to be active within the Laboratory for Virtual Augmentation and Simulation for Surgery and Therapy (the VASST Lab.) at the Robarts, where he and his team work closely with physicians and surgeons towards the development of minimally-invasive procedures for multiple organ systems. He has mentored more than 100 trainees and authored over 400 publications, including three books as editor. He was a founding member of the Society for Medical Image Computing and Computer Assisted Interventions (MICCAI) and has acted as General Chair or Program Chair for three of its annual meetings.

Awards

2019. *Canadian Organization of Medical Physicists (COMP) Gold Medal*. This is the highest honour that COMP bestows on one of its members in recognition of an outstanding career as a medical physicist who has worked mainly in Canada.

2018 *Fellow of the Royal Society of Canada*, for contributions to Medical Imaging and Image-guided surgical procedures.

2014 *Enduring Impact Award, MICCAI Society*, for contributions to MICCAI, Medical Imaging and computer-assisted interventions.

2013 *Fellow of the Canadian Organization of Medical Physicists (FCOMP)* in recognition of "his significant contribution to the organization and to the field of Medical Physics in Canada",

2011 *Fellow of the Institute of Electrical and Electronic Engineers* for Contributions to Medical Imaging and Image-guided Surgery.

11

Gorilla in the RF Cage and Other Recollections

Stephen R. Thomas

1 The Preparatory Highway

As is most often the case, performance in the later stages of one's career is intrinsically linked to experiences gained earlier on. There were two major phases in my days before medical physics that merit some mention with regard to how they shaped my professional perspective. These were (1) two years in the Peace Corps, and (2) many years as a graduate student in solid state physics. An old adage perhaps, of which I'm a believer, states that personality traits as entwined with world outlook aren't innate but are developed through the course of challenging individual events.

1.1 The Peace Corps Years

Following graduation from Williams College (Williamstown, MA) in 1963 with a BA in physics, I attended graduate school at Purdue University (West Lafayette, IN) for two semesters. A combination of circumstances including disillusionment following the assassination of John F. Kennedy in November 63 convinced me that I needed a break from academic pursuits. Thus, in

S. R. Thomas (✉)
University of Cincinnati Medical Center, Cincinnati, Ohio, USA
e-mail: steve.thomas@uc.edu

search of a "sabbatical", I applied to the Peace Corps with the specific request for an assignment in Africa. It all worked out as I headed to Ghana, West Africa in the Fall of 1964 as part of the fifth contingent sent to that country. As an historical note, the first group of Peace Corps volunteers ever to leave the United States under this new international program started by President Kennedy went to Ghana in 1961. Our responsibility (unlike that in many other host countries) was to teach science and mathematics in the Ghanaian secondary school system so, in that sense, I maintained some ties with my collegiate past. My institution, The West Africa Secondary School, was in the capital city of Accra on the "Gold Coast". Because the Ghanaian regime wanted to minimize the influence of any single foreign government (although it had a strong leaning toward the East with an anti-US press), I found myself part of an eclectic gathering of young expatriate teachers that included Russian, East German, British, and Canadian along with the Americans. As the Ghanaian school system was shaped by the legacy of British colonial rule, the structure was that of the United Kingdom with all courses taught in English. The students were receptive to all of us, enthusiastic learners, and a pleasure to teach (Fig. 1).

The immersion in Ghanaian culture provided an ultimate international experience. Social activities included lively parties with our Ghanaian hosts

Fig. 1 West Africa Secondary School students with me during a physics laboratory class (ca. 1965)

drinking the local Star beer and dancing to highlife music as well as rocking to the Beatles (Western rock and roll was quite popular with all nationalities). However, I will relate one sobering, eye-opening real-world reality at that time involving the East German teacher at my school. He, his wife and 2 youngest children shared a house with us within walking distance of the classrooms. They lived in the second story allowing many close and enjoyable personable interactions. After a while, we became aware by inference (not direct statement) that their oldest son who remained in East Germany was being held to "guarantee" that the family would return when the tour was over. This provided the somber realization that freedoms we took as a matter of accord were not available in all societies.

We took advantage of the school year vacation schedule to travel over many parts of the continent. In a number of regions, the politico-social kettle was at the point of pre-boil. The following provides a snapshot of some of the adventures: Pound wagons (so-called for the British sterling price per passenger) from Accra to Ouagadougou (capital of then Upper Volta); fish truck and bush hopper prop plane to Timbuktu, Mali; coal train to Kano and Benin, Nigeria (burned out vehicles signaled the beginning of the Biafran war); thumb solicited rides to the Zimbabwe ruins (Southern Rhodesia immediately before the declaration of independence by Ian Smith—the stress was evident); land rovers through the Tanzanian wildebeest paradise (prior to AIDS awareness although it must have been lurking somewhere); bus transit through the rift valley territory to Asmara (single soldier convoy riding night time shotgun foretold of the deepening Ethiopian-Eritrean strife); prop-plane flight to Khartoum (although the upstream Nile irrigation wheels continued to turn, Sudan political transitions would soon draw the attention of Egypt).

An additional word or two concerning the challenges of hitchhiking through East Africa. Rides were scarce to come by as traffic on the dirt roads in the bush was sporadic to just above nil. At one point being left off on the main track by a logger who was headed up to his camp some 50 miles (80 km) away, we received the admonition "If you get stuck here over night, watch out for the hyenas". Fortunately, the next ride did come along to further us on the way. At one junction point on the Zambia-Malawi border, we were faced with the decision of going South or North. Per reports from some source, strife and insurrection dominated the situation South, so North we went. Eventually, we did reach our interim destination, Dares Salaam, Tanzania, on the Indian Ocean.

In the middle of my second year, I was witness to one of the first of soon-to-be-all-too-common coup d'états both in-country and across the continent.

To provide a bit of historical background, *Kwame Nkrumah* was the President of Ghana at that time and the dominant political figure. He got his start as Prime Minister in 1952 and continued in that position after Ghana gained independence from Great Britain in 1957. In 1960, a new constitution was approved, and Nkrumah was elected President. Nkrumah became a cult figure and was addressed officially as Ossogeffo which means "redeemer" in the local language. When the Ghanaians approved a constitutional amendment in 1964 that made Ghana a one-party state, Nkrumah was designated President-for-Life. Under Nkrumah, Ghana was a socialist country. The Vietnam War was raging. Ghanaian politics and media were basically anti-West, anti-USA. However, that official, political sentiment did not permeate down to the general public. The average Ghanaian embraced the Western way of life and welcomed the company of American expatriates. Finally, on Feb 24, 1966, deep-seated irritation with the Ossogeffo persona resulted in Nkrumah being deposed by the National Liberation Council, an arm of the military. That day, radio news reports highlighted the rejoicing in the streets. As Peace Corps volunteers were not allowed to own cars, a British friend drove us into the centre of Accra to get a feel for the mood of the populous described as anti-Soviet, pro-Western. I recall the dark surge of self-authorized crowds led by an intoxicated, uniformed military individual surrounding our car as we drove away from the city following a trunk search and being glad we were with a British driver rather than Russian. (Another example, perhaps, of curiosity needing to be tempered to avoid undesirable consequences.) I am happy to recount that the Coup was essentially bloodless. Nkrumah was out of Ghana at the time and forced to spend the rest of his life in state exile, dying of natural causes in 1972. To wrap up this reminiscence, by the end of the tour, I felt that I had contributed to the scientific development of a few young minds. Without doubt, the experience had been beneficial toward expansion of my world view as well as personal introspective stability. Figure 2 is a letter from Lyndon B. Johnson acknowledging my contributions as a volunteer in the Peace Corps.

1.2   The Graduate School Years

The return after the Peace Corps was directly back to graduate school (Purdue University) in the fall of 1966. Often asked if I had any regrets in taking two years off from my professional track to spend time teaching in Africa, the answer—an unequivocal 'No!'. Following the requisite course sequence and various qualifying exams, it was time to choose a major professor and targeted area of research. At this juncture a momentous decision had to be

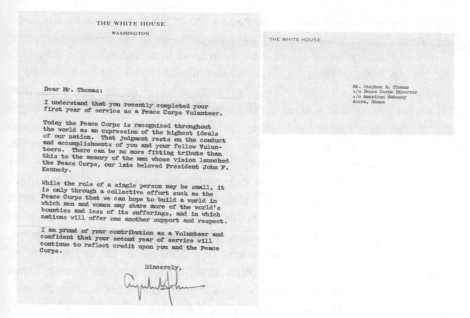

THE WHITE HOUSE
WASHINGTON

THE WHITE HOUSE

Mr. Stephen R. Thomas
c/o Peace Corps Director
c/o American Embassy
Accra, Ghana

Dear Mr. Thomas:

I understand that you recently completed your first year of service as a Peace Corps Volunteer.

Today the Peace Corps is recognized throughout the world as an expression of the highest ideals of our nation. That judgment rests on the conduct and accomplishments of you and your fellow Volunteers. There can be no more fitting tribute than this to the memory of the man whose vision launched the Peace Corps, our late beloved President John F. Kennedy.

While the role of a single person may be small, it is only through a collective effort such as the Peace Corps that we can hope to build a world in which men and women may share more of the world's bounties and less of its sufferings, and in which nations will offer one another support and respect.

I am proud of your contribution as a Volunteer and confident that your second year of service will continue to reflect credit upon you and the Peace Corps.

Sincerely,

Fig. 2 Letter received from the White House signed by president Lyndon B. Johnson following my first year as a peace corps volunteer at the West Africa Secondary School, Accra, Ghana (Summer 1965)

made. Input from students a few years ahead of you was critical. Through a series of consultations, I ended up focusing on a professor in solid state physics who was infamously known for chewing up his graduate students before eventually (hopefully) setting them free. Hsu-Yun (HY) Fan was born in Shanghai, China but earned a doctorate in electrical engineering from the Massachusetts Institute of Technology (MIT) in 1937. Before being hired by the Purdue University Department of Physics in 1948, he was an assistant professor at the National Tsing Hua University in Beijing, China. Of note, the final phase of the Chinese civil war took place between 1945 and 1949 and was referred to as the Chinese Communist Revolution. By 1951, Dr. Fan had received the rank of full professor and from 1963 until retirement in 1978, he was the Duncan Distinguished Professor of Physics. He was world renowned as a pioneer in semiconductor solid state physics. Professor Fan, though small in physical stature, was a stern task master who demanded excellence in performance. As I recall, he had a scar toward the back of his right cheek, never smiled around the office or laboratory, and, ironically, wore what many today would refer to as John Lennon type glasses (as opposed to those of Elton John). Finally, after much deliberation, I decided that I would be lacking in manly attributes if I did not sign on with HY Fan. With a deep breath, I made the commitment to follow the more tortuous path and took the plunge.

Thus began 5 years of relentless research—a true scientific odyssey. My topic was far infrared emission from semiconductors and involved the use of a Michaelson interferometer. I'll explain some of these terms in brief. As part of the electromagnetic spectrum, the infrared (or heat waves) have longer wavelengths than visible light while the far infrared goes out even beyond that (0.015–1 mm). Special detectors are required to record their presence. The interferometer is an instrument designed to split a beam of light (for me, far infrared radiation) into two perpendicular pathways and bring them back together so as to observe their interference upon recombining. Mirrors were located at both ends of the pathways with one of them controlled to move in tiny, precise increments. Thus, in the course of a sequence of steps, the signal recorded by the detector and digitized would be in the form of an interferogram. It could be decoded into the far infrared spectra I was investigating by the mathematical process known as Fourier transformation. My bench-top interferometer was a cube approximately nine inches on a side as contrasted to the Laser Interferometer Gravitational-wave Observatory (LIGO) with its four-kilometer-long perpendicular arms that was used in 2015 for the historic, landmark detection of two black holes merging.

Semiconductors, as the name implies, are materials with the ability to conduct electricity in between the extremes of conducting metals and non-conducting insulators. Common, better known ones include the elements silicon (Si) and germanium (Ge) (popularized in transistors) but there are others in the form of exotic compounds such as gallium arsenide (GaAs) or cadmium telluride (CdTe). One can manipulate the electrical properties of semiconductors by the deliberate addition of atomic impurities in a process known as "doping". This was the key to my research. Under the proper conditions (extremely cold temperatures), electrons in the semiconductor could be removed from the doped energy levels through the action of an applied electric field. During the transition back to the original level, they would release a small amount of energy in the form of far infrared radiation. Specific analysis of this radiation was the theme of my thesis. The game was to document experimentally the sharp peaks in the far infrared spectra and to assign the impurity energy levels responsible for this radiation according to theoretical models. So, the labour continued for years—a mixture of dedicated necessity as well as love. The challenges included: designing and assembling the equipment; preparing the semiconductor samples and specialized detectors; assembling the data acquisition hardware and data analysis software programs; correlating the resultant data output with theoretical expectations (perhaps the most demanding task). Of note is the fact that the data were in the form of punched IBM cards then dutifully carried in trays to the campus computer center main frame unit. With luck, results might be available a day or so later—no such thing as rapid turnaround! (Personal computers were not even on the fringe of the imagination in the late 1960s and early 1970s.) Oh, the desperation if the run was rejected due to one or more of my own programming errors. On a weekly basis, with trepidation and a stiffened upper lip, I would shuffle down the hall into Professor Fan's office to update him regarding progress and discuss results. Most often, I would emerge beaten but with renewed determination to forge onward in spite of the bruised ego. The experiments themselves were laborious. Data acquisition was slow with each run requiring several hours and up to six runs per session. I learned early on, given the multitude of things that could go wrong, if the equipment was running smoothly, don't stop, keep going. Thus, quite often the experiments continued overnight into the next morning. Perhaps this is a good point to mention a definite bright period in my existence at that time. I met my wife, Ingrid Erika (nee Storbeck from West Germany) and we were married in 1970. She had been recruited to come to the United States and work for a German professor in the life sciences. (Her paycheck was significantly more than my meager grad student stipend … but, by no means was that the reason we wed!) It was Ingrid who sustained me through the overnight sessions by shuttling McDonald's Big Macs into the lab. While on the subject of food, I do recall a dinner gathering at the Fan's home graciously hosted by his wife, Li Nien (Manya). Upon meeting Ingrid and conversing with her at length for the first time, Dr. Fan made a comment to the effect: "I sure wish Ingrid was my graduate student rather than you". Oh well! All in the family!

My experiments were conducted at very low, cold temperatures. The samples and detectors were housed separately in glass Dewars which are, in essence, scientific thermos bottles held at liquid helium temperature (~ minus 267 °C or 4.2-degree Kelvin) (Fig. 3). The Dewars were attached to the interferometer via an interconnected housing configuration. An integral part of the process was maintaining a vacuum independently within the Dewars and the interferometer. At the end of the experimental runs, when all components were back at room temperature, the vacuum would be released in a prescribed sequence—interferometer first and then the Dewars. After one particularly long run, I released the vacuum and heard a muffled implosion. It was with stunned awareness that I realized I had inadvertently reversed the order and a crystal quartz window on one of the Dewars blew off through the interferometer like a missile wreaking havoc by destroying the beam splitter, one mirror, and the window itself. Upon recovering from the initial shock,

Fig. 3 Room 186 of my laboratory in the Purdue physics building. The sample Dewar can be seen rising up vertically in front of my right shoulder. The sample, held at liquid helium temperature, would be in the bottom end and optically coupled to the interferometer. The detector Dewar is not in place in this photo but the holding collar, visible on the aluminum support table top, indicates its position. The interferometer cube itself is under the support table but essentially obscured by various other components (ca. 1971)

knowing I had to face the consequences, I marched down to Dr. Fan's office (as it was daytime now) to report this disaster that had occurred due to my negligence. Talk about visions of the Cross! Well, much to my great surprise, Dr. Fan was extremely understanding and supportive. Essentially, he said, pick up the pieces, we've got the funds to put the equipment back together again, and let's move forward with the business of scientific research. I wasn't crucified but rather re-invigorated through his encouragement to forge ahead. Most assuredly, I took two lessons to heart from that experience: (1) you should never underestimate the importance of sequence; and, (2) even the most feared adversary might turn around and express some compassion and encouragement when sorely needed.

Indeed, all ended well with the Ph.D. awarded in 1973. But I must admit, when I thumb through my thesis now, I am amazed that I ever produced such a tome. I conclude I must have been a lot smarter back then!

2   Transition into Medical Physics

Following receipt of the Ph.D., I spent one year at Kentucky Wesleyan, a small college in Owensboro, KY, as the 2nd man in a two-man physics department. Due to economic circumstances, that faculty position was dissolved in the Spring of 1974. As it turned out, this unexpected development made all the difference. Perhaps a perfect example of an event not forecasted and seemingly undesired at the time, but a game changer that, in retrospect, turned out to be eminently positive. I caught wind of a Post-Doctoral position in Medical Physics at the University of Cincinnati. It was with great excitement and anticipation of tremendous opportunities that we headed off to the "Queen City" to embark on this new career.

2.1   Of Mentors and Fishing Buddies

There was a lot to absorb upon entering this new realm of medical physics. My direct supervisor and mentor was James G. Kereiakes, Ph.D. Director of the Division of Medical Physics, Department of Radiology, University of Cincinnati, College of Medicine. He had at the time, and would continue to assemble, a distinguished list of credentials that included President of the American Association of Physicists in Medicine (AAPM), Coolidge Award of the AAPM, and Gold Medal of the Radiological Society of North America (RSNA) to mention only a few. Dr. K, as he was referred to around the Department, was nationally known for the quality of his training programs and output of productive medical physicists. I found myself in extremely capable hands. He enthusiastically ushered me into the Medical Center environment.

My initial assignments were within the Division of Nuclear Medicine that was directed by Eugene L. Saenger, MD. Nuclear medicine was the first Division within Radiology (and perhaps the entire Medical Center) to introduce the use of computers into diagnostic clinical medicine. Those were "power horse" machines in 1974 with a maximum RAM of 64 kB (four 16 k boards), cathode ray tube (CRT) monochrome monitor, removable hard disk, and paper tape punch.

Eugene Saenger (Gene) was one of the MDs who truly embraced physicists with the expectation that they would apply their expertise to advance medical research. There was an appreciation for the aptitudes brought in by the physicist including proficiency in analytical techniques. Most importantly, he was willing to provide the requisite resources as required to proceed.

Of note, he was instrumental in creating the position within the Department for Jim Kereiakes upon his joining the faculty in 1959. Gene received many accolades in his career including being awarded the George Charles de Hevesy Nuclear Pioneer Award for outstanding achievement from the Society of Nuclear Medicine (1987) and gold medals from the RSNA (1993) and the American Roentgen Ray Society (ARRS) (1998).

At one point in the late 1980s, Gene learned that I had been a trout fisherman in my youth. With exuberance he invited me as his guest to his fly fishing club about 100 miles (160 km) north of Cincinnati (Zanesfield Rod and Gun Club, Zanesfield, Ohio. Now, and as long as I knew it, ... no guns!). We drove up together and, upon arrival, I pulled out my fly rod, reel, and gear from the old days probably last used in the late 1950s. He grabbed them out of my hands, took a critical look and declared them less than fit for use. I was handed an extra rod and reel that he had with him that I used for the rest of the day. Fish were visible in the Club streams and, before the first cast, I thought the outing would be a piece of cake. However, that day I didn't catch a single trout while the others were reeling them in! Then and there, I realized I had a lot to learn concerning the art of fly fishing and decided to approach that educational objective with dedication. Since becoming a member of the club in 2000, Ingrid and I enjoy spending time there with reasonable success on landing the rainbows.

2.2   Humour in Medicine

Benjamin Felson, MD was the Director of the Department of Radiology at the University of Cincinnati from 1951–1973. Today the Chairman of the Department holds the Benjamin Felson Endowed Chair. Dr. Felson (Ben) was internationally recognized and acclaimed as the best-known radiologist of his time. Renowned as a chest radiologist, he created a number of the now classic diagnostic radiologic signs aiding in identification of various pulmonary disorders. He excelled in teaching (receiving the "Golden Apple" award eight times for outstanding teaching at the Cincinnati Medical School). His many awards included Gold Medals from the RSNA and the American College of Radiology (ACR). For many years, he was editor of the journal *Seminars in Roentgenology*. Ben Felson also had a far-reaching reputation as a humorist which was evident in his lecture presentations. He did, however, have limits and was not a proponent of malicious humour. In his wisdom, he has been quoted as saying "Racial, religious, sexist, nationalistic, or political anecdotes often sound as flat as a bassoonist with a hernia and are best avoided".

One of his short stories appeared in *Seminars in Roentgenology* in 1976 under the title "My Most Unforgettable Patient". This was a highly entertaining and delightful account of a late-night x-ray room adventure at Cincinnati General Hospital trying to obtain images of an uncooperative cow. In reaction to one of the many regrettable occurrences during that session, as colleagues and technicians maneuvered with mops in hand, Ben marveled at the volumetric capacity of the bovine bladder. Later, he published a book containing this and a number of his collective writings entitled "Humor in Medicine—and Other Topics" (Publisher RHA, 1989).

Ben was extremely competitive. That characteristic came out in spades on the tennis court at the annual Radiology Department picnic. If you were to be his partner, you had better be determined to win the set! Jerome F. Wiot, MD took over the reigns as Chair after Ben stepped down in 1973 but I do believe Ben continued to think of himself as Chief. Many times, when in a meeting in Jerry Wiot's office, Ben would burst in unannounced with some issue. Jerry would simply respond accordingly and shake his head patiently.

Another luminary within the University of Cincinnati Medical Center of the same vintage as Ben was Charles M. Barrett, MD. He was a pioneer in radiation oncology at the University while at the same time active in the role of president and chairman of two Cincinnati based insurance companies. In 1987, he was honoured with the distinction of being named a "Greatest living Cincinnatian" by the Chamber of Commerce. Dr. Barrett (Charlie) was one of those individuals who remembered your name after an initial introduction—a very impressive attribute to me as a young faculty member. He served on the Board of Trustees of the University for 20 years and was President of the Board for the 9 years before he retired. He was further honoured by having his name on the Charles M. Barrett Cancer Center which opened as a new facility as part of the Medical Center in 1989. In the 1970s and 80s it was common for the major medical companies to host department functions at the RSNA annual meeting in Chicago. At one such dinner sponsored by the Eastman Kodak Company, quite by chance, I was seated at a table with both Ben and Charlie. At one point for some reason, the conversation touched upon their respective portraits displayed on walls in various parts of the Medical Center. The discussion centered upon whose rendering garnered the most acclaim. I happened to have viewed both of them. Ben's was a detailed, lavish oil work very much after the style of the European masters. Charlie's was more abstract. I don't believe it was a watercolour, but perhaps of that genre. Charlie was lamenting how much better the presentation of Ben was than that of himself. Not to deny the truth in this assertation,

Ben leaned back, squinted with his mischievous smile and said: "Well, that's because I paid a lot more for my portrait than you did for yours!" Case closed!

3   Research Endeavors

3.1   Development of Quantitative Conjugate-View Counting Techniques

Within the Division of Nuclear Medicine, Harry R. Maxon, MD was treating patients with Paget's Disease of the bone. This disease, found in a small percentage of the elderly, is characterized by excessive production of structurally abnormal bone. In its initial phase, there is excess resorption of bone followed by an intense phase of osteoblastic response (i.e., abnormal new bone formation). When advanced, this disease can be highly painful and debilitating. Diagnosis and follow-up after therapy involve injecting the patient with a radiopharmaceutical (a pharmaceutical drug containing a radioactive isotope) that seeks areas of active bone growth. Following the radiopharmaceutical administration, the patient undergoes a nuclear medicine scanning procedure that provides images of the affected bones wherein abnormal regions (lesions) would show up as "hot spots" (representing high levels of radioactivity because the radiopharmaceutical is attracted to these lesions). Ideally, for a patient responding to therapy, the "hot spots" would decrease in intensity with time as the bone returned to normal. The difficulty on follow-up was to determine precisely the decrease (or not) of radioactivity in a given lesion as an indication of the degree of healing (or not) that had taken place. This challenge provided me the opportunity to develop quantitative imaging techniques enabling determination of radioactivity in a patient lesion at a given time. The method was known as *conjugate view imaging* where "conjugate" refers to the protocol of taking images of the affected area from diametrically opposite sides (i.e., 180° apart). I derived a series of equations that described the radiation emitted from lesions within the body under this geometry. In brief, the technique involves calibration of the imaging instrumentation, selected initial measurements on the patient, digital acquisition of lesion image data, and application of the equations allowing calculation of the lesion radioactivity. The results of this research were published in 1976. (Of additional note, in August 1976, we enthusiastically welcomed our daughter, Kirstin Erika, into the world.) This paper marked a defining moment in my career as it instilled in me the importance of multidisciplinary collaborations and the scope of accomplishments possible with the partnership between physician and medical physicist. Much of my continuing research within nuclear medicine focused on quantitative techniques as applied to in vivo radionuclide dosimetry including diagnosis and therapeutic follow-up of patients with well-differentiated thyroid cancer treated with radioiodine (iodine-131). The above-mentioned initial publication led to my involvement and subsequent long association (30 years) with the Medical Internal Radiation Dose (MIRD) Committee of the Society of Nuclear Medicine and Molecular Imaging.

3.2   Magnetic Resonance Imaging

Surprisingly to some perhaps, magnetic resonance imaging (MRI) came along right at the heels of computerized tomography imaging (CT). A general impression often is that MRI emerged decades after the CT scanner was introduced as a clinical instrument in the early 1970s. However, toward the end of the 70s, not long after Paul Lauterbur demonstrated the feasibility of using MR projections acquired under the presence of a magnetic field gradient to produce an image (1973), rudimentary MR head images were being circulated by some of the medical equipment manufacturers. In presenting these topics as part of the radiology resident physics lecture series, it was always fun to show the initial low contrast, low spatial resolution CT and MR images from the 70s followed by their transitional evolution into the exquisite quality images of the 90s and beyond. My foray into MRI took place in 1980 when we brought an electromagnet donated by Procter & Gamble (P&G) into my lab in the Medical Sciences building. It had been used by the P&G scientists for classical nuclear magnetic resonance (NMR) investigations of materials. The magnet had a maximum field strength of 1.4 tesla (T) (for comparison, the much weaker Earth's magnetic field is ~ 0.00003 T) and was unique in that it had a relatively large space of 5.7 cm between the magnet poles allowing investigation of objects larger than the typical 2–4 mm sample size in classical NMR experiments. Our intention was to transform this system into an imaging unit for small subjects such as rat pups and mice; however, as will be seen, we got waylaid by a much more extensive MRI project.

Given that my graduate school background was in solid state physics and did not include a component of NMR research, a requisite action was to reach out for competent professional collaborators. With little delay, I found a willing colleague within the University Department of Chemistry who was eager to participate and provide the technical expertise, Jerome L. Ackerman, Ph.D. With Jerry and some of his chemistry graduate students, we entered into another phase of productive interdisciplinary partnership. Work began in earnest on the small geometry, high field MRI system. However, on a 1981 trip to the Philips International headquarters in Eindhoven, Netherlands, ostensively to evaluate their high-end CT unit, I was given an advance look at their whole-body MRI system under development and brought back a preliminary image of a human head. I enthusiastically showed that image around the Radiology Department and recommended that the Chair consider acquiring a whole-body MRI unit to be on the leading edge of this new technology. As an alternative, I had become aware of a whole-body magnet that was available at a reasonable price ($125,000) from an Ohio manufacturer of radiological imaging equipment and offered to build a whole-body MR imaging unit in-house if the administration was willing to pick up the challenge and purchase that magnet. Upon deliberative consideration, rather than commit millions to one of the first commercial units and fight through the certificate-of-need process, my offer was accepted by Jerry Wiot and Gene Saenger. Funds were appropriated to acquire the magnet which was an Oxford Instruments (England) 6-coil resistive 0.15 T system with an 80 cm bore. In 1982, the unit was installed in a lab several doors from my office in the Medical Center … thus, the quest began. Following that initial investment by the Department, over the next several years, approximately one million dollars went into the project primarily through local and national grants. The Division of Medical Physics team included me as director, Larry Busse, Ph.D., Ron Pratt, Ph.D. and RC Samaratunga, Ph.D. Together, with input from outside expertise, we configured and assembled the requisite components: radiofrequency (RF) coils (they transmit the RF power into the subject following a prescribed pulse sequence and receive the return signal); gradient coils (they provide the superposition of weak magnetic fields on top of the main magnetic field to enable image encoding and slice selection); RF shielded room (this is necessary to prevent outside sources of RF signals from interfering with the system RF data collection); patient/subject couch transport mechanism; computer software for controlling the various components, acquiring the signal data, image reconstruction and display. Figure 4 shows pictures of the installation.

3.3   Brief Tutorial on How MRI Works

At this point, I will pause and provide the briefest of tutorials regarding magnetic resonance imaging (MRI) and further describe the function of some of the components listed above. A basic physical characteristic of an individual proton (hydrogen nucleus) is that it behaves like a tiny magnet

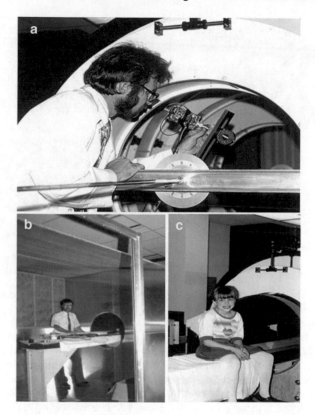

Fig. 4 **a** Shortly after installation of the magnet in 1982. I am checking the uniformity of the magnetic field as part of the process known as "shimming". **b** The radiofrequency (RF) shielded room enclosure under construction. When complete, it constituted the "RF cage". **c** Daughter, Kirstin, age 8, seated on the patient couch in front of the magnet (April 1985)

having a north and south pole with that property given the name *magnetic dipole moment*. That the human body contains a lot of water is a well-known fact (approximately 60% overall with the brain, as an example of a specific organ, composed of ~ 73% water). Thus, the two hydrogen atoms in each water molecule present the possibility of harnessing the magnetic field energy. However, under normal conditions, the individual proton magnets are not aligned; that is, they are pointed in all directions with no net magnetization (no overall, combined magnetic effect). If we should place the body in a strong external magnetic field, the individual magnetic dipole moments begin to line up and will exhibit a net magnetization. Another essential physical property is that the individual proton magnetic dipole moment will rotate around the external magnetic field in a process known as precession (Fig. 5a). Visualize the rotation of a spinning top around the vertical direction or an

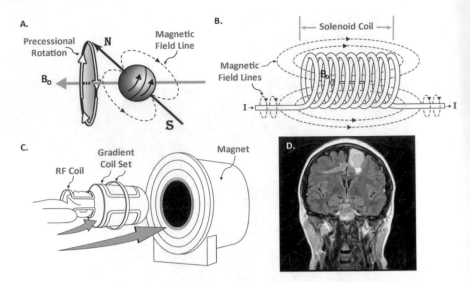

Fig. 5 **a** A proton with its intrinsic spin produces a magnetic field and acts as a small bar magnet with a north and south pole (magnetic dipole moment). In the presence of an external magnetic field, **Bo**, the magnetic dipole moment will rotate around the magnetic field with a motion known as precession. **b** Concept of the solenoid coil magnet: an electrical current, **I**, flowing through a wire conductor produces a magnetic field. If the wire is wound as a solenoid coil, a uniform magnetic field, **Bo**, is established inside the centre (arrow) where the patient would be located for imaging. **c** The MRI component "nest" configuration: the receiving RF coil is closest to the patient (here illustrated as a head coil for brain imaging). The patient with the RF coil goes inside the gradient coil set (partial representation), all of which are positioned in the middle of the magnet. The patient would be transported into the magnet on a motorized bed assembly (not shown). **d** An MRI of the brain. The arrow points to the abnormal area (tumour) that appears bright in contrast relative to the normal brain tissues

active gyroscope around a defined axis—both are examples of precession. The rate of this rotation or the frequency of precession is directly proportional to the magnetic field strength which means that the precession is faster for stronger magnetic fields (e.g., double the field strength, double the precession rate, etc.). For hydrogen, the precession rate is 42.6 million cycles (revolutions) per second at a one tesla (T) field strength written as 42.6 MHz/T where M stands for mega (million) and Hz (Hertz) for cycle per second. This value is unique to hydrogen and will be different for other nuclei that possess a magnetic dipole moment (not all nuclei do, but for example, in the case of carbon-13, the value is 10.7 MHz/T; nitrogen-14, 3.1 MHz/T; fluorine-19, 40.1 MHz/T). These frequencies are in what is known as the radiofrequency (RF) range. In the 1930s and 1940s, it was demonstrated that if an RF pulse of just the precise frequency and right orientation were incident on a nucleus with a magnetic dipole moment residing in a static magnetic field, an energy absorption process would take place. For hydrogen at one T, the required RF input would be 42.6 MHz. This phenomenon is known as resonance, hence the term nuclear magnetic resonance (NMR). As an historical aside, in the early 1980s as human body imaging goals using NMR were being realized, the term "nuclear" was dropped in favor of magnetic resonance imaging (MRI). The intent was not to confuse the public regarding standard nuclear medicine procedures involving radioisotopes nor conjure up thoughts associated with cold war apocalypse. Following absorption of the short input RF pulse, the nuclei will return to their undisturbed state and, in the process, re-emit energy at that same frequency. This is the output RF signal that can be detected with a properly tuned RF coil in the vicinity. This receiver RF coil may be the same as, or different from, the transmitting RF coil that sends the RF pulse into the subject.

The whole-body magnets used in MRI may be of various types (including permanent magnets); however, the most common ones used clinically employ superconductors. It is a property of physics that an electric current passing through a wire conductor will establish a magnetic field around that wire. If the wire is wound as a solenoid coil, a relatively uniform magnetic field will be set up in the centre of the coil (Fig. 5b). Superconductivity is that magical property where at sufficiently low temperatures, the resistivity of the conductor disappears (drops to zero thus no resistive heating) and any current introduced into a continuous loop under those conditions will continue to flow indefinitely. Liquid helium at ~ minus 267 °C or 4.2-degree Kelvin is used to cool the magnet coils. (Refer back to the graduate research section above for a different use of liquid helium.) Considerable resources were employed by manufacturers early on to develop large superconducting magnets capable of accommodating human bodies within the room temperature bore. Imaging units with magnetic field strengths from 0.5 to 10.5 T have been produced; however, the most typical clinical MRI systems operate at 1.5 T.

A final component to couple with the RF and magnet coils is the gradient coil set (x, y, and z). These are constructed according to a specific design configuration such that the current passing through the coils produces a weak magnetic field that will be superimposed on the much stronger main magnetic field. The gradient coils can be controlled to provide fields along the orthogonal coordinates (x, y, z) or in any random direction. One of the major functions of these gradient fields is to provide spatial encoding of the received RF signal. This allows formation of the clinical image following a computational process applied to the received RF data (2-dimensional

Fourier transform). They also enable selection of the precise location of the image slice within the body. This coil "triumvirate" may be envisioned as a nest configuration with the RF coils innermost (nearest the patient), followed by the gradient coils, and, finally, surrounded by the main magnet coils (Fig. 5c).

Subtle differences in the way water is incorporated within tissue affect the characteristics of the RF signal received. Manipulation of the input RF and gradient pulse sequences determines the image contrast (for example, the ability to distinguish diseased tissue from normal tissue) (Fig. 5d). At the present time, the clinical diagnostic power of MRI is enormous. We have the capability of diagnosing abnormalities of the brain, spine and most parts of the body; injuries of the joints; structure and function of the heart (cardiac imaging); functional activity of the brain (fMRI); blood flow (MR angiography); the chemical composition of tissues (spectroscopy); along with other sophisticated diagnostic imaging protocols.

Development takes time (indeed we had other responsibilities within the Division and Department that did not permit us the luxury of directing our full attention on the MRI unit) but by 1985 image quality was at a significant level to attract the attention of our chief neuro-radiologist, Robert Lukin, MD. Although the unit was intended for non-human research, we performed clinical brain scans on a number of patients in late 1985 just before the commercial MRI system was up and running in the Medical Center. Thus, the first patients to receive an MRI scan in the city of Cincinnati did so using the Division of Medical Physics unit (Fig. 6). Through the years, our research employed a menagerie of animals including mice, rats, rabbits, and pigs. A primary focus involved investigations of the blood substitute perfluorocarbon compounds (PFCs) in vivo (i.e., when the compounds are introduced into a living system). These materials have the ability to hold and transport oxygen. Our MRI system could be tuned to be responsive to stable fluorine-19 (F-19) in addition to the standard hydrogen imaging. (As stated above, it's the hydrogen of the abundant water molecule in the human body that provides the basis for 'ordinary' clinical MR images.) We published a number of articles showing how PFC F-19 MRI could be used in animal models as a reporter of the oxygen concentration in lungs and other organs.

3.4   Gorilla in the RF Cage

One of our more memorable "patients" arrived by the way of the world-renowned Cincinnati Zoo in January 1991. An infant gorilla was exhibiting neurological symptoms that were causing his handlers concern. Through

Fig. 6 Division of Medical Physics MRI system development team: (L to R) R.C. Sama-ratunga, Larry Busse, Ron Pratt, me. Following the first patient imaging in April 1985

some avenue of which I am not exactly sure, they became aware of our research facility at the Medical Center and requested time on the scanner for the youngster. Happy to oblige, we spent an eventful morning imaging the cute baby gorilla in diapers. His name was Kubatiza but, for some reason unknown to me, they called him Milt (Fig. 7). His introduction to the MRI unit went smoothly and he appeared to be comfortable within the radiofrequency (RF) enclosure surrounding the scanner. Caution had to be exercised with administration of the sedative all of which was in the capable hands of the zoo personnel in attendance. The session was a success with the good news being that no recognizable abnormality was noted in the young primate brain. In follow-up on the progression of this pediatric patient, it was reported that he had developed normally with no further issues. Shortly after this imaging session, he went off to reside at the San Diego Zoo and then years later to the Topeka Zoo (Kansas).

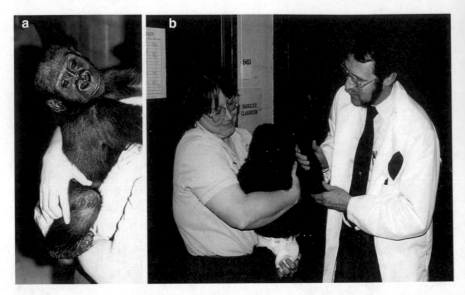

Fig. 7 Our pediatric patient from the Cincinnati Zoo, Kubatiza. **a** Close up. **b** In-between the primate nursery keeper and me (Jan 1991)

4   Mentors, Advisors, Professional Colleagues, and Friends

I want to conclude with some thoughts reinforcing a theme that has been interlaced throughout the above recollections. As an axiom experienced by many that bears value in re-emphasis, guidance received, and direction provided by mentors and advisors play a critical role in determining the outcome of one's career. From HY Fan to JG Kereiakes and others mentioned above (along with many other colleagues not named), the value of these individuals having confidence, putting trust in me, and supporting innovative new initiatives was of paramount importance (Fig. 8). I was allowed to spread my wings and attempt new ventures. One would hope that all professional environments would provide this opportunity for those qualified and that individuals in positions of leadership would promote this objective. Additionally, the value of assembling a dedicated, talented team with appropriate expertise for the projects at hand cannot be overstated. It is the team, more than the individual, that ends up making the difference. Through the years, my list of professional colleagues continued to expand significantly. As my career progressed, it was remarkable how those professional relationships evolved into personal friendships. Cherish the interactions at all levels!

Fig. 8 A radiology department function in Spring 1995 (L to R): Jerry Wiot, me, Jim Kereiakes, Gene Saenger

Stephen R. Thomas is Professor Emeritus of Radiology, Medical Physics, University of Cincinnati College of Medicine, Cincinnati, Ohio, USA. He has a BA in physics from Williams College, Williamstown, MA (1963), a 2-year Peace Corps tour in Ghana, West Africa (1964–66), and a Ph.D. in solid state physics from Purdue University, West Lafayette, IN (1973). His medical physics training started in 1974 at the University of Cincinnati. While progressing to full professor in 1987, he served as Director of the Division of Medical Physics, Department of Radiology from 1991 to 1998. He became Professor Emeritus in 2000. His research included nuclear medicine dosimetry and magnetic resonance imaging (MRI) with an emphasis on flourine-19 techniques. He had a number of National Institutes of Health grants and authored 110 publications. He had significant involvement in professional societies including: American Association of Physicists in Medicine (AAPM—President, 1997); Radiological Society of North America (RSNA); Society of Nuclear Medicine and Molecular Imaging (SNMMI); International Society of Magnetic Resonance in Medicine (ISMRM—Interim Board member facilitating the merger of the 2 major MRI societies in 1993); The American Board of Radiology (ABR). He has fellowships in the AAPM, ISMRM, and The American College of Radiology (ACR).

Awards

2012: William D. Coolidge Gold Medal Award, American Association of Physicists in Medicine (AAPM): The highest honor bestowed on a member by the AAPM in recognition of an eminent career in medical physics.

2012: Gold Medal, Radiological Society of North America (RSNA) The highest award given by the Society.

2009: Loevinger-Berman Award, Society of Nuclear Medicine and Molecular Imaging (SNMMI): For excellence in internal dosimetry.

2009: Life-Time Service Award, American Board of Radiology (ABR).

1991: Award of Merit (article co-author), Journal of Nuclear Medicine.

12

The Physicist and the Flea Market: At the Birth of Modern Image-Guided Radiotherapy

Marcel van Herk

1   The Flea Market and My Youth

I have always loved the 'Waterlooplein' flea market in my birth town of Amsterdam. Electronics was my oldest hobby, taking radios and alarm clocks apart as young as seven years, and demonstrating how a transistor works to the classroom at 11. I became interested in computers when as a 12 year old boy I read (and re-read many times) a 1967 Life magazine educational handout article explaining the IBM 360 computer in simple terms with beautiful pictures (by Croskinsky), such as a toy-train representation of the data flow in a computer.

At age 13, I attempted to repair a vintage TV I had found in the garbage and asked for help in a small repair shop. In response to my report of a bright red glowing vacuum tube, the shop owner knew exactly which component I needed to replace. When I reported back success the next week, he invited me to work for him as junior radio repair man. I did this a few years and learned a lot about electronics. I also found that electronics repair is an excellent training for life, physics, and software engineering. Being able to deduce where a fault lies by stepwise deduction makes all the difference for any project in medical physics, a skill that has helped me during my entire

M. van Herk (✉)
University of Manchester and Christie NHS Foundation Trust, Manchester, UK
e-mail: marcel.vanherk@manchester.ac.uk

career. In addition, electronics is unforgiving; if you do it wrong likely the parts will burn out, so you have to work very carefully and check and double check yourself all the time.

At the same age, I had started hunting computer parts and designing my own computers, that were partially functional but never finalised. One of the first things I did was to design and build a completely functional relay-based full adder (a circuit that can add two 4-bit binary numbers), soldered together while listening to Black Sabbath's Iron Man in the living room, not totally to my mother's liking due to the music and the spilled solder on the carpet. The parts I used were small electromechanical relays from 1950 punched card sorting machines, acquired cheaply (Fig. 1).

Pink Floyd and Supertramp were playing during more ambitious projects, such as an 18-bit arithmetic circuit I built in 1975 from recycled boards made

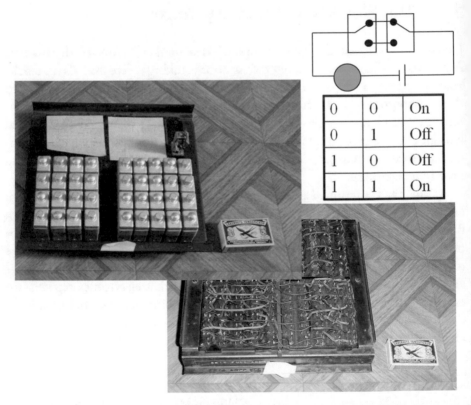

0	0	On
0	1	Off
1	0	Off
1	1	On

Fig. 1 Relay-based computer circuit made by the author at age 13 in 1973. It successfully added two 4-bit binary numbers. The inset shows the implementation of a so-called exclusive OR, the green light only turns on when the state of both relay switches differs. The small relays that are used in the actual circuit are recycled from discarded Bull punched card processing systems from the mid-1950s

in 1964 by Univac. I hoped to construct my own complete computer from it, but it got stuck at the assembly of a memory circuit. I had acquired the required parts, but they were not from the same manufacturer and connecting them was very tricky. Working on such systems without money tends to make oneself resourceful. Richard Feynman actually makes a similar observation in his book "Surely You're Joking, Mr Feynman!"

In parallel, I started learning the computer language "Algol60" and programming (this was an initiative of one of the math teachers at high school), and we connected to the University mainframe using an old tele-type in the school corridor, reading our programs on punched tape. With a friend, I also started sneaking into the University computer hall to write programs such as one that could calculate pi to 300 digits on their CDC 6600 computer.

In 1977, I acquired a recycled Intel 8080 CPU board (an early 8 bit micro-computer), and by the time I graduated from high school in 1978, I had an actual working computer based on this board combined with some (expensive) memory parts that my father had given me for my birthday (my stepmother called these a box of nails, and they found my birthday lists quite boring, for instance "could I have four pieces of the 2114 chip please?"). He also had given me a classic Tektronix 465 oscilloscope that he found in the inventory of a company that had gone bust. As chief engineer in a company installing heating and air conditioning, he was in contact with lots of small companies. I had written all code for my computer from scratch in assembly code (at the bit level) and input it by hand with recycled switches (16 for address and eight for data), exactly like in vintage computer systems such as described in the 1967 Life article. I used a broken old paper tape puncher that I had received from my brother's girlfriend's father as output. It came without any of the original electronics; so, I designed and constructed that myself, an early lesson in reverse engineering. One of my first serious programs solved the so-called Eight Queens Problem (how you can place eight queens on a chessboard without them being able to beat each other) and the output had all 92 solutions. My version of the program wrote each chessboard as eight characters on the paper tape, with a queen represented as a hole.

2 Beyond High School

After high school, I chose to study physics, sort of by default, because computer science was not yet taught as a separate science in Amsterdam. Because physics is such a wide field, I think it teaches problem solving more than specific topics, and I think it served my career much better than computer science would—if you can do multiple things you are much better off. But, of course, some of the physics material connected well with my later work, in particular experimental physics.

The computer I had built so far had no permanent memory and had to remain turned on all the time. During my study I extended my home computer with a chip programmer, such that its 'operating system' could be stored permanently. The chip contained the software that allowed the computer to be programmed using a keyboard and display from parts of a recycled calculator. This extension allowed me to actually turn my computer off without spending 30 min to get it up and running again. This came in handy, because for instance, at her first visit to my room, my now wife accidently short circuited the input keyboard by placing the naked electronics on a nail that I had left lying around, forcing me to re-enter the program from a hand-written listing. Later I added a recycled 1960 IBM punched card keyboard as input and an ancient Tektronix oscilloscope as output and used an old reel-to-reel tape recorder as the storage medium, creating something resembling a complete personal computer. Of course, I programmed everything myself, including the tools needed to actually program the computer.

I demonstrated this setup (transporting it on the back of my bicycle to the University, a very scary 20 min trip that the computer survived) as a digital techniques practical in 1980 to my professor, Prof Muller—scoring the highest mark. I think that building my own computer and its software was a very important experience, and it stimulated me later to develop all my software from the ground up. Knowing what is inside helps oneself not be intimidated by modern computer technology, realising that the underlying technology is the same and a lot of the hypes are just marketing bravado.

Apart from the physics courses, my study entailed quite a few computing courses such as numerical methods, and to my surprise punched cards were still to be used by the students to input code into the university's main computer. However, my study pretty much reached a standstill after my candidacy in 1981 (for graduate studies), mainly because attendance was not obligatory, and I had fun things to do at home—like programming and extending my home built computer until late at night. My home computer had grown with an outdated Olivetti Teletype as printer that I had begged from the computer science department when they were thrown out. They gave me a pair which was very useful because they kept breaking; so, I had a collection of spare parts.

At that time my computer (Fig. 2) also contained a flea-market bought

Fig. 2 My home-built computer in its 1984 state. The terminal (manu-factured by ICL around 1967) on the right has been donated to the computer history museum in Bletchley Park and is actually on display there (https://mpembed.com/show/?details=1&ga=UA-32443811-1&hdir=2&m=pq6v8g MmxMw&mdir=3&minimap=2&minimaptags=0 search for tag "Mainframes like the ICL 2966"). The Olivetti teletype shown bottom left broke down every few months and has long been discarded. I still have the computer, but the rubber wheels in the cassette drive have dissolved, so unfortunately it is not functional—still a restoration job to do! Photograph made by Els Couenberg

digital cassette tape drive (that was software controlled and could automatically locate the code I asked to load). I also bought an ancient ICL computer terminal at the flea market. I had trouble to get that to work due to lack of documentation (I had asked ICL but they did not want to help); so, I ripped out most of the electronics and replaced it by my own. Coincidentally, my later employer had exactly the same terminal (also acquired secondhand) for their first computerized radiation treatment planning system but did get the manuals—ICL had not refused to give it to them. Apparently, working in cancer research opens doors that remain closed for mere mortal students.

In mid-1982, I decided that I'd better get on with my studies, so I decided to do my practical work first, which at that time was a period of 1–1.25 years on one project. Options were offered at several labs and I visited all projects on offer, including one at the Netherlands Cancer Institute (Nederlands Kanker Instituut, NKI, also known as the Antoni van Leeuwenhoek

(AVL) Ziekenhuis). Medical Physicist Harm Meertens proposed to build a digital x-ray camera to verify radiation treatments. At that time most medical imaging was analogue and such a camera would replace the expensive x-ray films that were used to check whether the patient was in the right place during treatment, by comparing the film to a reference x-ray (Fig. 3). It sounded like fun, and I was quite charmed by the idea of doing research in medicine. But what really made me decide to work there was that during a brief tour of the department, I met engineer Jan de Gans. He was building his own microprocessor-based equipment and that was a chance not to miss—I had never seen so much hardware in my life! I believe that choice of first assignment was extremely important—it linked my passion and hobby with my research, and it offered the right environment to make it flourish. This gave me a head-start because I was one of few that combined physics and computer science knowledge in the radiotherapy domain at that time.

At my first meeting at the Netherlands Cancer Institute, I had asked Jan de Gans if he had some software that I could use (remember that before that I had written everything for my home PC from scratch, and there was no such thing as the Internet to download software) and he gave me an audio cassette tape with a standalone Basic interpreter for the 8080 processor. He did not expect me to be able to do anything with it because the tape format was not described at all, and even if I would be able to read it, the code would still have to be modified to run on my hardware, not an easy task without source code or documentation. But I liked the challenge and really wanted that software (it would make it much easier to program my computer) so I reverse engineered (hacked) the format and built the electronics and software required to read it.

Three months later, I started in NKI with a short term project to use digital filtering to enhance x-ray images used to verify radiotherapy treatments, which in those days were made with photographic film (Fig. 3). Jan was quite surprised when I told him that I had managed to read the tape he gave me before and had the Basic interpreter running on my home-built computer. At work I used the PDP11 computer of the department and wrote software to enhance images with Harm. The films were scanned with Jan's in-house built film scanner, based on the same microprocessor I used at home.

I also discovered that NKI's cellar had a treasure trove of old equipment for use at work and at home. From that collection, an ancient medical blood tester provided the 19 in. rack housing for my home computer, visible in Fig. 2, and an old analogue Hewlett Packard ECG recorder was connected as a graphics printer. Of course, all were interfaced with home built hardware and software. At regular intervals, Jan and myself, as we had become close

(a)

(b)

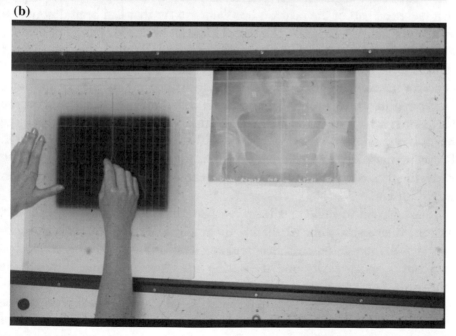

Fig. 3 Medical imaging before the 1980s was mostly analogue. **a** The 'hat stand' next to the patient contains an x-ray film, that when developed, allows evaluation of the treatment beam position relative to the anatomy. Such films were called 'portal' films. **b** Physician Joos Lebesque is comparing a portal film to a diagnostic x-ray image acting as reference. The contrast of portal films is very poor and film development takes several minutes; therefore, they were not widely used and electronic alternatives were needed

mates, went to the flea market, combining a search there with a nice lunch and discussions about new technology and things we would develop.

3   Developing a Detector for Image-Guided Radiation Therapy (IGRT)

After this period, I started my main project in the hospital, which was to actually build the high energy x-ray camera, aiming to instantly and electronically verify the treatment. With film, which had to be developed, such instant verification was not possible. Others worked on video-camera based systems, but we considered these too bulky. I first talked with high energy physics researchers at my University but the systems they used were built on counting interactions of photons with the detector. In radiation therapy this is not possible, because the number of particles per second used for treatment is much too high. For my project, I therefore needed to measure the bulk radiation intensity, which is commonly done with so called ionisation chambers, requiring a high voltage. The cellar provided a 1000 V power supply that had been part of the neutron irradiator system that had been decommissioned just before I arrived. (The use of neutron beams for radiation therapy was only done in a very few places in the world. While it was considered to have some clinical advantages, the results showed side effects that were of concern; hence, it went out of favour.) Jan helped me to wire it up, but I was scared to switch it on—for good reason, because when Jan toggled the ON switch it blew up with an almighty bang. We had forgotten to check the mains input voltage required which was 110 V, not the 220 V we put onto it. Recycling materials for a research project is a good idea, but make sure you check them carefully! Or at least have somebody else nearby to toggle the on switch for you.

I next started to build small arrays of so-called ionisation chambers and measure their signal while irradiating them with the cobalt-60 patient irradiator at NKI. Ionisation chambers measure irradiation by putting an electric field across an air cavity and measuring the current flowing through the ionised air. But since there is a lot of ambient air beyond the chamber volume, and I had failed to shield the high voltage and signal wire, this prototype picked up much more signal from the ambient air than from the volume of the tiny air chamber. By this time, I was well versed in electronics, but was unaware of the complications of working with radiation; measuring radiation signals requires very sensitive equipment, and you have to consider electrical shielding right from the start. We needed thousands of small ionisation chambers and I was worried that the signal would be too low. So, in parallel, I tested other principles, even going as far as trying some of old magnetic core computer memories (from the flea market) to measure radiation. However, there was no measurable signal; no demagnetization occurred. I guess in hindsight this is logical since core memories are still in military use because they can resist the electromagnetic pulse from a nuclear blast; a bit of medical radiation has negligible effect.

The first two-dimensional ion chamber prototype had 900 ion chambers (900 pixels) at 2.54 mm pitch that needed a switched high voltage as input and had a low current as output. The flea market came to the rescue and I bought an old telephone switchboard circuit full of reed relays (magnet controlled switches that are encapsulated in glass tubes), that I configured as high voltage switches and as readout circuit (technically it used 30 capacitors as current integrators that were read out with the relays) (Fig. 4). The whole system was controlled by the departments' classical PDP11 computer through very long cables and took about one minute to scan. The control software was written in Fortran by technical assistant Joos Weeda.

The images where very noisy, and Harm Meertens proposed to use liquid, which had been investigated in Sweden for super small ionisation chambers, and after constructing a liquid-tight detector it turned out to make a huge difference. The signal was about 100 times higher, and the images were much sharper, actually much more than we had expected. We used isooctane, a sort of pure gasoline. Luckily none of our detectors ever caught fire, even though later detectors would be powered up to 500 V. At first, we did not understand why the images were so much better. However, it turned out that ions formed by the irradiation in the liquid (like ozone in the irradiated air) actually stayed around so long that the readout process just picked up these long-lived ions, providing a form of memory that greatly improved image quality. At therapeutic dose rates, the ion life-time was actually on the order of 1 s, creating a useable detector. At lower dose rates, the detector would have been too slow, and at higher dose rates much faster electronics would have been needed. By chance we had hit a sweet spot!

Having some success, some more funding was forthcoming and during the last stages of my master's project I learned circuit board design, optically transferring a hand 'taped' design (symbols were on transfer stickers and connected by opaque black adhesive tape) and etching it all in a toilet space that had been repurposed to contain a small dark room, a photo transfer unit, and a fume hood. This version of the camera (Fig. 5) used semiconductor components, using consumer parts rather than dedicated precision parts to

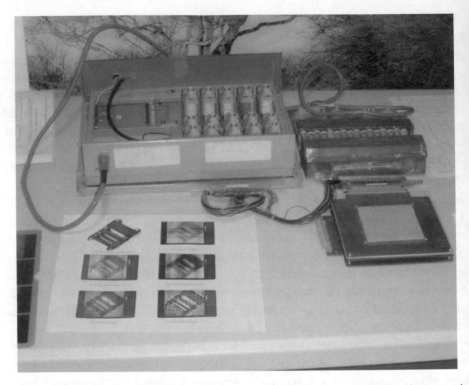

Fig. 4 Flea-market component based x-ray camera, built mid-1983. Photograph made during my 'farewell from NKI' exposition in 2015. The board on the right is the ion chamber array using drilled holes in a plastic plate as enclosures for the air, forming the detectors. The unit top left is a recycled (flea market) switch unit from a telephone exchange, rewired as high voltage switches. The unit top left measures the tiny output signals. The acquired images of the 'bar' phantom and the phantom itself are shown on the lower left

reduce costs. My home computer was used to simulate and design some of the electronics. This detector was much faster, so that the PDP11 became the limiting factor, allowing a fastest scan time of 20 s. This was way too slow, because a clinically useful detector had to be at least 16 times bigger and would require a scan time of 5 min. But the detector had to take an image in seconds, not minutes. Otherwise, the treatment delivery would be complete before an image had been made (a typical irradiation time was 20 s per beam). A better computer system was therefore required.

After I handed in my master's thesis, I was offered a Ph.D. position at NKI, but I had finished none of my major exams and did not formally have my master's degree yet—because I had skipped virtually all lectures. So, I decided to ask Harm for a half year delay, to give me time take eight major and really heavy exams, such as quantum mechanics, thermodynamics, relativity theory

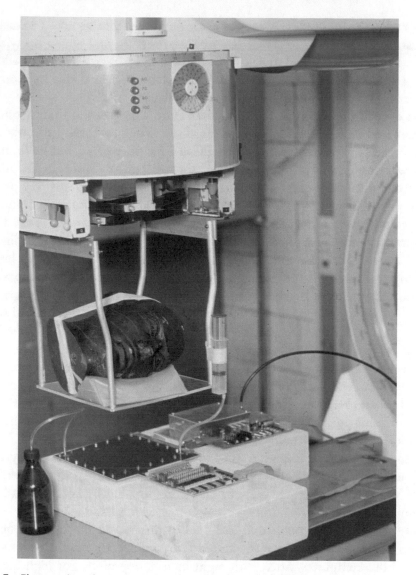

Fig. 5 First semiconductor-based prototype of a liquid filled x-ray camera (1984). The central detector plate is filled with liquid and scanned using high voltage switches and a heavily shielded amplifier set, in the metal box. The object to be imaged is the head of the so called Alderson-Rando phantom, a wax head around an actual human skull. Irradiation was by a cobalt radiation treatment machine; its opening is shown at the top

and such, scheduling three weeks to learn each one from scratch. I managed to do so, even having to revisit my first year's math, and ended up with a Cum Laude degree. I had to study efficiently and learn fast. This showed me that I could perform well under pressure. I later perfected this method by preparing lectures up to minutes before they had to be given.

In mid-1985, my Ph.D. research could start. The goal was to develop the camera and its software for actual clinical use. The first thing I did was ask Harm to buy me a just released IBM PC-AT for the project, but he thought this was too expensive and we could not get one. So instead, Jan and I built our own computer—experience enough. In hindsight this was penny wise, pound foolish, building all the computer parts straight for the standard PC would have saved us maybe a year of development later on, as the later commercial version of the system would be using such a PC. Jan built the 8086 processor board, and I designed electronics such as a control board for high voltage switching (with its own microprocessor), and a board to measure the detector signals at very high speed. This board was built around a very expensive converter chip and, to our shock, the first one arrived with bent pins as it had been packed in a cheap envelope and somebody had stood on it. Luckily, we were able to exchange it for free, and the replacement chip came better packaged! Designing and building all (quite complex) hardware was still done by hand—producing the circuit boards in the repurposed toilet, everything from design, laying out, etching, drilling each hole, and soldering the many connections being my task. Here my experience in computer design, electronics hobby, and perseverance were paying off. I also developed a graphics display card, and an interface card for a floppy disk and hard-drive. The graphics driver I wrote was later adapted to the standard IBM VGA card and formed part of Nucletrons' first brachytherapy treatment planning system. They paid me with a plotter, which I still have.

The big detector would have 128×128 pixels, and the detector board layout was done using a specially coded Fortran program on one of Jan's in-house built computers, fitted with eight inch floppy disks. For high quality drawing, I fitted a steel ink tip pen to a mechanical plotter that was the regular output device for planning systems at that time and plotted the layout on matte film. Plotting the board with 128 electrodes of 30 cm long took many hours, and the steel pen lost one mm steel after each plot, so it could not be reused!

In early 1986, all hardware was operational, and software had to be written. Because one deals with images with many thousands of pixels, the software had to be fast. I wrote the main software in the C programming language combined with hand-coded assembly routines. At that time the C compilers were not so good at optimising the machine instructions, and my hand-coded routines were actually much faster. My experience from my self-built computer at home came in handy again. Later the C compilers got much better and nowadays it is almost impossible to do better job by hand. I also developed fast image processing algorithms and if you do an internet search for the "van Herk algorithm" you will find an algorithm that has nothing to do with radiotherapy but everything with image processing. Programming was not always fun—I remember staring at one line of code for days—it wouldn't work! Turned out there as a spurious semicolon after an 'if' statement so the code that should be controlled by the 'if' would always be executed even if it was not allowed. The first main program used a horrible scripting code similar to the control language of text editors in use at the time (e.g., take four images and display them was written as 4{$im;di$}). This software was good enough to test the detector but really stifled our attempts to go into large scale clinical practice. The lesson here is that spending more time on design pays off later. I also learned that creating (and timely discarding of) a failure gives a lot of valuable experience.

4 Image-Guided Radiation Therapy (IGRT) in Practice

In May 1986, the system was used the first time to image actual patients (Figs. 6 and 7). The detector was placed on a cardboard box under to the patient by my supervisor Harm Meertens, and long cables led to the cellar where I was operating the computer system. Next the imaging hardware had to be updated, the prototype was built in a 10 kg steel box and not very ergonomic. Engineer Leo de Mooij convinced me that carbon fibre was the future, and we ordered rolls of the very expensive fabric on our account. Using a self-built vacuum table, he constructed a beautiful housing made of hard plastic foam embedded in carbon fibre fabric in resin. I redesigned the electronics to be further miniaturised (using surface mounted devices, with tiny extender boards produced in-house to create a full control board of 25 cm long and only 4 cm wide that fit beside the detector). The completed detector only weighed 5 kg, and then it was ready to be tested clinically on a larger scale. However, the 'simple' task of mounting the detector on the gantry turned out to be quite complex since the long support arm would bend a lot. Several mechanisms were constructed, with the most successful one being a holder made out of a honeycomb material used in the aerospace industry that was very rigid but only weighed a few kg. Later, Leo de Mooij made

276   M. van Herk

(a)

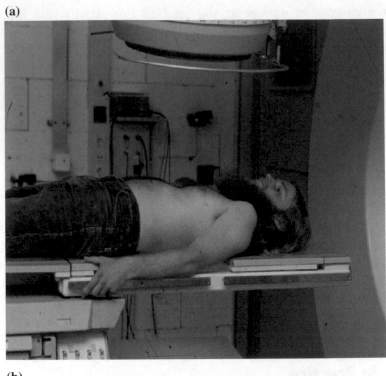

(b)

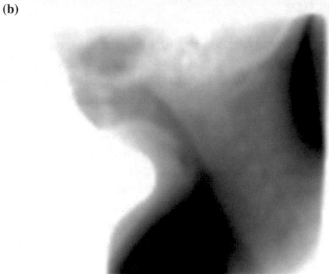

Fig. 6 a Setup of detector, with me as photo model (the detector is just visible in the lower part of the photo), and **b** first patient image taken in May 1986. It is a lateral head and neck irradiation. Air cavities and bony anatomy are well visualized in spite of the low resolution of the detector (128 × 128 pixels with 2.54 mm pixel size)

(a)

(b)

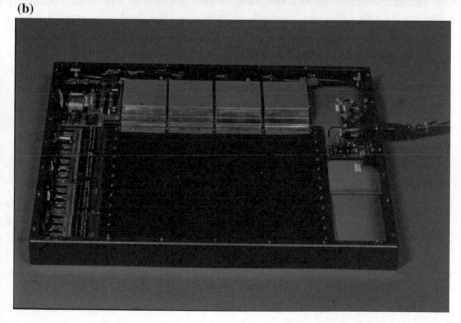

Fig. 7 **a** Control computer and **b** hardware of the first large-size liquid-filled x-ray camera. Apart from the CPU board built by Jan de Gans, all electronics and software were designed and built by me in the first year of my Ph.D. For good fortune, each detector we built had at least one flea market component. The detector was very heavy though because of its steel housing and could not be used routinely yet

his fortune by applying such materials to develop patient immobilisation and support structures (Fig. 8).

We negotiated licensing our electronic portal imaging device (EPID) design to Philips Medical (which would later become Elekta's external beam radiotherapy division), but they had just agreed to work with Rotterdam and Leiden on their recently completed video-camera based system and were not interested. In 1988, however, we managed to patent the detector and license it to ABB (Asea Brown-Boveri), with their main office in Baden, Switzerland. This is a huge company, but in the medical field they were just a small but innovative player. We agreed to license without any upfront payment, accepting just a percentage license fee. If you are ever in such a position, this is a very good idea. Once the percentage is agreed you can keep on innovating and sharing without any renegotiation as adding value to the design automatically leads to more revenue.

In 1988 and 1989, ABB constructed their system, further miniaturizing the components and increasing the resolution to 256×256 pixels. However, they had not listened to my warnings about shielding. The first microchip amplifiers that they created were wired such that inputs of one amplifier were placed right next to outputs of another amplifier, and the result was interference between all 32 channels in each microcircuit. They had to fix the design by micro-surgery on the ceramic modules, because they had not listened to the advice of a young Ph.D. student. Once their system worked, the higher resolution turned out to be a mixed blessing, requiring longer imaging acquisition and processing time, and software methods were actually added to lower the resolution if needed.

Around the same time Jurrien Bijhold and Kenneth Gilhuijs joined us for a master's project and later as Ph.D.s to work on automated image analysis. But it became clear that the core software design was very inadequate. Trying to fix one bug led to the others popping up elsewhere, while programming applications with the two letter style scripting languages proved to be very challenging. At home, in the evenings (in 1989), I wrote my own computer language interpreter for the system. We did not have a TV and internet did not exist yet, so I had time enough! I decided that this was needed because the PC operating system at the time (DOS) was limited to 640 Kbyte of memory. With a few 256×256 images in memory, there would very limited space for the software, and commercial software (e.g., for databases) was just too big, and would therefore reduce the processing speed. After I had convinced the team to adopt it as new core software in 1989, it was christened 'QUIRT' as the name of a European (EU) project that we just started—'QUality assurance and Imaging in RadioTherapy. This system was

(a)

(b)

Fig. 8 a Construction of housing of the first clinical electronic-portal imaging device (EPID) by Leo de Mooij and Peter de Groot in 1987. It was likely the first use of carbon fiber in radiotherapy—De Mooij went on to start a very successful company producing radiotherapy aids produced from this material. **b** This camera and the QUIRT software in clinical use in 1994 to verify lung shielding in total body irradiation, here shown from the control room

set out to be very modular, and modules were written for tasks such as image acquisition, image processing, display, database operation, and user interface design. Under the QUIRT project we installed electronic-portal imaging device (EPID) systems in Leuven, Belgium, Florence, Italy and Dijon, France to have the first international experience. Shipping the systems was done in the trunk of the project manager's car, driving all the way from Amsterdam with me in the passenger seat. We were never stopped at the border, but if we would have been, we may have had some problems!

The QUIRT software also became part of the first ABB release of the detector. The new software structure turned out be very robust, mainly because I dictated full independence of the modules; they would only come together in the scripts. In 1992, the portal imaging software was finalised. It was better than commercial systems, with fully automated image analysis built-in. The software was still in active clinical use in NKI in 2015, 25 years after its inception.

ABB's medical division was small. However, their hardware and software had impressed Varian, who then acquired ABB's radiotherapy division, making ABB's office their imaging headquarter. All of sudden our EPID was part of the world's largest radiotherapy manufacturer! They went on to develop robotic arms (Fig. 9), and with us designed an improved detector. We decided to prototype all hardware in Amsterdam and the Mark II, as Varian built it—is pretty much as designed by engineer Albert van Dalen in my group. All in all, over 1000 units were sold, and the licensing income for NKI funded the research group for many years, maintaining the same license sharing percentage agreed with ABB.

Around the time I decided to leave the group in 2015, NKI asked me to help replace my database code, which had been working faultlessly for over 25 years, for a commercial database, which I kindly refused. The underlying QUIRT software is also still active in research software (in particular in my new group in Manchester) albeit no longer in active development, and parts of it are still used in a system for tumour and normal tissue delineation training worldwide. One should realize that medical software is very long lived; once it has proven its worth, it can be used for decades. It is therefore very important to use stable and simple technology for it. Many computer technologies that have dominated the market at one stage, including big name software systems, have been outdated long since. The QUIRT software was written in C and can still be built and run on the most modern computers, just because it is based on simple and stable technology.

In 1988, I received the ESTRO-VARIAN award for this work, and in 1989, the Antoni van Leeuwenhoek award. When I left the group in 2015, I

(a)

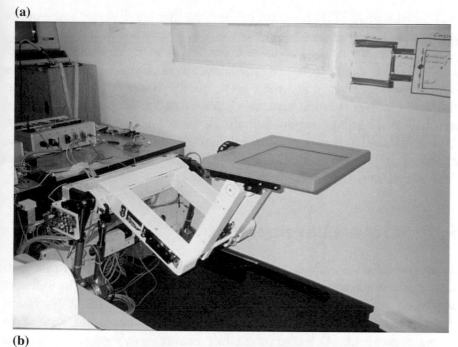

(b)

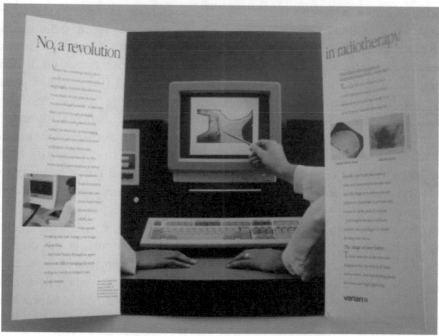

Fig. 9 **a** Prototype robotic arm with the EPID in the former ABB factory in Baden, Switzerland. At that time, it had become Varian's imaging division. **b** Sales brochure from the EPID from Varian. The specifications of the system were directly copied from the ABB sales brochure including typos

established a small EPID museum in the NKI that is still there. Some of the prototypes are displayed in my current office in Manchester.

I think that the largest impact of the worldwide implementation of electronic portal imaging was to raise the awareness of clinicians of uncertainties in the treatment chain, and I went on to quantify these (e.g., organ motion and inter-observer variation between clinicians) and develop a 'margin recipe' to calculate the safety margins that should be applied around the target to strike a balance between the risk of treating too much normal tissue and the risk of missing the target. However, the 2D nature of the x-ray images limited their application, and 3D methods were needed.

5   Software Development

So far you have seen that a small group in Amsterdam was able to develop hardware and software that reached departments world-wide. The software was based on 2D images. After completing my Ph.D. in 1992, I was strongly advised to do a post-doctoral fellowship abroad and wanted to get experience with 3D image processing. Hanne Kooy's group in Harvard Medical School in Boston was looking for somebody to develop a method for combining different patient scans for their high precision radiotherapy system. Harry Bartelink, the then head of the Radiation Oncology department in Amsterdam, advised me to go to Memorial Sloan Kettering in New York, but after visiting both groups, I preferred Boston. I liked the city and the job! They were using the AVS visual programming system, and they wanted me to develop the software with it. AVS is brilliantly designed. It is programmed by creating a visual data flow diagram and scans would be passed automatically from one processing block to the next (Fig. 10). Each scan could be 2- or 3-dimensional, and importantly had coordinates stored with the scan (for instance the size and thickness of each scan slice in cm). This means that if you write a software block that changes an image, you can process the coordinates at the same time, making it easy to keep images and coordinates consistent. In many software packages (such as the original QUIRT software that I talked about above, or even modern systems such as Python) this is not the case, and it is a very common error found even in medical packages for images to be a half pixel shifted, for instance after reducing the resolution.

However, when I started developing for the AVS system in 1992, I found that even though the images contained coordinates, many built-in blocks did not actually use them. Therefore, I was, in hindsight luckily, forced to write

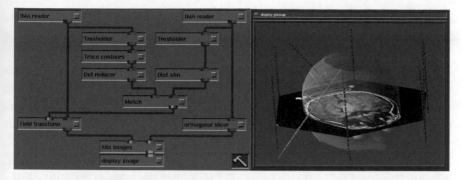

Fig. 10 Visual programming in the AVS system that I used in Boston. Each block that has its name start with a capital was written by me. Left: The network drawn on the display is the program that combines a computerized tomography (CT) and magnetic resonance imaging (MRI) scan and displays the result. The right hand image shows such a combined CT and MRI scan, embedded in the Radionics planning system. The blue shape is the tumour that is more easily visualised in MRI but more accurately targeted with CT. (The blue, green, red lines provide the "Z-display" and provide the relative location of each cross-sectional image.)

most modules myself. I wrote a 3D image registration algorithm based on the same algorithm we had developed for 2D images, and even though it was relatively simple, it performed very well. In 1993, the algorithm was licensed to Radionics, I think the first commercial program of its kind, and parts of it were in use at least until 2014, when I received a last (small) licensing fee cheque. My stay in Boston should have been a very happy one, but during this period, engineer and friend Jan de Gans died due to metastases of cancer of unknown origin, casting a very dark spell over that period.

Unfortunately, AVS was very expensive and required expensive Unix work-stations to run. But because I wrote my own AVS modules, it was not hard to connect these to QUIRT without the AVS software—something I did in the evenings in Boston on my first laptop that I had bought just before I travelled there (with a 386 processor, much less powerful than the workstations at that time). Using extra software, the 640 kB DOS limit could be overcome, and a new software architecture was born. Back in Amsterdam, this became the cornerstone for our work. However, the lack of modern user interface possi-bilities of DOS became problematic and the need for true 32-bit programs made us switch over to Microsoft Windows around 1994. I created a module for the QUIRT system to display windows, buttons and such and wrote a program to allow translation of apps defined in the free Windows design kit into a QUIRT program. Visual QUIRT was born. I coded a few viewers but was disheartened by the many controls that were needed, for instance four sliders would be required to change brightness and contrast of the two scans loaded into an image registration program (Fig. 11).

I therefore decided to hide these controls in a novel strange design with active regions hidden behind the display surfaces. Some of my colleagues thought I was crazy, but users easily adapted to it—and I wish I had them patented the idea—it is quite similar to modern smartphone apps where you swipe to perform certain actions.

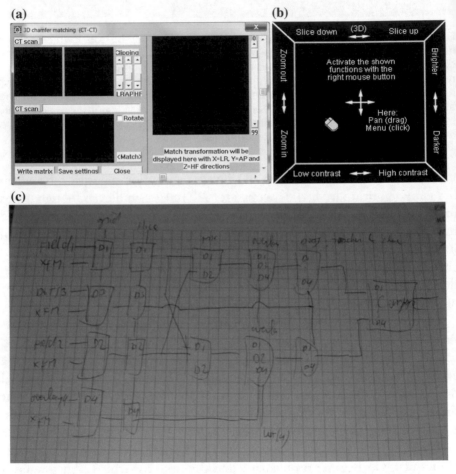

Fig. 11 **a** Early visual QUIRT program to combine two scans (1995). This was used by investigators at the Memorial Sloan Kettering hospital to study prostate motion. Note the many sliders that were needed for controlling the display. **b** In subsequent versions the controls were hidden behind the viewer panes, clearing up a lot of screen space. **c** Original viewer design drawn in 1994. This viewer design is in use up to the present in Elekta's image guidance system and (independently coded) in Aquilabs' web-based delineation teaching software

As before, I dictated that AVS modules should be independent, causing the core software to be easy to maintain and reliable. I had many fights with some colleagues who wanted to break the modularity to avoid duplicated code, but I am still convinced that modularity is much more important than code duplication. Also, after many struggles, I convinced my colleagues to adapt a data-flow based code design that greatly simplified coding. In a way, it is similar to the processing blocks in the AVS system, where a change in one scan automatically causes a recalculation of those and only those scans and data that depend on it. In computer programming, this is called the 'make' algorithm, and I believe I was one of the first to apply it to data processing. This greatly improved the reliability and simplicity of large chunks of the software.

6   What I Did During My Christmas Holidays

In the year 2000, NKI signed a principle collaboration agreement with Elekta, after a long process of negotiating with both Varian and Elekta, the main vendors of radiotherapy equipment. One of the main factors in the decision was the behavior of the contract lawyers—it was difficult to negotiate an agreement with Varian because every time there was a request for one change, many changes were made elsewhere in the documents. Elekta was more straightforward and that just tipped the balance. Around the same time, I had also started working with David Jaffray on software for image-guided radiotherapy. David Jaffray and John Wong (see John's Chap. 7) had proposed to combine a radiotherapy machine with a CT scanner around 1998 and got Elekta interested. When Elekta built four prototype cone beam CT (CBCT) guided treatment machines in 2002–2003, NKI got one of these. It was delivered without any image guidance software though; so, we had to write our own.

Medical devices require regulatory approval before they can be used clinically. Only two days after the US Food and Drug Administration (FDA) approval for imaging arrived (June 2003), we took our first patient scan. We had most components from my Boston working and some (slow) CT reconstruction code I had written in 1992. Up to that time, CBCT was expected to be impractical because of the long computing times. Over the Christmas holiday in 2002–2003, I accelerated it using the fact that vertical lines are not deformed due to perspective or rotation, and by organising the memory such that these vertical lines ran along the fastest memory axis, the reconstruction time was brought down to about 30 s. Combined with the viewer, registration and knowledge of the clinical workflow, we (myself, Jan-Jakob Sonke and Peter Remeijer) coded a full image-guidance system that was introduced clinically in-house in January 2004, making us the first hospital to introduce CBCT based image-guidance clinically. This meant that a scan could be taken of the patient on the radiation treatment machine, each day prior to the treatment beam being turned on. Thus, the tumour and the normal tissues could be localised to ensure accurate dose delivery in a way that was not possible before.

Elekta was struggling to develop their own solution, and having seen our work and its clinical application, agreed to license it. Because the QUIRT system was too big, and we had integrated it in Delphi (visual Pascal), I decided to create a QUIRT-less version, combining AVS modules with just Delphi code. However, this meant the entire viewer (which had grown to be quite an extensive component) had to be re-programmed. This was mostly done by me during another Christmas holiday (2003–2004) and first released to Elekta in April 2004. Our software design turned out to be very reliable with only a small percentage of issues attributable to our components in the XVI system. The lesson in this is that many clinical software systems were developed by hospitals and later transferred to industry. Nowadays FDA regulations make this much more difficult, limiting the rate of development.

7   From Two-Dimensions to Three-Dimensions

We anticipated that the transition from 2D to 3D would be seen as complicated; so, we decided to reduce functionality as much as possible (Fig. 12). The software release to Elekta that would be commercialised was delivered in December 2004 (it fitted on a single 1.44 MB floppy disk) and received FDA clearance in June 2005. We had expected that only a few units would be sold, but it was a big hit. All of a sudden, virtually all linear accelerators (linacs) were sold equipped with CBCT, and to date all Elekta users are running the software that I and my colleagues wrote in that frantic year. The modular software design had paid off, and the strange viewer design is now commonplace at thousands of clinics; it is still highly appreciated as it reduces screen clutter and limits mouse motion for many actions.

I received the European Society for Radiotherapy and Oncology (ESTRO) Breur award in 2004 for this and other work on organ motion and safety margins.

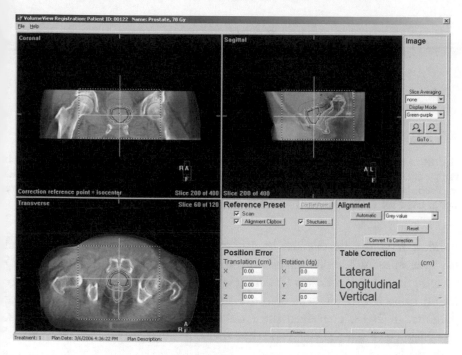

Fig. 12 Software for image-guided radiotherapy as developed by me and colleagues and released by Elekta in 2005. The display panes show the CBCT scan in green overlaid with the planning CT in purple. By design, the amount of functionality was limited for ease of use. My 'strange' viewer design with hidden controls is an integral part of the system. All calculations are controlled by the 'Make' algorithm, greatly limiting the amount of code needed for the user interface

8   Four-Dimensional Imaging

In 2001, I presented several papers at the ESTRO meeting in Seville, Spain on behalf of American colleagues who could not travel due to the 9/11 attacks. One of them was Gig Mageras' lecture on respiratory gating that showed so called, four-dimensional CT (4DCT) for the first time. The fourth dimension refers to time. I decided to implement 4DCT in the NKI and the flea market and cellar came in handy again. I combined a small temperature sensitive resistor from my flea market collection with a microcontroller that Jan had designed 10 years earlier and that I had kept (I had actually kept all of his stuff by moving into his room after he died) (Fig. 13).

This 4D work would be inspiration for the development of 4D CBCT, first developed and coded by Jan-Jakob Sonke and Lambert Zijp in 2003, and later redeveloped by myself to make it work in parallel with the image acquisition—allowing it to be used clinically in practice—over the 2006 summer

Fig. 13 Left: Temperature sensor (the point at the end of the black wire is a small temperature sensor that detects the patients breathing), and (centre and right) microcontroller built from recycled parts that was used to provide the respiratory signal used for NKI's first 4DCT and PET acquisitions

holiday. It would be released by Elekta in 2009 and has been in clinical use ever since. While physicians were initially skeptical of using 3D and 4D imaging in a daily workflow (before, only 2D was used), the workflows were found to be highly efficient, and are now standard for the vast majority of patients. In 2018, I received the Amsterdam Impact Award for this work, and Elekta estimated at that time that over 6 million people had been treated with our system.

9   Move to Manchester

In 2007, Harry Bartelink retired. I had made the mistake not to formalise my role as research lead in Amsterdam, and with clinical physics becoming more and more formalised, I found it harder and harder to develop new sytems and make an impact on clinical practice. In 2013, I had a sabbatical in London, UK at University College London (UCL), working on different subjects. After I returned to the NKI, my position was furter weakened and this severely affected my output and joy in the work. In 2014, I was asked by Manchester University and the Christie NHS Foundation Trust to consider moving there and, after deliberating and negotiating for over a half a year, I finally decided to leave for Manchester at the end of 2014. Starting there with a new group allowed me to refocus my research, and I started working on "big data"—using images from thousands of patients to try and learn how to treat future patients better.

10   Big Data and IGRT

Radiotherapy is an evidence-based cancer treatment process with precise technology and methods. However, many clinical and biological factors are poorly known. The piece of work which changed my research focus came, as is often the case, from a very unexpected result. In the literature, a difference in bowel filling was found to have an effect on patient outcomes (failure-free survival) in radiotherapy of prostate cancer. The hypothesis was that bowel filling influences the position of the prostate, and the prostate subsequently could be missed by the treatment beams. However, modelling by a member of my team (Marnix Witte) was unable to confirm this hypothesis, unless it was assumed that the majority of tumour cells were outside the prostate. Therefore, Marnix Witte and I decided, in 2010, for the first time, to link the entire 3D radiation dose distribution with clinical outcome. This technique highlighted a region a few cm from the prostate where a higher dose led to better outcome in advanced stage prostate cancer patients. This was the first instance that a retrospective method could identify that dose away from the targeted tumour volume had an impact on the success of treatment. This method is now our workhorse for clinical dose-response studies and I have many students in Manchester that use it. We have, for instance, identified the base of the heart as a sensitive region, and located the anatomy responsible for several side effects. These findings are currently being translated into clinical trials.

Recently, Corinne Johnson-Hart analysed a large cohort of lung cancer patients to link small residual beam placement errors with patient survival. In clinical practice, often action thresholds are used, not correcting small errors that are considered clinically insignificant. To our surprise, such a link was found, and was attributed to a higher than expected sensitivity of a specific region of the heart to radiation, measurable because these residual errors after IGRT are totally random. This work received the best abstract award (from > 2000 abstracts) at the ESTRO meeting in 2018. This finding has led cancer centres to re-assess their clinical practice.

11   Computer Restoration

Digital imaging in medicine only became possible with the advent of affordable yet powerful computers. Fifty year ago, two former employees of Digital Equipment Corporation delivered such a computer—the Data General (DG) Nova. Hounsfield and colleagues, who were developing the first CT scanner

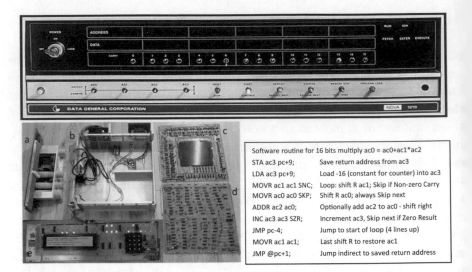

Fig. 14 Top: Photo of an original Nova front panel [Wikipedia]. Lower left: Components of my Nova computer. **a** Power supply; **b** Casing with transformer and fan; **c** Core memory (cores are under the black cover); **d** CPU; **e** (Inset) Restored front panel with modern display and keypad, an Arduino Nano processor sits on the right with USB connector to connect it to the Internet. Lower right: Original software multiplication routine

with EMI, chose this computer to drive their machine. I could recently acquire one of these machines on eBay—the modern version of the flea market, although the computer I could buy was incomplete (Fig. 14). In the past few years, I have successfully restored this Nova 1210 computer and I am now writing CT reconstruction code for it based on my experience with CBCT (see https://www.youtube.com/watch?v=n7GDlJNpB8Y). Hopefully, this work will help future generations understand more about the workings of computers and image processing—the older hardware is simpler to understand yet quite similar to modern computers.

12   Conclusions

Having an interest in flea markets has taught me to recycle, and study and repair many devices. I sincerely believe this experience has helped me enormously to successfully develop medical imaging systems that are in wide use today. To be able to develop such systems, you have to have perseverance, trust in your own ideas, and of course have a good knowledge of the field. While originally this approach was applied to hardware, later I applied it to software. In medical software, you should be very careful to use stable technology, because its lifetime is much longer than consumer products. Therefore, it is important to think through the design, and maintain sufficiently in control such that key design principles are not forgotten. Nowadays, however, it is hard is to combine engineering principles with the requirements of modern research, where each student chooses the latest tools that are being developed at a fantastic speed. Finally, old stuff can come in handy as I showed in our development of 4DCT, but it is also true for data. Treatment data of patients treated decades ago contain valuable clues for future patients and should be saved for many years with utmost care.

Professor Marcel van Herk Active in radiotherapy since 1982, he is Chair in Radiotherapy Physics at the University of Manchester. His group improves accuracy of radiotherapy in all aspects. He studied physics at the University of Amsterdam and started his career at the Netherlands Cancer Institute. There he developed an electronic portal imaging device, marketed by Varian between 1990 and 2000. In 1992, he worked briefly on 3D image registration at Harvard Medical School, and developed a system licensed to Radionics. Later work includes quantification of organ motion and determining physician observer variation. His method for margin determination forms the basis for margin guidelines. His group developed the software for the Elekta cone-beam CT image guidance system integrated in medical linear accelerators and is used clinically worldwide. This system uses patented methods for 4D image guidance. Recent work includes big data methods correlating details of the treatment with clinical outcome, showing, for instance, that irradiation of the base of the heart leads to worse survival. For the past 15 years, he held a part-time Professor appointment at the University of Amsterdam, working amongst others at optical imaging such as optical coherence tomography. His hobbies include electronics, computers, nature walks and swimming. He is married to Els Couenberg and has two children. Els is acknowledged for her help in preparing this manuscript.

Awards

2021. The Royal College of Radiologists (RCR) Honorary Fellowship, recognizing his career and the outstanding work with RCR on contouring training.

2021. Turing Fellowship, University of Manchester, given to a leading UK researcher, in recognition of his work on machine learning and big data methods.

2018. Amsterdam Impact Award for innovative ideas with a societal and/or commercial impact, specifically for the clinical impact of cone-beam CT-guided radiotherapy.

2004. ESTRO Breur Award is the highest honour that can be conferred on an ESTRO member and is awarded in recognition of the major contribution made by the winner to European

Radiotherapy, in this case for the development of margin recip
and CBCT guided radiotherapy. (ESTRO = European Society fo
Radiotherapy and Oncology)

1988. ESTRO Varian Award, awarded to a radiotherapy professiona
for research in the field of radiobiology, radiation physics, clinica
radiotherapy or radiation technology, during the ESTRO Annua
Congress, specifically for the development of a liquid-filled porta
imaging device.

Part IV

Medical Physics: More than Protection of the Public

"*There is also hope that even in these days of increasing specialization there is unity in human experience.*"

Allan McLeod Cormack, 1979 Nobel Prize winner in Physiology "for the development of computer assisted tomography."

13

The Scariest Day of My Life

Cari Borrás

1   Background

The year was 1989, sometime at the beginning of March. I had been the Regional Advisor on Radiological Health at the Pan American Health Organization/World Health Organization (PAHO/WHO)[1] for less than one year. I wasn't sure when I joined the organization what my exact job was going to be. I had assumed that activities would be similar to those carried out at the Bureau of Radiological Health (later on called the Center for Devices and Radiological Health) of the U.S. Food and Drug Administration (FDA). So, I was expecting a lot of medical physics and obviously, some radiation safety tasks. But I always assumed that I was going to be working in the medical field, PAHO/WHO's mandate being health.

I learned that there had been a radiation accident with an industrial irradiator in San Salvador, El Salvador (ELS), exposing staff to high doses of

---

[1] The Pan American Health Organization (PAHO) was founded in 1902. In 1949, after the World Health Organization (WHO) was created, it agreed—without losing its independence—to also serve as the WHO Regional Office for the Americas.

C. Borrás (✉)
Radiological Physics and Health Services, Washington, D.C., USA
e-mail: cariborras@starpower.net

gamma rays from cobalt-60 sources. I was excited to accompany an International Atomic Energy Agency (IAEA) expert mission to assess the situation. However, I have to confess, I had no idea how an industrial irradiator functioned. So, I was looking forward to working with the IAEA expert team consisting of two physicians from the Radiological Emergencies and Assistance Center/Training Site (REAC/TS) and a health physicist from the Oak Ridge Associated Universities (ORAU), both in Oak Ridge, Tennessee. It so happened that REAC/TS was also a PAHO/WHO Collaborating Center for Radiological Emergencies and I had already established contact with its director shortly after I started working at PAHO.

At the end of February 1989, the Minister of Labour and Social Security of ELS had requested the IAEA "for medical assistance with radiological emergency in El Salvador", and informed the Minister of Health of the accident. On March 2nd, the Minister of Health asked PAHO/WHO for a mission to provide advice on the situation and an IAEA fax to PAHO confirmed that I was going to go with the REAC/TS team, planning to fly on March 2nd. I got permission to travel to ELS and I reserved a flight from Washington to San Salvador for March 3rd.

However, I had to cancel my travel plans. I got an urgent call from REAC/TS that the Department of State had forbidden the mission experts to travel to ELS because of the civil war there. I had no choice. I had to go. I had been invited by the Minister of Health of ELS and PAHO/WHO was governed by the Ministers of Health of the Americas. The Minister, de facto, was my boss. I had to obey. Furthermore, I was a Spanish citizen; the directives of the U.S. Department for governmental employees (REAC/TS is part of the U.S. Department of Energy) which are advisory for US citizens, were not applicable to me, who, in any case, was an international civil servant, PAHO/WHO being an international organization.

2   What is an Industrial Irradiator?

To be honest, I don't remember how I learned about industrial irradiators. But, somehow, I did. I phoned the manufacturer, a Canadian company, and asked as many technical questions as I could. I also had a lot of telephone conversations with the ORAU health physicist, who also had an extensive conversation with the Canadians. He learned that two Canadian technical personnel had traveled to ELS February 12–18 at the request of the president of the Salvadorean factory, who told them the irradiator was not working; it seemed that some of the cobalt sources did not return to the safe position n the pool. The Canadians realized what the problem was, and seeing the run-down status of the machine, they disabled it completely for future use, thus ensuring the workers' safety. Had they known that on February 5th a serious radiation overexposure had occurred, they would have provided very much needed advice on how to proceed.

However, at the time of their visit, no one in ELS had realized that three workers had been overexposed. They were sick, but their sickness was attributed to food poisoning. And they certainly did not tell anyone what had happened, being concerned about potential work reprisals.

An industrial irradiator is a machine that irradiates products in order to sterilize them. There arc several types. The irradiator in ELS was of the "pool type". A generic diagram of this type of "pool" irradiator, also called "panoramic wet storage irradiator" is shown in Fig. 1a. The irradiation was provided by cobalt-60 sources, the same type of radioactive sources that are used in radiation therapy for cancer treatment. But their geometry and activity (the amount of radioactive material) are very different. In an irradiator, the cobalt-60 slugs are encapsulated and lodged in the so-called "source pencils" which are placed one next to the other on a "source module". There may be several modules (the ELS irradiator had two) in a rectangular "source rack", which is held in the vertical position. The assembly of these parts is shown in Fig. 1b.

When not in operation, the rack is kept in a pool of water that shields the radiation emitted by the radioactive sources and is located below the irradiator room. The functioning mechanism is simple. The source rack is attached to a cable that using a hydraulic system holds it over the water in the irradiation position and returns it to the storage water pool by gravity. The tote boxes containing the products to be sterilized enter through a small door into the irradiation room on a conveyor belt, similar to the one we see in the airport to transport luggage, and they change positions automatically by a system of valves and pistons. When the irradiation time is completed, they leave the irradiation room though a small exit door.

The time each container has to be irradiated is calculated and depends on the activity of the sources and the degree of sterilization required. Complete sterilization is checked with a dosimeter—in ELS it was a plastic material that changed colour when irradiated. There are other devices ("dosimeters") that can measure radiation by the charge produced in a chamber filled with air (ionization chamber). When radiation interacts with the chamber, it produces an electrical signal that can be collected and displayed. Some of these radiation detectors are the so-called Geiger counters. These are very sensitive—very little radiation produces a big electrical signal which can

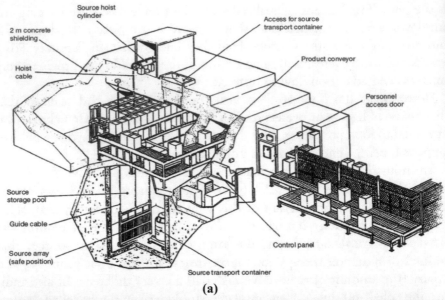

(a)

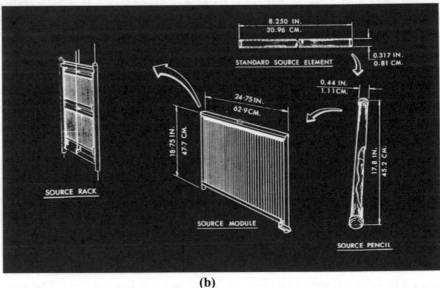

(b)

Fig. 1   a Category IV gamma irradiation facility: panoramic wet source storage irradiator. With permission from: International Atomic Energy Agency, Radiation Safety of Gamma, Electron and X Ray Irradiation Facilities, IAEA Safety Standards Series No. SSG-8, IAEA, Vienna (2010). **b** Schematic diagram showing the build-up of a typical cobalt source rack from slugs, pencils, and modules (courtesy of Nordion, International Inc.)

be read immediately on the device itself. All facilities using ionizing radiation have them to check environmental radiation levels. They can alert people of a potential radiation problem by triggering an audio alarm when a radiation level is considered dangerous. The irradiator electrical system has embedded safety features, some of which are activated by the signal produced by Geiger-type detectors.

I may not have known about irradiators, but I did know about radiation hazards, and in fact, if workers had been able to enter the irradiation room with the source rack above the water, their radiation doses must have been very large. The cobalt-60 sources at the time had an activity of about four times what a cobalt radiation therapy machine for cancer treatment usually has, and the radiation field sizes are much larger as the tote boxes containing the material to be sterilized can be quite big.

I learned that irradiators have safety interlocks designed to prevent the source from being raised when personnel are in the irradiation room and to prevent access when there are abnormal radiation levels in the room. Other automatic safety features lower the source and shut down the irradiator in the event of an electromechanical malfunction or when power is cut off, something I did not know was occurring very often in San Salvador.

Although not all the safety systems were functioning at the time of the accident, there were enough to prevent an accidental irradiation. The fact that they were bypassed by one operator indicates how resourceful the employee was and how well he understood what was needed for the machine to function. It also reflects how little he knew about radiation hazards, especially from radioactive sources. It is not my intention to describe the details of the accident. They were published by the IAEA in a report that is accessible and freely available on the web.[2]

The more I learned about irradiators, the more I feared I couldn't provide proper technical assistance on my own. And what about my own safety? As long as the sources were in the water pool, no significant radiation was present in the irradiation room. But how did I know that there were no loose sources around the factory?

And what about the civil war? How dangerous was ELS going to be?

---

[2] International Atomic Energy Agency. The Radiological Accident in San Salvador: A report prepared by IAEA in cooperation with PAHO/WHO. IAEA. Vienna 1990. https://www.iaea.org/publications/3718/the-radiological-accident-in-san-salvador.

3   There is a Civil War in El Salvador

The ELS civil war had started in 1979; so, at the time of the radiological accident, the country had been ravaged for 10 years. I was born in Barcelona, Spain a few years after our civil war had ended, but the tragic events my family went through were very much alive in their minds. My uncle and my father actually fought on opposites sides because of their convictions, a common phenomenon in civil wars.

Coming from Franco's Spain, and having joined as a college student a militant group against the dictator, I was very much against the repressive regimes that were controlling Central America in those days –with the open or clandestine support of the U.S. But I couldn't agree with communist parties, as I truly believed—and still do—in having a democracy where political parties with different opinions can alternate in the government, as long as this is the desire of the country's citizens, who elect them. In ELS, the forces fighting the government were a coalition of left-wing groups named collectively the Farabundo Martí National Liberation Front (FMLN) which was joined by the communist party of El Salvador. And in fact, by 1989, the country's governance was divided almost equally between the government and the FMLN, whose army guerrilla fighters were very well organized and had weapons and money. When I arrived in El Salvador I found out that the country was in the midst of election preparations. What I did not realize then is that the government had agreed to allow the FLMN to participate in electoral politics. That started peace talks, which eventually broke up, and the war would continue until 1992, with over 75,000 Salvadoreans killed in the conflict, and a larger number having emigrated.

One may think that the fact that there was a war would have very little to do with a radiation accident in a factory, but, in my opinion, it was the main contributing factor. For years, the machine had been without proper maintenance; the radioactive sources had not been changed since their installation in 1975, and the operators who had initially been trained not just in the operation of the irradiator but also in their safety, had been replaced over the years, leaving verbal recommendations to the newcomers, who received no training in radiation protection.

The sources had so little activity, that sterilization times were too long, and the process was very inefficient. To discuss their replacement, in 1981, a representative of the Canadian firm that built the irradiator traveled to ELS. Unfortunately, however, he never left the airport because of the worsening civil war situation and returned back to Canada on the next plane. Since then, contact between the factory management and the Canadians had just been over the phone, and the sources were never replaced. Had the security of the Canadians been ensured, the plant would have had the added benefit of very much needed radiation safety training. The war impeded it.

At PAHO, we were "non-political" (our job contract had a requirement never to side with a political party). We provided technical cooperation to any member state that requested it, from Fidel Castro in Cuba to Pinochet in Chile. In fact, the PAHO Director at the time, Dr. Carlyle Guerra de Macedo, had an initiative for the whole Central American region: "Salud, Puente para la Paz" (Health, Bridge to Peace) that hoped PAHO-sponsored health activities would facilitate peace in the long run.

But I had read and been horrified about the killing of Archbishop Oscar Romero (made a saint of the Catholic Church by Pope Francis in 2018) and I was aware that brutal acts had been and could be perpetrated.

Yet the country's political situation did not worry me. If PAHO, which was accustomed to all these kinds of situations, was asking me to go, it must have been OK, no?

4   Home Preparations and Travel to El Salvador

No, my concern was technical. Not only did I lack the knowledge about irradiators, I had no instrumentation, for example a Geiger counter, that could help me assess the radiation risk. All I had was a device that could measure radiation around x-ray machines and thus did not have the configuration nor the tools to measure the high energy radiation emitted by cobalt-60. I remember looking at home for an acrylic plate that I could put in front of my detector to see if that way I could detect the cobalt-60 radiation, which is much more penetrating than the x-rays from diagnostic radiographic units. But I did not have any calibration for this (improvised) arrangement, nor did I know whether it would even work at all; the expected levels of radiation were unknown, and suspected of being beyond the range for which my device had been designed and calibrated.

Worse still, I did not even have a personal dosimeter, something that all radiation workers are supposed to have. But I was not a radiation worker; I was an international civil servant, working in a building with no access to radiation sources or radiation detection instrumentation. I had been in Washington DC for less than a year now and had not made any contact with medical physics colleagues in neighboring hospitals.

REAC/TS-ORAU came to my help. Not only did I speak often on the phone with the ORAU health physicist there learning as much as possible about what to search for and what precautions to take, but he also sent me an electronic personal dosimeter that would trigger an alarm if the radiation level was too high. The device came in time for me to travel to ELS. I will be forever indebted to him. A pity he won't be able to read this; he passed on in 2019.

I flew to San Salvador on March 7, via Houston. My plane departed from National Airport, and since in early March Washington DC can be very icy, I remembered the plane that in 1982, upon departing from that airport, crashed into Washington DC's 14th Street Bridge and plunged into the icy Potomac River in the midst of a blinding snowstorm—we were told the reason was that its wings had not been properly de-iced (the TV image of the volunteer who helped so many people to safety at the cost of his own life is still vivid in my memory—I often wondered whether I would have ever been capable of such a sacrifice and decided probably not ...). I had learned that after de-icing, a plane has 10 min to depart to be safe. I had a window seat. I remember watching the de-icing operation and then looking nervously at my watch while clutching in my handbag the personal dosimeter that ORAU-REAC/TS had lent me. I don't remember whether we departed within the 10-min period, but nothing happened to my flight, and we arrived in Houston safely. I do not remember either whether I had to change planes there, but I do remember that, as we departed Houston for San Salvador, the pilot announced that a volcano in Costa Rica had erupted and that it could affect our trip. Somehow, it didn't. But the warning certainly contributed to make my flight quite memorable.

I arrived safely in San Salvador, as planned. As I greeted the PAHO driver who met me at the airport, I reminded myself that I had to go to ELS not just because my "boss" had told me to go, but also because of a moral imperative and the fact that, yes, I was/I am a "radiation expert".

5   How Three Workers Were Accidentally Irradiated

After an introductory meeting with the PAHO/WHO Representative (PWR) at the PAHO-WHO San Salvador Office (PAHO is a decentralized organization) to discuss with him my work plan, I went to the industrial irradiation facility. I learned that the plant manufactured and sterilized surgical and medical equipment, such as blood bags, for Central America. For the sterilization processes it used either the cobalt-60 irradiator or an autoclave. A view of the main building is shown in Fig. 2.

The responsibility of the irradiator plant was under a maintenance manager and a technical advisor. The latter was an industrial technical engineer who had joined the factory in May 1988. The rest of the workers had been with the company from three months to five years. An electrical engineer supervised the technicians. There were three irradiator operators and five general maintenance workers, some of whom were electricians. I interviewed most of them on a one-to-one basis in an empty office of the factory. The workers were asked whether the conversation could be taped. They all replied negatively. They were very nervous and apprehensive but were put at ease when I told them that I was a radiation worker like themselves, whose purpose was to ascertain the cause and circumstances of the accident in order to establish accurate dosimetry that could impact the treatment success of the irradiated coworkers and to determine whether the design of the irradiator required modifications to prevent similar accidents in the future.

In theory, the responsibility of the irradiator operators was to run the irradiator; the responsibility of the general maintenance technicians, except that of the refrigeration technician, was to repair the electromechanical components when they failed. In practice, the maintenance technicians could—and

Fig. 2 Street view of the factory where the irradiator was located

did—run the irradiator, and the operators knew how to fix the irradiator—and did. This was necessary since the irradiation plant was operating 24 h a day. All the workers seemed very secure about the operation of the irradiator, and they all explained in the same manner the methods used for the irradiation process, including the significance of the green, white and red lights in the control panel, shown in Fig. 3.

Although they all had seen cables on the floor (see Fig. 4), no one in the factory knew what was supposed to be attached to them. Thus, they could not imagine that, once upon a time, a Geiger counter had been attached, a detector that was an integral part of the door safety interlock.

The circuit was supposed to register background radiation pulses that allowed the door to open when the key was inserted in the keyhole.

The factory workers, however, had learned that pushing the circuit start button quickly several times allowed them to open the door—not realizing the resulting pulses mimicked those the Geiger counter would have produced when measuring background radiation. They were all convinced that this was "the method" of opening the door to the irradiation room. And they were certain—and in that they were correct—that, for the door to be opened, all the lights on the control panel had to be green.

Unfortunately, they did not know the status of the lights on the fateful night of February 5th nor what the irradiated workers did. On February 6th,

Fig. 3 Irradiator control panel

Fig. 4 Cables that should have been connected to a Geiger counter

when the next operator came to continue the sterilization process, he found the door open and some totes in disarray. He put everything back in order and continued working, as if nothing had happened; the machine seemed to be OK. In the following days though, the irradiator stopped again, a recurrent problem the irradiator suffered, usually due to the power shortages in ELS.

I learned that in the period February 6–10, some workers had entered the irradiation room when the Geiger counter showed that the radiation level was too high. I identified 13 individuals who could have been exposed to radiation and arranged to have their blood samples taken and sent to another PAHO/WHO Collaborating Center in Buenos Aires, Argentina for cytogenetic analysis. This is a biological dosimetry method: by counting the number of chromosomal aberrations in the blood, one can estimate the radiation dose received by an irradiated individual. I did consider sending the samples to REAC/TS, but I was concerned the blood samples could get stuck in US customs. To send them to Argentina was easy: shipments from a PAHO office to another PAHO office went by diplomatic courier. Although the results showed that four individuals had received measurable doses, none were significant. At least this was good news!

I had asked all the workers about radiation safety, since I felt they did not understand why I was so worried about possible deaths. All the workers acknowledged that entering the irradiator plant could be life threatening, but in a country with ten years of civil war, "when over 500 people died every week", the perception of danger had lost its meaning.

6   The United Nations Cannot Guarantee My Safety: I Have to Leave San Salvador

I took copious notes of all my activities and made measurements in the room trying to figure out where the irradiated workers had been at the time of their overexposure. For some of the measurements, I was helped by a nuclear medicine physicist working in a San Salvador Hospital; he was the only medical physicist in the country. I even took a detailed video of how the irradiator was supposed to function; on it I described the different areas and components. All the documents were filed in my office at PAHO and when I retired, I took them to my house and stored them in my basement, only to access them for the writing of this chapter.

The one issue I could not figure out was how the door of the irradiation room was opened. The safety circuits should have prevented it. We considered using the entry through the door allowing the entry of the containers for irradiation, and we even forced the door using several methods. But nothing seemed plausible. It became evident I had to talk to the irradiated victims.

At the time of my visit, the two most exposed victims had been moved to a hospital in Mexico for specialized treatment of radiation exposure. The least irradiated one was still in a San Salvador hospital, and I went to speak with him there. From him, I learned that on the night of February 5th the three workers had climbed to the irradiator second tier and from there had freed the source rack which had become stuck in the irradiation position. He did not know how the jam had occurred nor could he tell me how the operator had gained access to the irradiation room. All he could say was that he and another worker (from different departments in the factory) were asked to help him and they did. By the time they arrived at the irradiator, the door to the irradiating room was open, so the two "helpers" followed the operator into the room, having been assured that there was no radiation risk as the "electrical power to the irradiator had been cut off for two hours", by which time "there should have not been any residual radiation". I do not recall whether I told him at this point that the source rack contained radioactive material, which was always emitting radiation, not like x-rays that get turned off when the power is off. Probably I did not, as he was very scared, and was going to be transferred to Mexico to get treatment for his irradiation at the same hospital where the other irradiated workers were.

Clearly, I had to go to Mexico to interview them if I wanted to know more. This was not straightforward; the Mexico PWR was hesitant to allow me to see the radiation victims. I was a physicist, "what does a physicist have to do with patient treatment?"

Eventually, on March 14, I flew to Mexico City and was allowed to visit them in the hospital. There I learned the sequence of events and could correlate their description of positions within the irradiator with the measurements I had taken in San Salvador. This was needed to perform a radiation dose reconstruction and compare it with the medical radiation effects now visible, such as moist desquamation of the skin on their legs. The IAEA-PAHO/WHO publication contains all the medical details. Suffice to say that the main operator died from the radiation exposure six months after the accident and one of the helpers required amputation of his two legs. The other helper fared better; he went back to the factory to work.

By the time I arrived in Mexico, the workers had already been visited by the REAC/TS-ORAU team consisting of two physicians from REAC/TS and the health physicist from ORAU. From March 8–12, while I was in ELS, they had evaluated their medical condition and made recommendations for further treatment. I was in constant contact with the ORAU physicist, sharing with him the information and measurements I was gathering at the irradiator plant.

On March 12, I was ordered by the United Nations to leave ELS; they could not guarantee my safety any longer—the war was now too dangerous. So, I made plans to fly to Mexico. The day before my departure I received an important request from the ORAU health physicist. "What is the distance from the ceiling to the place where the workers had stood to lower the source rack into the water?" This was a measurement I had not taken, but it was essential, because it could tell whether the workers had been standing straight or were in a bent position, thus offering more area of their bodies exposed to the radiation.

Without any hesitancy, I asked the PAHO driver to take me to the factory for one last measurement.

As we got into the car, the driver was very attentive to the surrounding sounds. "Shall we go this way?" "No, I hear too much gunfire". "Let's try this other way". "No, no, I can hear the bullets shooting". "This way". "That way"…

Eventually we made it safely and I measured a distance I will never forget: 1.5 m. One of the workers was 1.59 m high. He had to bend.

Fig. 5 Location on the irradiator 2nd tier where the workers stood to extricate the jammed source rack and lowered it to the water pool

Figure 5 shows where in the irradiator the workers stood and Fig. 6 is a diagram of a potential configuration of one the worker's position, drawn by the REAC/TS-ORAU team.

The driver with his wonderful sense of danger took me back to my hotel and we agreed that he would pick me up the next morning to drive me to the airport. I had been asked by the airline to arrive by 5 a.m.

That evening I organized all my data and decided to put all my notes in my carry-on bag together with my video camera, my still camera, and the instruments I had taken with me. I have to say that I was leaving ELS thinking that at least one of the safety interlocks of the irradiation room had failed. How otherwise could the operator have entered the room and seen that the rack was stuck by five totes being in the place of four due to a piston malfunction?

I had partial answers in Mexico City. There, I talked to the irradiated operator in the hospital where he was being treated and I also toured a similar irradiator with an excellent physicist who apprised me of what I still did not know about the machine.

I gathered additional information on May 25–26, when all of us who had been working on the ELS accident met in Nashville at the REAC/TS site: the Canadian representatives who had gone to ELS at the end of February,

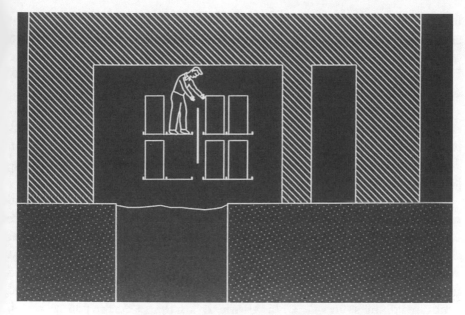

Fig. 6 Diagram showing a potential position of a worker freeing the stuck source rack, relative to the irradiator ceiling and the tote boxes

the mission team members who had gone to Mexico, the director of the Argentina Collaborating Center on Radiation Emergencies who had facilitated the cytogenetic dosimetry of the other potentially irradiated workers and I. We compared notes, clarified issues and decided to write a report. The report was not going to be written as planned. The IAEA convened a meeting in San Salvador in July 1989 and in 1990 a report of the San Salvador accident was published by the IAEA, albeit "in cooperation with PAHO/WHO".

7   Writing a Report of the Radiation Accident

The meeting convened by the IAEA about the ELS radiation accident took place on July 24–28, 1989, in the offices of the United Nations Development Program (UNDP) in San Salvador. The meeting objectives were to compile and analyze the findings and to identify the lessons learned to avoid a similar occurrence in any other part of the world.

The meeting was attended by three IAEA staff members, representatives from the Ministries of Labor and Health, the Salvadorean nuclear medicine physicist who had helped me in some of the measurements, a local radiation oncologist who had examined the irradiated workers, several people from the factory, the two Canadians who had gone to ELS at the end of February and three PAHO/WHO staff members: one from the PAHO/WHO-ELS office, another one from the Regional Emergency and Disaster Program and I. Again, no REAC/TS-ORAU persons were able to participate due to a delay in the U.S. Department of State travel authorization.

At that meeting, I shared with the IAEA the information gathered in my early March trip to San Salvador. But I also used the opportunity to go back to the factory with the Canadians to review the situation of the irradiator. They showed me that all the lights at the control panel were functioning properly. There had been no malfunction as I had assumed.

The "Danger High Radiation Area" illuminated sign above the door had not been operating for an undetermined length of time, but the red "source up" light, the green "source down" light and the source transit alarm that alerts the operator of an abnormal incident, were all operating correctly. When I interviewed the irradiator operator in Mexico City, I had learned that when he came back from a coffee break, the alarm was ringing and continued doing so in spite of his efforts to "fix things" at the control panel. He had then climbed to the irradiator roof, where he found out that the cable was not under tension, confirming that the source rack was jammed in an intermediate position. I understood how he tried unsuccessfully to release air pressure from the source hoist mechanism. What I had not understood is that he then pulled the source cable completely out of the source hoist cylinder, bringing together the two hoist sheaves, thus activating the "source down" microswitch. That was brilliant. The alarm stopped and when he returned to the control panel, he saw the "source down" light illuminated, allowing him to open the door by their "usual" method. A pity that when he entered the room, he did not carry the Geiger meter with him... He thought that because he had turned the power off for two hours, any "lingering" radiation was gone.

Figure 7a through c show the hoist mechanism, the cable and sheave assembly, the locked fence the operator had to open to access the roof, the ladder he had to climb, and the top of the irradiator as it was left by the Canadians during their late February visit when they disabled the irradiator for safety precautions.

Sometime in September a draft report of the accident, written by the IAEA staff, came to PAHO for my review. When I read it, I was very concerned. It seemed very insulting to the ELS governmental authorities and to the factory management. I had lived in the United States 23 years already and was mindful of liability issues. So, trying to protect PAHO/WHO, I sent the report, already with my technical comments, to our legal department for

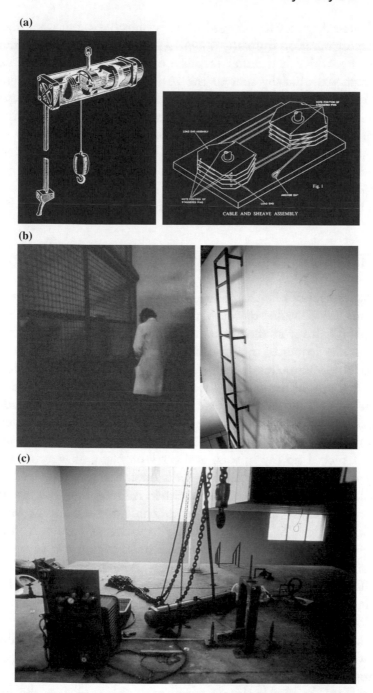

Fig. 7   **a** Diagrams of the hoist mechanism and the cable and sheave assembly (Courtesy of Nordion International Inc.). **b** Access to irradiator roof: locked fence and climbing ladder. **c** Top of the irradiator roof, as left by the Canadians

review. I guess they may have agreed with my objections. But what they did to "save" the organization from any potential legal problem, was to remove any credit to the work I had done. To make matters worse, when the redacted draft report was officially sent to the IAEA, it went with a sentence that read "We are very sorry for taking so long in reviewing the draft report and to study Doctor Borrás observations, but we should be particularly careful when we are dealing with a report on a subject matter foreign to our area of institutional expertise." To which, the IAEA replied: "We understand your reservations in 'dealing with a report on a subject foreign to [your] area of institutional expertise'".

Did I say somewhere that I was a "radiation expert"?

8   The Scariest Day Of My Life

I did not know any of this the morning of March 14, when I was going to fly from San Salvador to Mexico City. I was picked up from my hotel by the PAHO driver at the agreed time, and I arrived at the ELS Airport at 5:00 a.m., as requested by the airline. The airport, built in the late 1970s, was a modern structure located at 42 km (26 miles) from San Salvador, in the south-central area of the country, bordering the Pacific Ocean. I don't know about the present, but in 1989 there were no other buildings around. It was quite dark when we arrived, but all the lights inside the airport were on. I thanked the driver and told him he could leave me, as I assumed everything was OK. I picked up my suitcase and carry-on bag and went to a door to enter the building. The door did not open. Assuming that door was locked, I went to another door; it did not open either. And another one. Eventually I tried every door in the airport. I had to conclude that the airport was closed. I got a bit angry. Why did they tell me to arrive at 5:00 a.m. if nobody was going to be around? So, what could I do? I sat on top of my suitcase by one of the doors and looked around me. Facing the airport, there were some hills, I guessed part of the coastal range. They were not very high and were quite distant from me, but they looked majestic against the dim light of dawn. I could start seeing the outline of the mountains—very beautiful… I thought nobody being around, the place would be silent, but no, I could hear the sound of marching feet. Uno, dos, uno, dos, uno, dos…. My goodness! Some troops were training behind those hills. And it didn't matter whether they were from the El Salvadorean army or whether they were the guerrillas. A civil war is a civil war. Both sides are brutal. The sound continued for a long time as the sky was getting lighter and lighter. Uno, dos, uno, dos, uno, dos... I kept looking inside the airport to see if there was any movement. But no, everything was very quiet. Clearly, either no one was there or, if anyone was inside, they were not ready to check-in passengers. So, I kept looking around. And then, against the outline of the mountains, I could see a man coming down the hill. He had no suitcase with him. He was carrying nothing. And so, I was really afraid. Was he one of the people marching? He did not seem to wear any uniform. Was he a bandit? I had been alerted by the ELS PAHO/WHO office that crime had flared up during the civil war. So, yes, probably, he was a bandit. And I certainly was an easy target to him. A middle-aged woman alone in a totally isolated place. I really got very, very scared. I kept looking at the man coming down the mountain; as he was getting closer to me, I was getting scarier. How could I convince him not to attack me? Because, clearly, he was going to take my suitcase, my carry-on and all my belongings and then... He would kill me. This was very clear. He would kill me. I had to devise a plan. A plan that would save me. What could I say to him? To tell him that he could take my suitcase but not my carry-on and to promise I would not identify him, did not seem convincing at all. So, I went through my handbag and took out my United Nations laissez-passer (a UN diplomatic travel document which can be used like a passport in connection with official travel duties) and my Spanish passport.

I thought I would try to explain to him that I was an international civil servant originally from Spain—"look, look, see, I work for the United Nations and I am a Spaniard" and that I had been in the country trying to figure out how a radiation accident had occurred in order to prevent other radiation accidents to occur in the future. At that point, I took even my notes from my carry-on and was planning to show them to him and say "it is very, very important that these papers arrive in Mexico. I am going to Mexico to see the irradiated victims. These notes can prevent their death. I need to get them to my organization, the Pan American Health Organization. We are a health organization. We want to help your country. We are helping your country. We have done a lot of things for your country. It is very important that I get back from this mission". But I have to confess that clutching my passports in one hand and my papers in the other one, I was not convinced that I had the power to dissuade him from his intentions. He would kill me. I had no exit. The dawn was giving way to the sunrise. I could see the man very clearly now. And he was getting closer and closer to me. I was getting more and more scared. I have never been so scared in my life!

And then, something happened. A truck full of passengers arrived at the airport. Several people came out of the truck and lined up near me by one of the doors, waiting for the airport to open. By the time that I looked back at the mountains, the man had disappeared. I was saved!

9   'Aftermath'

How little did I know that the ELS radiation accident was going to take a toll not just on my professional life but on my personal life as well.

I was born in Barcelona, Spain, on February 18th during a snowfall, a most unusual precipitation in my hometown. To celebrate the event, I always try to see snow on my birthday—it has become a yearly ritual. So, February 15–20, 1989, just three weeks before my trip to ELS, I took some days off from work and I went to Lake Tahoe with a friend of mine to do some cross-country skiing, a sport I am lousy at, but I loved passionately. Lake Tahoe is my favorite spot in California, and I went there often during the 14 years I worked in San Francisco. The friend with whom I went this time was a relatively recent friend. Although he had become a widower and was still mourning the passing of his wife, we had a wonderful time. Lake Tahoe was glorious, the snow was deep and dry, the sun, bright and warm.

I wrote to my friend just before I left to ELS (there was no email in 1989), and while I was there, I shared with him my fears and concerns, what I was doing and what I had accomplished. It was like writing a diary, only that it was addressed to him, and yes, it was intense, as I am. I felt good writing it, I did not feel alone. When I was back in DC, I mailed to him the long letter I had written. I don't remember when his reply came. Probably the letter is saved somewhere in my basement where all my history is kept. But I remember what he wrote. He was very sorry, but he felt he could not respond to my "intensity" the way that he thought I was expecting him to respond. So, he was terminating our relationship… I may be passionate and romantic, but I don't linger in the past. Yes, it hurt. But one has to move on… And I did. Except that when I think about El Salvador I also think about my lost friend.

Cari (Caridad) Borrás obtained a Doctor of Science degree from the University of Barcelona, having done her thesis research at Thomas Jefferson University in Philadelphia, PA, as a Fulbright scholar. The American Board of Radiology certified her in Radiological Physics and the American Board of Medical Physics in Medical Health Physics. She has worked as a radiological physicist in: Barcelona, Spain; Philadelphia, PA; San Francisco, CA; Recife, Brazil; and Washington DC, where for 15 years she ran the Radiological Health Program of the Pan American/World Health Organization, and where she currently holds an adjunct faculty position at The George Washington University School of Medicine and Health Sciences and works as a consultant. She served in several committees of the American Association of Physicists in Medicine (AAPM), American College of Radiology (ACR), Health Physics Society (HPS), International Organization for Medical Physics (IOMP), International Union for Physical and Engineering Sciences in Medicine (IUPESM) and European Federation of Organisations for Medical Physics. She has organized and/or participated in more than 300 international courses/workshops, written over 100 publications and edited two books on radiology services and radiation safety. She is a Fellow of the ACR, AAPM, IOMP, HPS and IUPESM.

Awards

2018. International Day of Medical Physics (IDMP 2018). Award for *"promoting medical physics to a larger audience and highlighting the contributions medical physicists make for patient care."* Presented by the International Organization for Medical Physics at the 24th International Conference on Medical Physics, Santiago, Chile, 2019.

2013. AAPM Edith H. Quimby Lifetime Achievement Award. This award recognizes AAPM members whose careers have been notable based on *"significant scientific achievement in medical physics, or significant influence on the professional development of the careers of other medical physicists, or leadership in national and/or international organizations, and active participation in the AAPM"*. Presented at the 55th American Association of Physicists in Medicine Annual Meeting & Exhibition, Indianapolis, IN, USA.

2013. Selected by the IOMP as one out of 50 medical physicists *"who have made an outstanding contribution to the advancement of medical physics over the last 50 years."* This recognition was part of the IOMP's 50th anniversary and it was presented at the 20th International Conference of Medical Physics, Brighton, UK.

2012. IUPESM Award of Merit. It is the highest honor the IUPESM bestows to a medical physicist and a biomedical engineer for *"outstanding achievements in physical and engineering sciences in medicine"*. Presented at the IUPESM World Congress on Medical Physics and Biomedical Engineering, Beijing, China.

2003. Sociedad Española de Física Médica (SEFM) Gold Medal. This is the highest honor bestowed to one of its members *"in recognition of their contributions to medical physics"*. Presented at the XIV SEFM Congress, Vigo, Spain.

14

The Goiânia Accident: A Candid and Personal Experience

Carlos E. de Almeida

1 Introduction

Generally, radiation therapy is a very safe mode of cancer treatment. However, under unusual circumstances and through inappropriate activities, accidents can happen. This chapter is not intended to give the full picture of the tragic Goiânia accident in detail since there are already several reports specifically about this incident.[1] Instead, I shall present some of my observations and discuss the difficulties that arose during the first 40 days following the discovery of the incident when I was present on the scene and sharing the coordination of the operation.

It is important to emphasize that I respect all the victims and their families and by no means should these stories be interpreted as disrespect to any of them. It is also important to recognise the support from the local authorities to the field workers, as well as the dedication of the medical staff provided

---

[1] See for example: The Radiological Accident in Goiania. International Atomic Energy Agency (IAEA), Vienna, Austria, 1998. Available on-line at https://www-pub.iaea.org/mtcd/publications/pdf/pub815_web.pdf.

C. E. de Almeida (✉)
Universidade do Estado do Rio de Janeiro, Rio de Janeiro, Brazil
e-mail: cea71@yahoo.com.br

© The Author(s), under exclusive license to Springer Nature Switzerland AG 2022
J. Van Dyk (ed.), *True Tales of Medical Physics*,
https://doi.org/10.1007/978-3-030-91724-1_14

by many institutions, and the entire radiation protection staff of several institutions as part of the Comissão Nacional de Energia Nuclear (CNEN the National Nuclear Energy Commission of Brazil), who were tasked with different responsibilities whilst working in the contaminated areas.

2   Conventional Methods of External Beam Radiotherapy

The use of radiation to treat patients started a few months after the x-ray was discovered by Roentgen in 1895 and has been used since then. Several radiotherapy machine designs were developed using x-rays providing maximum peak energies of 400 kVp. However, with these energies the maximum dose in the patient occurs at the skin surface and for many years how the skin reacted was the biological indicator of how much radiation it could take. The skin reaction scale consisted of redness, slight and moderate desquamation, moist desquamation, and finally, tissue necrosis. Since physical measurement of radiation dose was not yet available, the interpretation of those indicators could vary significantly from physician to physician. Only several years later it was realised that the need to define some physical quantities that could serve as normalisation was crucial so that everyone could speak the same language and enable comparison of the clinical observations and results.

With the advent of nuclear reactors, it was possible to process nuclear waste and condense caesium-137, as an artificial by-product of the uranium nuclear reaction. It emits only one photon with a single energy of 662 keV and has a half-life of about 30 years. It could be produced in quantities sufficient to be inserted into a small capsule and placed in a shielded container, known as the radiation treatment head. The radiation beam is then collimated in a solid angle with a set of mechanically movable collimators made of high atomic number and high density to allow the collimated beam to reach the patient surface as well as tumours located up to 5 cm below the patient surface.

Caesium-137 is a soft, flexible, silvery-white metal that becomes liquid near room temperature, but easily bonds with chlorides to create a crystalline powder. It reacts with the environment like sodium chloride (table salt). Thus, it moves easily through air, it dissolves easily in water, and it binds strongly to soil and concrete. External exposure to large amounts of caesium-137 can cause radiation sickness and death.

Although the energy of the caesium-137 photons was higher than the ones produced by x-ray units of that time, the dose delivered was not enough to reach deep-seated tumours without unnecessarily irradiating the surrounding normal tissues to unacceptably high doses, including the skin. With the advent of research and the use of industrial nuclear reactors it was possible to apply the neutron activation principle to turn some materials such as the natural element cobalt-59, into a radioactive element, cobalt-60, which is an emitter of two photons with energies of 1.17 and 1.33 MeV, thus an average energy of 1.25 MeV. It has a half-life of 5.27 years.

With cobalt-60, the maximum dose deposition is typically around 5 mm below the skin surface, with the skin receiving about 30% of that dose. With this discovery of "skin sparing", physicians lost the qualitative concept of skin reaction as the maximum dose was no longer at the surface and a need for physical quantities became even more vital. The production of caesium sources for external beam radiotherapy was eventually discontinued, although some centres continued using it until it was possible to replace it with a cobalt-60 unit or a linear accelerator. An additional problem related to caesium-137 sources is its long half-life of close to 30 years. It requires appropriate decommissioning and radioactive waste disposal procedures when the sources have completed their clinical usage, especially since they have such a long half-life. If appropriate waste repositories are not available or disposal procedures are not implemented properly, there is the potential for a major radiation hazard.

It is probably fair to say that, from the time of the use of radioactive sources for radiation therapy, the role of medical physics professionals became invaluable in the area of radiation therapy.

In recent years, new machines replacing cobalt-60 and caesium-137 units have arisen using different types of beams including protons and electrons. In addition, computerized treatment planning systems have developed using highly sophisticated dose calculation methods. These dose calculation methods use imaging data from different imaging modalities to localize the tumour and normal tissues. Various imaging modalities can be used although the most often used is computerized tomography (CT). Additional tumour information can be obtained from positron-emission tomography (PET) and magnetic resonance imaging (MRI). The different imaging modalities are combined (fused) to obtain the best possible information about the location and make-up of the tumour. Modern radiotherapy has capabilities of controlling the beam precisely to account for patient motion as for example occurs in the thorax due to breathing. All of this technology requires sophisticated quality assurance procedures, in addition to oncology computer information systems. Together, these technologies ensure that the dose delivered to the patient is indeed the dose prescribed to the tumour by the radiation oncologist.

2.1   The Effects of Radiation on the Human Body

The use of radiation is very effective for diagnostics. Since radiation absorption is strongly related to the atomic number of the tissues, such as bone and air in the lungs, high contrast images can be produced with excellent quality for diagnosis. The effect of the radiation in the body is considered low risk if the requested exam is justified, if the equipment is well-calibrated, and if the patient positioning is adequate so as to avoid the need for exam repetition. In this case, the benefit outweighs the small risk involved, in analogy with the use of other medications.

In radiotherapy, the use of high doses of radiation is aimed at the tumour cells, which divide more rapidly than normal tissue cells. Cell division is considered the most sensitive phase of the cell cycle, and hence, radiotherapy is more effective in targeting the tumour cells. Simplistically, one may say that the repair mechanisms of the normal cell are more efficient than in the tumour cell. The dose fractionation, which is usually divided into daily treatment fractions given at five days per week, usually over multiple weeks, also allows the normal cells to repair a large portion of the sub-lethal damage.

The undesirable effects of high radiation exposure to the whole human body can be clinically expressed in different ways depending on the magnitude of the doses involved, the type of radiation, and the exposure time. For instance, the effects of whole-body, high radiation exposures are often described by different phases of the acute radiation syndrome:

- The *prodromal phase*, which occurs within hours after exposure and has signs of nausea and vomiting with the likely outcome of death.
- *Central nervous system* and *gastrointestinal* effects are seen within days with early symptoms being nausea and vomiting also with the possible outcome of death.
- The *haematological syndrome* affecting blood forming organs requires observation of a possible decrease of lymphocytes and platelets. These effects may require a bone-marrow transplant to replenish the bone-marrow cells that have been destroyed.

Sometimes biological dosimetry is used to estimate the radiation dose the individual was exposed to. Biological dosimetry can include chromosome analyses, where chromosomal abnormalities, consisting of the number of dicentrics, are measured and then correlated to a previously calibrated curve to determine the dose received by the individual.

The above describes some acute effects of whole-body exposure. However, for smaller radiation fields, other effects can occur as well, such as skin reactions. In addition, for both whole-body and partial-body exposures, there is the potential for longer term, late effects such as carcinogenesis, with potential cancer onsets occurring more than 5 years later.

In the case of the Goiânia accident, besides the external exposure, some of the individuals exposed were heavily contaminated internally with caesium-137 from their hands during eating. In this case, bioassay analysis of the faeces and urine played an important role in determining the level of contamination. Unfortunately, two individuals died because of this and all attempts to increase the excretion were unsuccessful.

The simulation of physical dosimetry was very difficult as there was uncertainty about each of the irradiation parameters, such as: the precise hour the radioactive source arrived in the homes of the individuals, the distances involved, the persons' posture (standing or squatting) during exposure, and the duration of the activities in the kitchen.

3   The Accident

In brief, a radiotherapy centre in Goiânia moved to a new premises and abandoned its old premises, leaving in place a caesium-137 treatment machine without informing the radiation licensing authority, as is normally required. The unit contained a 50.9 TBq (1375 curie) caesium-137 source, which still had 70% of its original activity. Two people who thought that the machine may have some scrap metal value took it home and tried to dismantle it resulting in a major radiation accident in September 1987. The consequences were grave due to the long interval (26 days) between the source removal and the identification of the first symptoms. Subsequently, many people were exposed and contaminated, leading to the death of several people as well as the contamination of a critical area of the city.

The description of the persons involved in the accident, in general, matched with the radiation doses measured in the afflicted areas and the clinical symptoms. This led to the mobilisation of medical assistance and a strategy for the decontamination of the areas to be defined. The first priorities were to provide care for the victims and exclude critical pathways in order to avoid further exposure of the population or contamination of new areas.

The corrective actions taken during the first 40 days after the discovery of the accident involved communicating the incident to the public and the press, implementing medical assistance specific for each patient, and the critical and open evaluation of the overall situation, which resulted in "lessons learned" that might be of benefit to us and hopefully, to others.

The picture of a caesium-137 teletherapy machine like the one used for years in the cancer centre is shown in Fig. 1.

After a legal dispute between the radiation oncology group and the hospital where the treatments were delivered, a teletherapy machine was left in an abandoned bunker without any external protection. The bunker was later vandalised, with the doors being removed, which left the machine exposed without any signs warning of the danger. Waste pickers living nearby found the machine and believed that they could sell it as scrap metal and thus decided to remove it using a wheelbarrow.

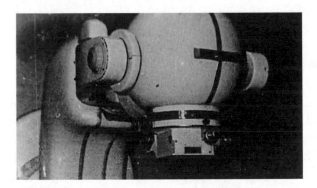

Fig. 1 A caesium-137 unit like the one in the accident

Fig. 2 The bunker remains after it was vandalised. Reproduced with permission from International Atomic Energy Agency, The Radiological Accident in Goiânia, IAEA, Vienna (1988)

Fig. 3 The remains of the treatment machine head. The inset on the left shows a schematic drawing of the source, which is usually 1.5–3 cm in diameter

Figure 2 shows a picture of the remains of the bunker, already vandalised, where the machine was abandoned still containing the source. It was intriguing though how a tiny wheelbarrow was able to withstand the weight of the head of the machine.

Figure 3 shows the remains of the treatment machine head completely destroyed and without the source. At this point, small pieces of the source had already been distributed to some friends of the waste pickers due to the interesting light blue colour emitted by the source.

Figure 4 shows a photo of the actual wheelbarrow used to transport the treatment source head to an outside area adjacent to the house of one of the eventual patients. The remains of the source in the area resulted in the death of his wife by acute radiation syndrome (as described previously) and his daughter experienced extremely severe internal exposure that also led to acute radiation syndrome and her death.

3.1   The Accident Notification

The Goiânia State Health Authority notified the CNEN of a possible radioactive source found inside a plastic bag, which was subsequently checked by a local physicist to measure if there was any radioactivity present in the object. The local physicist had to use a second scintillometer to measure the object, since the first had indicated a level off scale and he believed, for a moment,

Fig. 4 The wheelbarrow used to transport the treatment head of the caesium-137 machine

that the instrument was not functioning properly. At this point, some people exposed to the source were already being admitted to the Hospital for Tropical Diseases since the symptoms were not correlated to radiation exposure yet.

The Health Authority, who indicated that the occurrence was possibly related to an old caesium-137 source that was used for many years to treat cancer patients, contacted the owner of the source now working in a different place. The same day, a group from the CNEN was sent to Goiânia, followed by a larger number of workers in the following days.

In January 1987, I was back from a one-year post-doctoral position in Paris at the Bureau International Des Poids et Mesures (BIPM, International Bureau of Weights and Measures), sponsored by a Union for International Cancer Control (UICC) fellowship, where I had the privilege of working with Prof Andre Allisy. During that time, a consistent evaluation against the BIPM radiation dosimetry standards was carried out with two 1 cc graphite ionisation chambers, kindly provided by Klaus Dufschmit, with whom I became friends when we both served on the International Atomic Energy Agency's (IAEA) Secondary Standards Dosimetry Laboratory (SSDL) Committee. I must state how lucky I was to have met and been part of this committee, which has been made up of such prominent scientists like Harold Wyckoff, Bob Loevinger, and Andre Allisy. I learned a lot over those six years. Dr. Wyckoff, as a very kind and courteous person, once invited me to have lunch with him at the very prestigious Cosmos Club in Washington D.C., founded in 1878, as a private social club for men and women distinguished in science, literature, and the arts.

It was a normal day, and I was working tirelessly to finalise the facility design to implement the first Primary Standard for air kerma, in Brazil. Air kerma is a quantity used by medical physicists as part of the dose calibration chain related to radiation therapy machines and the dose delivered to patients. This involved the construction of an isolated room (with four walls, a glass ceiling, and a wooden floor placed over 20 cm thick Styrofoam), inside the existing bunker of a Picker cobalt-60 teletherapy unit. Through the creation of this isolated room, it was possible to reduce the temperature fluctuation to about 0.01 °C, as monitored by two thermal probes, one placed near the ionisation chamber and the other at an extreme position inside the room.

I was just about to leave when the Institute director, Mrs. Anamelia Mendonça, who replaced me when I left the BIPM, told me that an accident was reported with a caesium-137 source and a large area of the Goiânia city was contaminated. Initially, I was shocked.

My experience was mostly with radium-226 tubes and needles used with the Manchester rubber ovoid and tandem system used for gynaecological treatment from my time working in a cancer centre in Bahia, Brazil. I was aware of the possible leak of a radioactive radon gas (a radium-226 daughter product), from the source if they were bent, cracked or broken with a fissure; so, a check of each tube and needle was carried out periodically by placing them inside a test tube with cotton inside, and after several days, a measurement of the cotton was made with a Geiger counter.

When I was very young in the field, as a senior physics undergraduate student with very little formal training, I was a dedicated reader of the Johns' book, The Physics of Radiology, and the journals, Acta Radiologica and the American Journal of Roentgenology, known as the Yellow Journal because of the colour of its cover. My preceptor was a young radiation oncologist, Dr. Calmon, who gave me a tremendous incentive and books to read. My perception of the risks involved with the radium sources was vast and several protective measures were implemented in the hospital, such as the use and control of personal pen dosimeters and movable lead construction barriers to protect the nurses while they were cleaning and feeding patients during the 24 h of treatment with radioactive sources. Perhaps because of all of that, Dr. Hlasivec, an IAEA expert during an evaluation mission of the therapy centres in Brazil in late 1969, proposed my name to be granted a one-year fellowship abroad, as well as that of my boss for a 3-month period.

On July 1, 1970, I arrived at the MD Anderson Hospital in Houston Texas, with very poor knowledge of the English language. A wonderful and unforgettable person, Robert Shalek, the head of the Physics Department welcomed me and after 6 months he told me "you've come a long way" and that I should consider enrolling in the M.Sc. program. He promised to ask the IAEA for an additional year of extension of my fellowship. He did so, and the fellowship was extended. During my second year, I started working on my M.Sc. thesis with Peter Almond (see Chap. 3). The projected consisted of comparing electron beam data generated by a Siemens Betatron to that of a Sagittaire Linac. The betatron used scattering foils to flatten the beam, which caused some disadvantageous effect in the delivered dose distribution. The Sagittaire Linac used a beam scanning system to flatten the beam without affecting the energy of the beam. He was a most memorable mentor in several aspects. Not only did he show me how to do science, but he was an ethical person and very supportive during my academic life at the University of Texas. Under his guidance, I worked on my Ph.D. thesis, supported by the Brazilian Government, during which I used a Cerenkov detector to measure the energy and the spectrum of electron beams from different machines. Cerenkov radiation occurs when a particle passing through a material at a velocity greater than that at which light can travel through that material. Some people compare this to the sonic boom when an airplane is traveling through the air faster than sound waves can move through the air. As a result, we published several papers together and a kind and great relationship was established. In that period, I had the first contact with caesium-137 sources and the Fletcher after-loaded applicator for gynaecological treatments, but it never occurred to me to know the chemical composition of this type of source. A californium-152 neutron emitter source was also used at that time but discontinued later.

When Mrs. Anamelia Mendonça told me about the accident, I went back to the lab and after a search in some books (there was no Google at the time), I found that the chemical composition was caesium chloride, hence, dangerously soluble in water. I immediately told her: being water-soluble, in addition to the direct radiation exposure to people, the contamination could be even more serious, affecting not only the people nearby, but also the environment. She then asked me to go to Goiânia.

The next morning, I flew to Goiânia in the company of two doctors (Dr. Nelson Valverde and Dr. Alexandre Oliveira), expert professionals trained for the specific treatment of overexposed people, as well as two radiation protection colleagues.

Upon arrival we were immediately faced with multiple critical situations. These are summarized in point form below.

1. Patients with severe acute radiation syndrome were in need of immediate medical assistance. They were identified, externally decontaminated and were sent by an Air Force plane, which was fully prepared in order to avoid contamination of both the plane and staff, to the decontamination and medical facility at the Navy Hospital in Rio de Janeiro, accompanied by Dr. Valverde. Almost all of them died a few days later of massive internal radiation contamination, including the daughter and the wife of the waste picker.

2. The request by Dr. Oliveira at the Federal Hospital in Goiânia to have a ward area prepared immediately to provide medical assistance to the remaining patients was unfortunately delayed as the whole hospital staff was on full strike for different reasons. It was not an easy task to convince even part of the staff to come back, as they were afraid of being contaminated. Initial blood tests were almost impossible to carry out because again, the lab staff was afraid to handle the blood samples.

   I must say that this reaction was perfectly understandable, since the staff perception of the risks involved were not in their heads. The situation was indeed very new to them, so we had to work very hard to convince them. Slowly but steadily, we started to get some of the staff back the following day, and a large special care nursery ward was made available and isolated, with control points for access.

3. A large number of people were looking for the contamination check point located outside a soccer stadium, where at that moment, scintillation counters were being used (before a huge number of instruments were shipped from Rio de Janeiro). Figure 5 shows a picture of the long line of people waiting to be monitored during the first days after the accident was discovered.

4. The dressing and shower areas of the stadium were used to clean, with soap, people with surface contamination and to change their clothes. At that point, we decided to set a collection and control point at the exit of the stadium sewer system to monitor the potential level of contamination. The levels measured were not considered relevant at that moment.

5. By looking at the improvised managing area, I requested a city map, taping it on the wall and circling all areas identified as contaminated. In two days, we concluded that the major contaminated points were closely

Fig. 5 Long line of people in Goiânia waiting to be checked for radioactive contamination by caesium-137

located, with some exceptions, to the individuals that transported the small pieces of the source to their homes.

Figure 6 shows, on the left, the effect on the hand of a patient that took home a small piece of the source that was the size of a rice bean, and on the right, a gradually evolving lesion on the right leg of a patient that also took home a small piece in his pocket. The pictures were taken around the 30th day after the accident.

6. The major part of the source was still in a bag lying over a chair on the ground floor of the entrance of the Goiânia State Health Authority building in need of immediate shielding. Mr. Rozenthal and staff provided the shielding very diligently, as shown in Fig. 7.

7. The need to communicate with the public was put under heavy pressure by the government as well as the local, national, and international press. A telephone line was made available to the public, and every morning, either Mr. Rozenthal or I were on the news giving answers to all different

Fig. 6 The left figure shows the effect of radiation on the hand and the right shows a picture of a lesion on the right leg

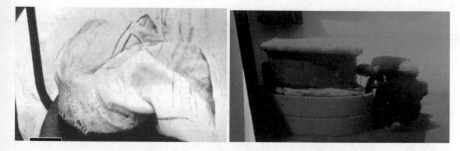

Fig. 7 On the left is a picture of the bag containing the source and, on the right the concrete container quickly built to house the remains of the source

types of questions, and at the end of the day, a press conference was carried out. We learned quickly that the best way to have the quality and contents of our messages preserved was through live messages on the radio and TV, since many times what was said and then written by the press for the daily paper was done in a rather bizarre way. This, of course, is understandable, since this event was extraordinarily unusual for most people.

8. A decision had to be made as to whether to evacuate the public living near the highly contaminated areas. With the help of the State Justice Department this was done in a very careful but legal way.

9. Our staff working in identifying the contaminated areas was not yet using the proper clothes; so, orange suits and black boots were quickly purchased and every day they were checked for contamination before they entered the central administration building. They were known in the city as the "yellow guys". Personal radiation monitoring film badges were provided to each staff member as well as a personal digital monitor with pre-set alarm to those working the contaminated areas.

10. The burial of the first victims a few days later was a special and delicate situation since the population strongly rejected the procedure, creating a local commotion. The dead had to be buried in a way as to prevent soil contamination, so it was not like a traditional burial. It had to be controlled, almost by police force, and by long and exhaustive conversations with the local leaders. Figure 8 shows the lead walled coffin used to bury the highly contaminated victims that had died with acute radiation syndrome.

11. The identification and isolation of the contaminated individuals had to be done quickly, together with checking and isolating their families, work colleagues, and friends with whom they had contact. All tests were done (blood, urine, faeces, and later by a body counter installed in Goiânia).

Fig. 8 A lead-walled coffin used to bury one of the highly contaminated victims

All these procedures followed the horizontal principle of finding the contaminated individuals, decontaminating, and promoting the social distance.

12. The identification of the bus and the driver that took a mother and son with the remains of the source to the local health station was an interesting one. Two days after our arrival we just had to stay at the bus stop and wait until the bus passed by. The bag with the source had been placed almost under the driver's seat. It was possible to identify the driver and take him to the hospital for evaluation and a gross dose estimate was done for his legs and gluteal muscles. Luckily, the bus route was only a few km and the doses were not a point of great concern at that moment.

13. The contamination of the circulating paper money in the local central bank was also an interesting investigation. Contamination of a huge number of notes was found, which were subsequently separated from the intact notes and considered as radioactive waste.

14. The CNEN headquarters and the Institute of Radiation Protection and Dosimetry (IRD) were informed to have the following ready and on alert: the labs, environmental bioassays, whole body counters, cytogenetics, and metrology. The last one to guarantee that all instruments were measuring the same value.

I was personally pleased that during the five years that I was the director of the IRD, we were able to install several labs to support the nuclear power plant program, and at that moment, they could be quickly activated to support the actions in Goiânia. It's interesting to note that, like a premonition perhaps, a few months before the accident, I had conducted a graduate course called Special Topics in Radiation Protection as part of the M.Sc. and Ph.D. academic programs, with more than 10 staff members from the IRD. A full discussion of the whole book edited by Karl Hubner and Shirley Fry called the *Medical Basis for Radiation Accident Preparedness* was made. This book describes a series of accidents by experienced professionals like C. Lusbaugh, Bob Ricks, H. Jammet, Eugene Saenger, E. Komarov, N. Rasmussen and others. In one of the chapters, the first case of child abuse by radiation exposure was described by V. Collins; the famous Houston case, where a father placed caesium-137 spheres under the testicles of his son to make him sterile. At the same time, the IRD had participated with a medical group to handle overexposed people.

15. To assess the severity of each patient, the clinical histories and description of their involvement were analysed and registered. Haematological counts (red and white blood cells and platelets) were performed, and a bit later faeces and urine were submitted to bioassay and cytogenetic studies to confirm the doses received. Several people received doses in a protracted fashion, but the severity of cutaneous (skin) reactions (erythema), the in vivo bioassays and the haematological profile, especially in terms of the lymphocytes and neutrophils, served as the initial basis to hospitalise the victims.

4 Brief Anecdotes

1. One day a physicist from Belgium came to the city of Goiânia and started to take measurements, passing information contradicting ours to the press. In the afternoon I received a wave of reporters questioning our measurements, saying that our data was wrong and the data from the Belgian scientist was right. I took a deep breath and said, "Please ask the gentleman to come and perform the measurement jointly with our instrument at the same point." The reporters still insisted, and with my patience wearing thin, I said, "Why do you believe in what he is saying and not in what I am saying? Is it because he is tall, blonde and has blue eyes?" After a moment of silence, they laughed and never touched on this point again. The gentleman disappeared soon after that.

2. Coincidentally, I can say that, with a metrological background, several of us made an effort to quickly set up a lab with some sources to carry out daily checks of our instruments, which was fundamental to ensure that we all were correctly measuring the exposure due to the contamination level.

3. Every day on the Morning News, the incident was being covered. I was sure that we had identified all the contaminated spots; however, some slightly contaminated people were still being identified. Unfortunately, the

newsroom is always looking for news, especially bad news, and so they began saying that a new spot had been detected. All I could think was 'Ok! Let's double-check, but I am almost certain that it is a false alarm or fake news", the expression used nowadays. And we were right almost 100% of the time!

4. One point that drew my concern was: at the house site where the source was opened, a contamination in the leaves of a mango tree was detected. I was intrigued by that, and called a biologist and asked, "How much time does it take for a tree to transfer soluble chemical compounds left in the ground to the leaves?" That was a difficult question! Later we found that the roof of the adjacent house was also contaminated, and soil evaporation could not be responsible for that. The best possible answer was that since a helicopter came very close to the scene for filming purposes, it most likely raised contaminated dust from the ground to the surrounding areas. Nothing could be done except to remove the tree and the adjacent house and treat it all as radioactive waste.

5. One day, the State Secretary of Transport, a great supporter of our work by providing staff, cars, trucks, and buses, called us and asked if we could gather the whole staff involved, about 100 people, and explain what radiation is, how it is detected, what protective measures to take, and so on. I went to a black board and for an hour I tried to cover the subject in the most understandable way possible. (That lecture happened just after the mother and daughter were buried in the local cemetery, under heavy protest by the population.) Several questions were asked but a middle-aged gentleman made the most unusual one. He asked me the following: "Professor, will there be a problem for the people buried next to the contaminated ones?" I took the chalk again and explained that they were buried inside a thick lead-walled coffin to avoid any soil or underground water contamination. He rose from his chair and told us loudly, "I am worried because my mother-in-law is buried right next to them." Under the circumstances I replied, "Sir, you are most likely the only person that is worried about the health of his mother-in-law even after her death." There was a general laugh in the audience. He rose again and said, "Professor, I am worried because she was always a very unlucky person."

6. The Call Centre received several calls every day with different requests and every single one was investigated; none was undervalued. They included complaints like: (a) My hair is falling out (she was using a different shampoo). (b) My fish are dying in the aquarium. (c) My skin is full of blisters (they were pimples). We tried to respond to every single call that could potentially be related to the accident.

That procedure increased the level of confidence of the population in the working team.

5   Final Remarks

The Goiânia accident was one of the most serious radiological accidents that had occurred up to that date. Radiation injury caused the death of four people and the contamination of some houses and small parts of the city soil. Radiological accidents are not very frequent, but this should not lead to complacency, and the public must feel confident that the authorities are capable of doing everything possible to avoid it or mitigate problems, should any arise. About 112,000 people were checked for contamination and out of this group, 129 people had been confirmed with different levels of internal contamination with doses smaller than 50 mSv. (50 mSv is the annual acceptable whole-body dose limit for radiation workers.) A thousand people were identified as having been exposed in a short time of doses greater than one year of natural background radiation. (Natural background radiation is about 2–3 mSv per year with significant populations receiving 10–20 mSv per year depending on their location.)

Carlos E. de Almeida   is Full Professor in Medical Physics at the State University of Rio de Janeiro (UERJ), former Director of the Instituto de Radioproteção e Dosimetria (IRD) (1980–1985), and a Senior Researcher IRD (1987–1991) and the National Cancer Institute (1991–1999). He obtained his M.Sc. and Ph.D. at the MD Anderson Hospital—University of Texas. He spent one year at the BIPM (1986), one year at the Institute Gustave Roussy (2007), and at the Institute Curie (2010). He has acted as a professional expert for the International Atomic Energy Agency (IAEA). He served on the IAEA-SSDL Scientific Committee and was strongly involved in the National Laboratory for Ionizing Radiation. He served as chairman on various committees of the International Organization of Medical Physics (IOMP). He was Chairman of the National Quality Assurance Program (PQRT) and Chairman of the National Training Program in Radiation Oncology for physicists, radiation oncologists, nurses and radiation technologist involving more than 100 teachers and over 1000 students. He was the mentor and Chairman of the first Accreditation Board in Medical Physics of the Brazilian Society in Medical Physics. He supervised 68 M.Sc. and 28 Ph.D. students. He has published over 200 peer reviewed journal articles, books and chapters.

Awards

2020. International Day of Medical Physics (IDMP) Award for promoting medical physics to a larger audience and highlighting the contributions of medical physicists make for patient care. Awarded by the International Organization for Medical Physics.

2014. The Gold Medal awarded by the Brazilian Society of Radiation Oncology. The highest medal awarded in recognition for his work on the development of radiation oncology in Brazil.

2013. Selected by the International Organization for Medical Physics as one of the 50 medical physicists "who made an outstanding contribution to the advancement of medical physics over the last 50 years". This recognition was given as part of IOMP's 50th anniversary.

2013. Fellow of the International Organization for Medical Physics. This recognition was given as part of IOMP's 50th anniversary.

2010. Henry Becquerel Medal awarded by the French Academy of Sciences and Letters of Languedoc in recognition of his contribution to Medical Physics.

15

True Tales from India

Arun Chougule

1 Little Known Profession of Medical Physicist

It was October 1982. I had completed my master's degree in physics (material science) and started a job as a junior lecturer in a science college near my native place of Miraj, India (Fig. 1). My roommate during my master's program (he had done a master's in chemistry) asked me whether I would be interested to work in a cobalt plant at Mission Hospital, Miraj, India as they were in need of a medical physicist. Until then, I did not know about the medical physics profession or what he meant by cobalt plant. However, my curiosity made me visit the hospital and the superintendent of the hospital introduced me to a radiologist. In those days the combined degree/diploma holders of radiology used to practice radiology and radiotherapy. During the discussions with him, the radiologist realised that I am neither qualified nor do I have any experience in the field and he found that I was not suitable for that medical physicist's job. At the time, I was not disappointed, but it did generate an interest in the profession, which was quite unknown, even in the university physics departments. At that time only the Bhabha

A. Chougule (✉)
Sawai Man Singh (SMS) Medical College and Hospital, Jaipur, India
e-mail: arunchougule11@gmail.com

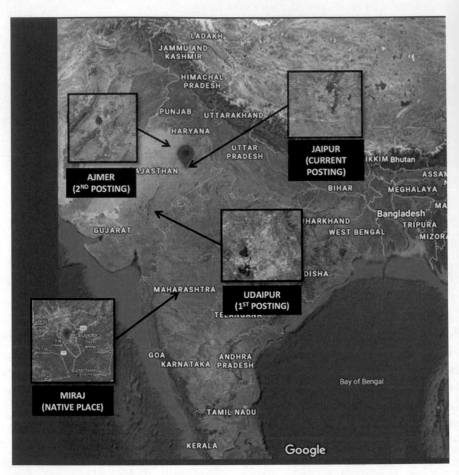

Fig. 1 Map of India showing my native/birthplace and places of posting from 1984 to 2021

Atomic Research Centre (BARC) in Mumbai was offering a one-year post-graduate diploma in Radiological and Hospital Physics (Dip.R.P.) with an intake capacity of 15 candidates per year (50% of the students were sponsored by their employers). In those days, the internet was not available in India (Internet services were launched in India on 15th August 1995 by Videsh Sanchar Nigam Limited) and the only way of communication was by letter writing through Indian postal services (no courier services were available at that time). In March 1983, while in the science college library, I came across an advertisement in "*Employment News*" magazine regarding the invitation of applications for the new batch of Dip.R.P. course starting in September 1983. I decided to apply; however, I was not sure of getting selected due to limited placements. I was called for an interview at BARC in the last week of August 1983. It was the first time that I saw the BARC institute and was so fascinated with the campus, the number of working scientists, and the facilities that I felt that I would be the luckiest person to get an opportunity to study and work here. I returned home and continued my junior lecturer job as I was not that confident that I would be selected.

2   Life Changing Decision

On 12 September 1983, I received a telegram from BARC indicating that I am selected for the course and that I need to join by 15 September 1983. I was happy but at the same time worried as I would have to resign the job and join the training program with a very small stipend. My parents were against my leaving my job to go for training. They were in the process of choosing a bride for me and get me married. (It is the tradition in my region and religion that parents choose the life partner for their children.) However, I decided to resign and join the training programme. It was a hard decision for me; however, when I reflect on it now, I feel I took a wise decision.

During the training, we were taught the subjects of radiation, dosimetry, radiobiology, anatomy and physiology, radiation protection, etc., by very experienced and eminent scientists from BARC. The clinical practical training was at the renowned Tata Memorial Hospital, which is the oldest and one of the major cancer treatment hospitals in Mumbai, India, founded in February 1941 by the Tata Trust. I had three months of field training at the Kidwai Institute of Oncology, Bangalore, India. While I had taken the decision of joining the course after resigning my job as lecturer, I continued to have uncertainty about future job prospects. The course coordinator, Dr. P. S. Iyer, who was a very senior scientist, always used to inspire us and remind us (the non-sponsored students) that only the outstanding top students in the course will get a good job. I was very dedicated and always tried to be in merit. My hard work paid off and I obtained the second top rank in the group.

Lesson to be learned: Life is not about waiting for the storm to pass but learning to dance in the rain.

3   Starting My Career as a Medical Physicist

In December 1983, the Atomic Energy Regulatory Board (AERB) was established as an autonomous competent authority to implement the Atomic

Energy Act of India. Ionising radiation such as x-ray, gamma-ray, or other radiations from radioisotopes could be hazardous to human health if not used appropriately. Therefore, its use for human welfare needs to be regulated. Dr. K. S. Parthasarathy was the first Secretary of the AERB and was looking for bright young medical physicists for appointments in the AERB. He was my chief, practical examiner for the final examination of Dip.R.P. in September 1984. He was impressed by my performance and knowledge and offered me a job at the AERB. However, I was more interested in academics and working as a clinical medical physicist in a hospital; therefore, I preferred to join an academic medical institute. Being the bright student of my batch, I was offered jobs in industry, hospitals, and medical colleges. I joined the RNT Medical College and Hospitals as assistant professor in Udaipur, Rajasthan on 17 December 1984 in the department of Radiology. My career started as a Medical Physicist and a Radiation Safety Officer. Dr. Gopiram Agarwal, Head of the Department of Radiology, was my first boss and mentor. In India, until the late 1980s, many medical colleges and institutes had Radiology Departments that have both diagnostic as well as radiation therapy activities. For the treatment of cancer patients by radiotherapy (cancer treatment is done by surgery, radiotherapy and chemotherapy depending on the type and stage of cancer), a Toshiba RCR-5 cobalt-60 teletherapy machine with a cobalt-60 source from Amersham, UK was installed in September 1984. (A teletherapy machine is a machine that treats with external beam radiation therapy. A telecobalt refers to an external beam treatment machine using gamma radiation from a cobalt-60 source.) It was the third cobalt-60 machine in the state of Rajasthan for a population of 55 million at that time. Cobalt-60 is an artificially produced radioisotope by irradiating cobalt-59 in a nuclear reactor. Cobalt-60 has half-life of 5.27 years and emits beta and gamma rays. This means that the activity is reduced by a factor of two in 5.27 years. The gamma rays of cobalt-60 are used to treat cancer patients.

The dose rate to treat patients is determined in gray per minute (Gy/min) where gray is the unit of radiation absorbed dose. When the activity goes down by a factor of 2 after 5.27 years, the dose rate to treat patients reduces to 50% of the original value. This has an impact on patient treatment time and the patient throughput on the treatment machine.

The cobalt-60 teletherapy machine installations at the Radiology Department of RNT Medical College, Udaipur has a remarkably interesting story. Being a medical college and public hospital, there was a strong interest in developing a cancer therapy facility in the region since there was none in existence, even in the private sector, at that time. Philanthropists, along with surgeons and radiologists formed the Udaipur Cancer Society in 1982 and started collecting donations to establish the cobalt therapy facility to cater to the need of poor cancer patients of the region. From the funds collected, a cobalt-60 bunker was built with the approval of BARC. (Before the establishment of AERB, BARC was the competent authority for radiation protection.) A Toshiba telecobalt machine was purchased through an M/S Amazon Consultant from Delhi in early 1984. A cobalt-60 source was imported and loaded in the machine in September 1984. The bunker is a heavily shielded treatment room to protect staff and patients from unnecessary exposure and consists of walls that are about 1.3–1.5 m thick (roughly 4–5 feet) made of high-density concrete. Two unfortunate things happened, resulting in a great problem for the project. The Jost Engineering Company from Mumbai, who were contracted to supply the cobalt-60 source from Amersham, UK, did not obtain the mandatory import license from BARC. The source container including the source landed at the Mumbai dockyard in May 1984 and was seized since there was no import license from a competent authority. There was a two-fold loss: (1) demurrage charges by port authorities, and (2) the source decay without any clinical usage. However, with the pursuance and undertaking from the company, the source container was released, and the source was allowed to be loaded in the treatment machine with the condition that it would not be used for treatment until the type approval tests (being the first Toshiba telecobalt machine in India), the acceptance tests, and the radiation protection survey were done by BARC (Fig. 2). The institute needed a medical physicist and a radiation safety officer desperately. That is how I landed in Udaipur in December 1984.

Another bad thing happened. The M/S Amazon Consultant committed fraud and cheated the authorities of SMS Medical College, Jaipur, Rajasthan, which is a state Government Medical College (where presently I am heading the Department of Radiological Physics), in supplying the Toshiba telecobalt machine (the second order in India) and was blacklisted and put into jail. As a result, this Toshiba telecobalt machine was never installed and remained nonfunctional, wasting public funding as well as leading to suffering for cancer patients not receiving the required treatment. As of today, the machine is lying dumped in the open (Fig. 3).

The situation made it exceedingly difficult for us to run the Toshiba teletherapy unit at RNT Medical College, Udaipur without any authorised service engineer and no spare parts. With this background on a difficult situation, I started my career as a Medical Physicist and Radiation Safety Officer in December 1984. I was only recently trained. The staff radiologist did not have any experience working with cobalt-60 teletherapy nor a linear accelerator for the treatment of cancer patients. He had a little experience treating

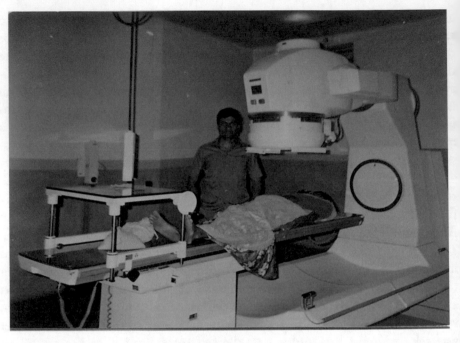

Fig. 2 The Toshiba telecobalt unit at Udaipur installed in 1984 and used for treatment until 1995

with deep x-ray therapy. Having some knowledge about the regulations about the safe use of ionising radiation in healthcare, we started fulfilling the mandatory requirements using a secondary standard dosimeter for calibrating the cobalt machine, survey meters for radiation protection measurements, and personnel monitoring with film badges for radiation workers. Ionising radiation is a dual edged sword: on the one hand, it can cure cancer; on the other hand, its inappropriate use can be hazardous. The radiation dose to the malignant tumour needs to be delivered with an accuracy of ± 5%; therefore, accurate measurement of radiation output at the normal treatment distance is necessary, requiring the use of various dosimeters. We, along with BARC experts, conducted the type approval tests, acceptance tests, radiation protection survey, and submitted the reports with request for permission to commission the telecobalt unit for patient treatment. With my communications and assurances, we got permission in the last week of April 1985 to commission the Toshiba telecobalt unit for cancer patients—a long-awaited dream. Being in the Chief Minister's region, the authorities wanted to inaugurate the facility by the hands of Chief Minister of Rajasthan. The date of 13 May 1985 was fixed for the inauguration. All preparations were done, and a rehearsal was completed on 12 May. We were ready for the inauguration on

Fig. 3 Photo taken on 20 April 2021 of the remnants (on right) of the Toshiba telecobalt machine at Jaipur, lying unused since 1984

13 May 1985. Unfortunately, the Chief Minister's mother expired early in the morning of 13 May 1985 and the inauguration function was cancelled. However, the inauguration could finally happen on 28 May 1985 and the long awaited teletherapy services could begin.

Lesson learned: The struggle you are in today is developing the strength you need for tomorrow.

4   Where There is a Will, There is a Way

I had a hard job, being the only medical physicist. Initially we were treating over 80 patients per day with radiographers who were not trained in radiotherapy but only had experience in diagnostic radiography. Slowly the patient numbers started increasing. To replace the old cobalt machine at Jaipur, a new Toshiba telecobalt machine was purchased but could not be installed due to the fraud and arrest of the Indian supplier as mentioned earlier. Due to the non-availability of a radiotherapy facility at Jaipur, only two cobalt machines were functional in the Rajasthan state for radiotherapy, one at Udaipur and another at Bikaner. The number of patients for radiotherapy, increased to over 200 per day on a single cobalt machine. By comparison, developed countries on average treat 20–40 patients per day on a teletherapy machine in an 8–10 h day; thus, the number of patients treated per day by us on a machine is unimaginable for them.

Treatments used to start at 6 a.m. and continued till 2 a.m. the next day morning, 20 h per day, representing a huge overload on the machine. As there was no authorised service engineer to maintain the machine, we received help from local industry and engineers. With support of philanthropists, the Toshiba telecobalt machine provided clinical service for over 11 years, without any major breakdowns despite the heavy workload, the non-availability of spare parts, and no authorised service agency. However, the cobalt-60 source decayed and the output at the patient became less than 50 cGy/min, the lower limit set by the AERB and, hence, we needed to stop treatments and replace the decayed source. Below 50 cGy/min, the time required to treat a patient will be about 8–10 min. As a result of the length of time, the accuracy of the treatment goes down due to the movement of the patient or the internal organs; hence, the regulatory authority has restricted the use of cobalt-60 therapy when the output is below 50 cGy/min.

5   Nightmare for the Cobalt-60 Source Change

The telecobalt machine in Udaipur was the only Toshiba machine functioning in India and the source container for transporting the new source and returning the decayed source was not available in India. The source transport container is a specially designed shielding container for radioactive sources and must be approved by the International Atomic Energy Agency (IAEA) for radiation safety. The first cobalt-60 source in India was imported from Amersham, UK and as per the AERB regulations needed to be sent back to the country of origin for its safe disposal. On contacting Amersham, UK for the source transport and disposal, they quoted an exorbitant cost, almost four times the cost of a new cobalt-60 teletherapy machine, perhaps due to many logistic issues.

Another stumbling block was related to the government rules for decommissioning and disposal of used equipment and their accessories. The rules mandated that the dismantling and disposal of old equipment should yield a minimum of 10% of the original purchase cost. Since the disposal of decayed radioactive sources was happening for the first time in the state, there was no provision to pay for the source disposal. We failed to convince the authorities of the need to dispose of the decayed source and that this would cost money. The issue did not get resolved. The treatments were stopped on this unit. A decision was made to purchase a new telecobalt unit; however, the existing telecobalt bunker was not vacated due to the non-decommissioning of the machine and the lack of source disposal. While it was decided to construct a new bunker, the problem was that there was not sufficient free space near the radiotherapy department to construct a bunker. I took up the challenge and showed that medical physicists are problem solvers. The shielding walls were expanded outside of the facility rather than inside the bunker, and room for a new cobalt machine was created within extremely limited space. Figure 4 shows the modified plan where the shielding is bulging outside the machine room as compared to standard plan (Fig. 5). This has helped in two ways: (1) it created a sufficient space inside the room, and (2) space is left between the existing building wall and the bunker so that the natural light and ventilation is not obstructed.

Lesson learned: If you are not willing to learn, no one can help you, but if you are determined to learn, no one can stop you.

6   Problems Do Not Come Alone

With funds received from the Government of India, it was decided to buy a new telecobalt unit. An order was placed to M/S UB Electronics, Hyderabad, in September 1994 to supply the ATC C/9 telecobalt unit, manufactured in the USA. The purchase order indicated that this purchase was exempt from the usual local municipal tax. In January 1995, the truck with the machine started from the Mumbai port on its way to Udaipur. When it reached the Udaipur city entry point, the municipal tax authorities stopped the truck and refused its entry into Udaipur unless the local municipal tax was paid. After waiting for four days at the entry point to the city, the transport company offloaded the machine boxes onto the side of the road and went back to Mumbai. The machine boxes were seized by the municipal tax authorities. The RNT Medical College authorities tried to convince the municipal authorities to release the machine, but to no avail. Then, I as the radiation safety officer intervened with a clever ploy. I told the authorities of the Municipal Corporation that the cobalt teletherapy machine head contains depleted uranium (since it is an excellent material for shielding

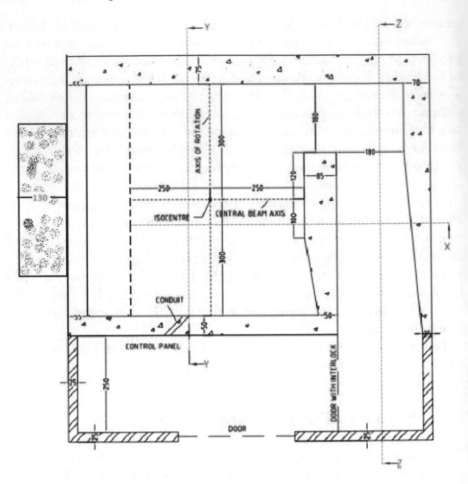

Fig. 4 Modified bunker plan. The bulging of the shielding on the left side is outside the bunker creating shielding, sufficient space inside the bunker, and saving natural light and ventilation to adjacent existing building

purposes because of its high density and its pyrophoric properties), which is a hazardous radioactive material and cannot be left on the roadside and that it needs to be secured in a cobalt machine bunker. Thus, it finally got to the RNT Medical College cobalt bunker site. The Medical College account authorities deducted the local municipal tax from the payment due to M/S UB electronics, and that amount was paid to the municipal tax authorities. Against this action the machine supplier went to court for justice. The case dragged on for 17 years and, finally, the Supreme Court of India ordered the RNT Medical College to pay the wrongfully deducted amount with 18% interest to the machine supplier. It was a difficult time for me as a

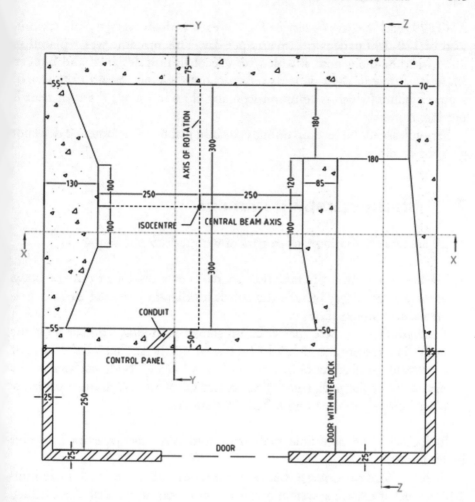

Fig. 5 Original bunker plan with bulging of shielding inside the room

medical physicist to keep up with the treatments and the machine mainte-nance. I sustained good cordial relations with the company and earned lot of appreciation from administration.

The initial cobalt-60 source loaded in 1995 needed replacement in 2004. Despite the ongoing court case, M/S UB Electronics cooperated and under-took the source replacement in December 2004. As the initial source was supplied by the Board of Radiation and Isotope Technology (BRIT), BARC, and the Government of India, disposal of the decayed source was done by BARC with very minimal charges. By this time, the amendment in the finance rules (instigated by the first incident of the cobalt source disposal problem in 1995) mandated payment for the decayed source disposal. The

ATC C/9 machine gave us service for 19 years, with an average patient workload of 120–140 patient treatments per day. This machine was replaced i
2014 with the indigenous Bhabhatron telecobalt unit designed and built ir
Mumbai. It has an eight-hour battery backup; therefore, even in the case of
a power failure (which is common in many cities in India), the treatment is
not interrupted.

Lesson learned: Never stop doing your best just because someone does not
give you credit.

7   Brachytherapy Treatment

There are two major modes of treatment with radiation therapy:

1. Teletherapy—where the radiation source is at a distance from the body.
   Sample technologies include the cobalt teletherapy machine, linear accelerator, and proton therapy.
2. Brachytherapy—where the radiation source is in close contact with the
   body. The radiation source can be placed on the surface of the tumour
   or patient (surface mould), or it can be placed in body cavities in the
   proximity of the malignancy (intracavitary), or the radiation sources are
   placed directly in the tumour bed (interstitial).

In localised and accessible malignant tumours, brachytherapy can give
better results.

In RNT Medical College, before I joined in 1984, they had a caesium-
137 manual afterloading system but it was not used. In manual afterloading,
sources are preloaded inside of hollow tubes prior to inserting the tubes in
the patient. After I joined, I put efforts into its use for patient treatment;
however, the source transport system to move sources from the hot lab to the
operating suite and the movable lead barriers for shielding the staff were not
available. The source transport system minimizes radiation exposures to the
staff. I prepared a sketch with the required lead thickness and space to hold
the caesium-137 sources and got the lead transport container constructed by
a local machine shop (Fig. 6). Figure 7 shows autoradiographs of the caesium-
137 sources. These are used as quality control of the activity distributions of
these sources.

Furthermore, the lead movable barrier available in the department to
protect staff had a thickness of 1.5 mm of lead-equivalent used in x-ray rooms
and was not of use in brachytherapy with caesium-137 since it has a gamma

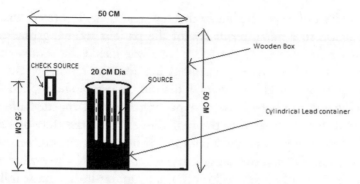

Fig. 6   Sketch diagram of lead transport container in wooden box

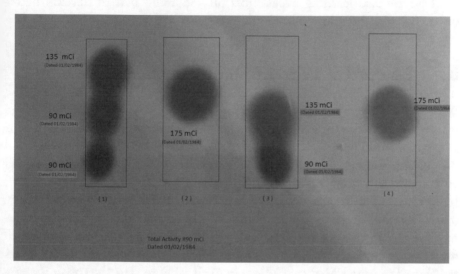

Fig. 7   Auto-radiographs of caesium-137 sources with a total activity of 890 mCi on 1 February 1984. An autoradiograph is an x-ray film that is exposed by placing the radioactive sources near, or on, the film

ray energy of 667 keV whereas diagnostic x-rays only have an energy range of 20–150 keV. I designed and prepared a lead barrier of 27 mm lead thickness. It took about four months to procure all these necessary radiation protection accessories. We then started manual afterloading brachytherapy, initially for gynaecological cancers (cancer of the cervix, vaginal cancer) and then for other malignancies as we gained experience. For radiation protection reasons, patients with caesium-137 sources in applicators in the body are isolated in special brachytherapy rooms until the applicators with the radioactive sources are taken out and placed in storage. No patient relative or attendant is allowed in the room while the brachytherapy treatment is in progress. Within a few days of the start of a specific brachytherapy treatment, I noticed, that despite the instruction to a daughter-in-law of the patient receiving brachytherapy treatment for her cancer of the cervix, she was inside the room taking care of her mother-in-law. When I scolded her for not following the instructions and putting herself at risk of radiation exposure, she replied "the risk of annoying the mother-in-law and putting her own married life at stake is a bigger danger compared to the radiation risk". In the state of Rajasthan, where the female literacy rate was below 20% and the social custom of Pardah or purdah (from Persian meaning "curtain", a burqa) is a prevalent religious and social practice of female seclusion, the daughter-in-law must follow each and every instruction of the mother-in-law.

These are also learning experiences for me and I had to develop better systems so that radiation safety is not compromised.

8   Reward of Working Sincerely and Positively

With my positive attitude and working with sincerity as a problem solver, involved in activities of non-government organizations in support of patient and patient attendants, I became extremely popular among the staff and the public. Though I was an Assistant Professor/Medical Physicist, I had an exceptionally good reputation with administration and local politicians. In 1993, a cobalt therapy machine was installed at Ajmer Medical College (another city in the state of Rajasthan—280 km from Udaipur). However, no medical physicists were available and, hence, AERB did not give permission for its use for patient treatments. Due to political pressure from Ajmer, I was transferred to Ajmer Medical College in April 1994 to start the cobalt therapy facility there. Since I was the only medical physicist at Udaipur, and upon my transfer to Ajmer, the AERB issued a notice to RNT Medical College to stop radiotherapy due to the absence of a qualified medical physicist. This news flared up like wildfire in the media and a politician from Udaipur demanded my transfer back to Udaipur. This helped to bring the role of the importance and the necessity of qualified medical physicists in radiotherapy to the limelight. The Rajasthan government, under pressure from both Ajmer and Udaipur, finally, posted me for 15 days a month, alternatively, in Ajmer and Udaipur till a new physicist was appointed. It took about six months to get through the bureaucracy for the appointment of a medical physicist at Ajmer and my transfer back to Udaipur as a fulltime medical physicist. This incident gave me, and medical physicists in general, a much-deserved prominence and recognition. Because of my positive problem-solving attitude, I was elected as Secretary of the Medical Teachers Association and assigned many administrative jobs in the institute.

Lesson learned: A simple rule for a happy and successful life is to never try to defeat anyone, but just try to win everyone.

9   Radiation Awareness Efforts

I was the first one, having some knowledge and training about radiation hazards and radiation safety, appointed in the Medical College, and the attached hospital. In the second month of my joining, I happened to visit the hospital trauma and emergency department and, to my surprise, I noticed that the conventional fluoroscopy x-ray machine was being used (rather misused) by interns and final-year graduate medical students just for fun, and even by nursing staff for unnecessary screening. They had no knowledge about radiation hazards. Immediately, I spoke with the superintendent of the hospital, and explained to him about radiation hazards of fluoroscopic screening in emergency departments by untrained and unauthorized staff, as well as the regulatory requirements with appropriate documents. He was convinced, and the fluoroscopy unit was moved to the radiology department in the main hospital.

This incident motivated me. I aggressively made efforts to teach and make the graduate and post-graduate medical students, the nursing students and staff, and the doctors and faculty from all non-radiation professional departments aware of radiation safety. Part of that included preparing posters and reference protocols. Even though it was a huge task for me, it had very good impact. Many radiographers and radiology post-graduates, who never used their personnel monitoring badges (dosimeters to record the radiation doses received while working in a radiation environment) received the monitors and started using them. Available radiation protection accessories like lead aprons, lead goggles, and lead barriers started to be used appropriately. Even private clinics with x-ray facilities started seeking my advice. This proactive approach not only created the radiation safety awareness but also helped publicize the usefulness of qualified medical physicists. The media attention by these efforts lead to my writing articles and briefs for newspapers about radiation applications in healthcare, the associated risks, and corresponding radiation protection considerations.

In 1990, a BBC report (by an antinuclear lobby group) regarding the Rajasthan Atomic Power Plant (RAPP) at Rawatbhata, about 200 km from Udaipur, indicated that radiation had caused damage to people in nearby villages, creating a panic and putting pressure on the government. I was included in a team of experts to investigate the issue. I was the only radiation expert in the team, others being doctors and social scientists. The team spent a few months in the area, did some measurements, and health checkups of people in the villages within 20 km around RAPP. The team found neither an increase in radiation levels as reported by the media, nor any health issues in comparison to other parts of the state, relating to radiation. This participation in the team encouraged me to plan a study about radiation exposure levels around RAPP and other parts of Rajasthan away from RAPP. The project was prepared, submitted to a government funding agency, and funded for three years. The research funding allowed me to appoint helping hands with a senior research fellow (SRF) and a junior research fellow (JRF). Another project was submitted and funded by AERB to study the radiation protection status of x-ray installations in Rajasthan.

In Warren Buffet's words "An idiot with a plan can beat a genius without any plan."

10   Struggle for Getting Ph.D. Guide and Enrollment

I joined an academic institution since I was interested in research and, therefore, I needed to earn a Ph.D. degree as a necessary requirement to be promoted to Associate Professor, and to be a successful academic. The stumbling block was to find a Ph.D. supervisor in medical physics since very few Ph.D. supervisors with experience in medical physics were available in those days. Furthermore, every university has its requirements for registration as a Ph.D. student. Some needed a minimum of three-years continuous presence with a supervisor there. While on the job, it was difficult to do this without any study leave, since this was not permitted in my institution due to no other medical physicists being available to take care of my duties. I was running from pillar-to-post to find a supervisor with a research question having relevance to clinical radiotherapy. After about three years of desperate efforts, I finally got a supervisor from BARC in Mumbai with the condition that the clinical research work will take place at RNT Medical College and that I spend at least four months every year with him at BARC. Finally, it involved seven years of research work, with great hurdles of shuttling between BARC, Mumbai and Udaipur, along with financial hardship and sacrifices of time, both on the job and the family front. However, the struggle paid off in the long term and the Ph.D. entitled me for promotion to Associate Professor, bypassing four other senior colleagues, and finally reaching a Senior Professor position. Hard work and sacrifices never go waste!

I would like to quote as the lesson learned: A blind man asked St. Anthony "can there be any thing worse than losing eyesight?" St. Anthony replied "Yes, losing your vision".

11 Golden Period of My Career

After putting in over 16 years of dedicated service at RNT Medical College, Udaipur, I was promoted to Associate Professor and transferred to SMS Medical College and Hospitals, in Jaipur in March 2000. The SMS Medical College and Hospitals is the oldest medical college and hospitals in the state and northern India (the hospital started in 1934 as Sawai Man Singh-King of Jaipur state and the medical college opened in 1947). It presently has about 5000 beds; however, at any given time over 7000 patients could be admitted, almost 140% capacity. Some patients are on the floors in the corridors. The yearly outdoor patient foot fall in 2019 (before the COVID-19 pandemic) was 6.4 million patients. This is a huge patient number and is completely unimaginable in advanced countries. Despite the extensive experience, it was very tough for me to adjust with such a huge workload with only one assistant medical physicist to support me. The two of us were managing daily with 8–10 brachytherapy procedures and treatments of about 350 cancer patients by teletherapy on three teletherapy units (two cobalt-60 machines and one linear accelerator, operating 15–16 h over two shifts). We had additional duties of teaching and training of radiotherapy and radio-diagnostic postgraduate students and radiography technology students. It was very difficult to convince the government bureaucrats to create additional posts of medical physicists. I had learned that it is your contacts and influence in power which work to get things done in government circles in India. Immediately after joining at Jaipur, I got involved with the Jaipur Cancer Relief Society (JCRS), a non-government organization (NGO) providing services in radiotherapy and helping patients. Most of the members were retired government officials and influential people of Jaipur. My experience of working with an NGO at Udaipur helped me join the JCRS and eventually become one of the executive members. Furthermore, I organized a National Conference of Medical Physicists in 2002 and invited the Health Minister and Health Secretary for the inaugural function. This brought medical physicists into the limelight and promoted the needs of expanding services of medical physicist, additional manpower, and resources. I understood the true meaning of "*Wood in a Wilderness*".

With the help of influential people in government circles, I went on requesting and getting additional posts of medical physicists. In 2010, I was promoted to full Professor in Medical Physics; however, there was no independent department of Medical Physics; we were working under the department of Radiotherapy. Further efforts worked on starting an independent medical physics department. Despite stiff resistance, especially from Radiotherapy and Radiology, an independent department of Medical Physics was created in 2013, a landmark achievement for the state and the profession.

Due to my involvement in various activities at the institute, I became an indispensable staff member for the institute administration and for the government. This has created more opportunities, additional responsibilities, and visibility for medical physics services. I was assigned the responsibility as a nodal officer of telemedicine services, member of the library committee, the ethics committee, the research review board, the purchase committee and so on. I learned how to turn the tides, and hammering when the iron is hot, to achieve the desired shape.

In 2006, a separate university, Rajasthan University of Health Sciences (RUHS) was established for health sciences. I was assigned many responsibilities, appointed member of the board of management, nodal officer for entrance examinations which created more opportunities to develop and get funds for the Department of Medical Physics. A separate building for Medical Physics was constructed with all the facilities like offices for all faculty members of the department, a lecture hall, demonstration rooms with dosimetry and quality assurance equipment, and a departmental library. It was appreciated by medical physicists from the country and abroad, and an attraction for young medical physicists wanting to join. Starting from a humble beginning in a not well-known profession, I rose to Pro-Vice Chancellor (Pro-Rector/Vice Rector in the Western world) of RUHS, a great and unimaginable achievement for a medical physicist in the given circumstances.

Simultaneously, I continued to regularly organize scientific meetings, workshops, and conferences at Jaipur, the Pink City of India, and a tourist attraction that people love to visit. Association with various NGOs working in cancer treatment gave opportunities to participate in cancer awareness programs, creating new facilities, writing two books in Hindi for the public, and the publication of a quarterly magazine.

12   Strong Dream to Travel Abroad

I came from a poor family background, the fifth child of a farmer father struggling to make ends meet. However, my father was a firm believer of educating his offspring despite the many hurdles and discouragement from close relatives. He admitted me in the first level of a government school when I turned seven years old. I had to walk to school over 6 km one-way in bare feet from our hut on the farm to the school. However, I was good at studies and obtained the third top position in the fourth-class examination in the entire district. This earned me scholarship, and, from that, I purchased, for the first time, sandals to wear on my feet. Our teachers were very devoted to their jobs and they used to pick up good students from the class and tutor them extra time without any additional remuneration. Teachers had all the freedoms to take the students to task or punish them without any interference from parents. When I passed my senior secondary board examinations (Xth class) with flying colors, my maternal uncle gifted me a secondhand bicycle. I started going to senior secondary school by bicycle, 8 km away from my home, and after school hours, I worked on the farm to help my parents.

While in a XIIth class annual function, a scientist from the National Chemical Laboratory (NCL), Pune, was invited as the chief guest. In his speech, he described the importance of science and his experience of a visit to France, a better life, and advances in science. I was impressed and started dreaming of getting a chance to visit abroad, which was rare for a lower/middle class family in those days. While at my job at Udaipur, a tourist place visited by many foreigners every year, I used to interact, discuss, and try to find out the possibilities of foreign travel—the cherished dream. The dream was fulfilled when I travelled abroad for the first time to Milan, Italy, on an International Cancer Technology Transfer (ICRETT), fellowship from the Union for International Cancer Control (UICC) at San Rafael Hospital in February 2001. For me, a vegetarian, with limited warm clothes in a severe winter was difficult, but the joy of the first foreign visit was much to cherish. At San Rafael, I was able to observe the planning and the treatment on a Gamma Knife, the use of a PET scanner, a cyclotron and a linear accelerator and I learned many things. GOD is great! While I dreamed of at least one foreign visit, he gave me so many after Milan, Italy, that I am the one in my family and my institute having many foreign visits. My basket is full, and I am grateful to my parents, teachers, mentors, well-wishers, and also my critics. As the great, ancient Indian philosopher, Chanakya, said, that for your success, critics are also as important as well-wishers.

13   Difficult Task to Convince the Administrators

In 2006, a decision was made to install a linear accelerator in the institute. Until that time, no linear accelerator was available anywhere in Rajasthan. Because of the growth of the hospital campus, there was no horizontal space to expand the radiotherapy department for a linear accelerator bunker. In a planning meeting for the linear accelerator bunker, the medical superintendent asked us to plan the bunker on the third floor of the existing radiotherapy ward. I pointed out that it is not possible to build a linear accelerator bunker on the third floor. He and his advisers kidded and mocked me. When flyovers can be constructed over bridges, swimming pools can be constructed on 20th floors, why can linear accelerator bunkers not be constructed on the third floor of a radiotherapy ward? He requested the preparation of a plan and the corresponding requirements to be submitted in 15 days without ifs ands or buts. There was no further discussion. I was disappointed by the attitude of the administration, but I had to follow orders. I convened a meeting with engineers of the public works department (PWD), the government department for the planning and execution of government buildings. I gave them the model layout of the linear accelerator bunker, along with the thicknesses of the required primary walls. None of the civil engineers of the PWD had any experience with planning or constructing a linear accelerator bunker. They were amazed to know that the wall thickness of the primary walls needed to be about 2.5 m thick of high-density concrete. I planned a meeting of the PWD engineers with the Varian Medical System linear accelerator project manager, along with the company's architect, who was the expert in linear accelerator bunker planning and implementation. During that meeting, the Varian architect questioned my wisdom of suggesting the construction of a linear accelerator bunker on the third floor, even though it was not my suggestion but that of the hospital administrator. During the meeting with hospital administrator, the Varian architect categorically told him and the advisers that it is not possible to have a linear accelerator bunker above the third-floor radiotherapy ward and asked for a space on the ground floor, which was not available. In a typical bureaucratic manner, the hospital administrator blamed us for being rigid and not being interested in having the linear accelerator. Of course, we were hurt. In search of a suitable place on campus, we found underground and ground floor parking areas under the neurosciences centre. I performed measurements of the space between the pillars and the height from the underground base to the first floor. The distance between pillars was 24 feet and the height from the base to the first floor was 21 feet. I spent lot of time trying to fit the linear accelerator bunker plan within the available space without damaging or shifting the pillars. Finally, it worked. I developed a plan with bulging primary walls outside, creating enough space inside the room within the confines of the pillars. The bulging roof of the bunker was also taken into account. The plan was very tight, but it worked. My experience of planning a cobalt bunker at Udaipur, in limited free space, taking out the internal shielding and bulging it outward, as described above, worked as a handy experience. The architect involved was very experienced with a positive attitude and prepared a plan within the available space. The plan was approved by the AERB; the bunker was constructed; the linear accelerator was installed and made operational in April 2009. I was appreciated and credited for making it happen without violating the regulations and not expanding horizontally. Medical physicists are problem solvers and people of action.

14   Campus in Jaipur

The Jaipur position has given me more opportunities and visibility, and further recognition of the contributions of medical physicists to healthcare, radiation protection, and education. By virtue of hard work, I was able to earn many fellowships and opportunities to travel abroad, to participate in training programs, conferences, and international medical physics professional organizations. For contributions to healthcare, research, education and establishing clinical services, I have been awarded the second highest award of medical sciences—the prestigious Fellow of the National Academy of Medical Sciences "FAMS" of India. I am the only clinical medical physicist getting the award since its establishment sixty years ago.

I am strong follower of Mahatma Gandhi and he said, "*Whatever you do will be insignificant, but it is very important that you do it.*"

I have thoroughly enjoyed being a medical physicist along with the opportunity to impact many lives and careers. I am proud to be a medical physicist—a healthcare professional. Being a clinical medical physicist in a medical college, I was involved in teaching, training, and health care. I consider it to consist of two very holistic jobs/professions (1) educator and (2) health care provider. I am fortunate to be part of both. Finally, I quote Ratan Tata, the great industrialist and inspirer, "none can destroy iron, but its own rust can; likewise, none can destroy a person, but his mindset can."

Therefore, always be positive and love your work.

Dr. Arun Chougule is the Senior Professor and Head of Department of Radiological Physics, SMS Medical College & Hospitals Jaipur, Ex. Pro Vice Chancellor, Rajasthan University of Health Sciences and Dean, Faculty of Paramedical Science, Jaipur, India. He has 37 years of professional and teaching experience in medical physics. He is considered a pioneer in experimental radiation dosimetry in India. He has held many significant positions in committees and organizations. He is the past President of the Association of Medical Physicist of India (AMPI) and currently the President of the Asia-Oceania Federation of Organization for Medical Physics (AFOMP) and Chair of the Education and Training Committee of the International Organization for Medical Physics (IOMP). He has more than 120 publications in national and international journals and has presented more than 350 papers in national and international conferences. He has authored two books. His research interests include radiobiology, experimental dosimetry, radiation safety and quality assurance in radiology. He has served as an expert to the International Atomic Energy Agency and has been a regular associate to the International Centre for Theoretical Physics. He has done work for the Radiation Safety Training programme of the Flemish Interuniversity Council (VLIR), Belgium.

Awards

2021. Awarded the prestigious Fellow of the Indian National Academy of Medical Sciences—FAMS, for contributions to healthcare, research, education and establishing clinical services.

2020. Awarded the Outstanding Medical Physicist Award from the Asia-Oceania Federation of Organizations for Medical Physics (AFOMP).

2019. Awarded outstanding medical faculty, SMS Medical College, Government of Rajasthan, India.

2016. International Day of Medical Physics (IDMP) Award for "promoting medical physics to a larger audience and highlighting the contributions medical physicists make for patient care." Awarded by the International Organization for Medical Physics.

2012. Prestigious Dr. Farukh Abdulla Sher—e-Kashmir best researcher award for 2011–12.

Part V

Medical Physics: More than Teaching

"Nothing in life is to be feared, it is only to be understood. Now is the time to understand more, so that we may fear less."

Marie Sklodowska-Curie, a pioneer of research on radioactivity and winner of two Nobel prizes in Physics (1903) and Chemistry (1911). She was the first woman to win a Nobel prize and the only woman to win the Nobel prize twice.

16

Introduction to 3D Medical Imaging: Of Mice and Men, Music and Mummies

Jerry J. Battista

1   Introduction

The need to see inside the human body for medical reasons probably dates back to prehistoric times perhaps when humans developed primitive tools and sustained some injuries. During the Stone Age, millions of years ago, let's imagine a hunter returning home after a hard day's work and complaining of a sharp pain in the abdomen. The dialogue below is purely fictional since coherent languages probably did not exist way back then.

> "Laali!!! I was hit by a flying arrow from our son Zakkal and I've got a major pain in my belly." "Don't worry Brac, I will find the cause. I can't really see much in the darkness of the cave here, so let's go outside into the sunlight. Now I see a sharp glittering stone embedded in your liver and will extract it carefully. Bite on this branch – 1,2,3." Brac grunts as Laali pulls away the arrow tip. "Does that feel better now, my dear?" "Oh yes, Laali. I don't know where we would be without solar-guided surgery".

The ability to peer inside the human body evolved millions of years later. On November 5, 1895, Wilhelm Conrad Röntgen discovered x-rays

J. J. Battista (✉)
Western University, London, ON, Canada
e-mail: j2b@uwo.ca

© The Author(s), under exclusive license to Springer Nature
Switzerland AG 2022
J. Van Dyk (ed.), *True Tales of Medical Physics*,
https://doi.org/10.1007/978-3-030-91724-1_16

produced by a cathode ray tube. He wasn't sure why these invisible rays caused a nearby phosphorescent plate to glow in the dark, but he soon realized that they were penetrating enough to cast fascinating shadows of intervening solid objects. Here is another imaginary dialogue between William and his spouse, Anna Bertha.

"Anna, Anna, you must come and see these amazing shadows! I discovered a new type of light that goes right through some solid objects. I have no idea what this radiation is, so let's name it "x", just like Professor Zimmerman used in algebra. Place your hand over this photographic plate and we will make the world's first radiograph of human bones. The appearance of your wedding ring will be most charming for generations to come. This is an entirely painless and harmless procedure, …as far as I know".

News of this remarkable discovery spread rapidly through newspapers around the world. Diagnostic and therapeutic applications of x-rays soon began to emerge internationally. Fifty years earlier, general anesthesia using inhaled ether had been demonstrated publicly for a tooth extraction by a Boston dentist, Dr. William Morton. X-ray imaging would later play a major role in planning more complex surgical procedures.

In the following sections, we will explore various types of medical imaging, with emphasis on x-ray transmission computed tomography (CT). Along the way, I will describe my own experiments with gamma (γ) ray scatter imaging performed during my Ph.D. studies (1973–77). I will use everyday language and analogies as much as possible to make this material comprehensible, although not necessarily comprehensive. This journey will cross over into non-human applications including images of Egyptian mummies, animals, old car parts, guitars, specimens retrieved from a seventeenth century sunken ship and a vehicle with hidden contraband that was detected at a border crossing.

1.1   What is Tomographic Imaging?

Our eyes infer the external shape of a three-dimensional (3D) object using ambient light that is absorbed, emitted, scattered, or reflected. We mentally reconstruct a face by looking at someone from different viewpoints and noting the reflections and shadows. Extending this concept to the *interior* of the human body requires much more penetrating rays that can pass right through the human body. This type of imaging was popularized by Superman's x-ray vision in science fiction.

Tomography is derived from the Greek word *tomos* meaning a "cut or slice". This imaging procedure aims to overcome the fundamental limitation of radiography where shadows of multiple structures overlay each other on the imaging plate. Tomography allows visualization of a set of individual thin slices in cross-sectional views, much like slices of bread. Using a series of abutted slices, picture elements (pixels) can be re-sorted to form other views, including 3D renderings. Early forms of tomography (known as focal plane tomography) were already developed in the 1930s and relied on radiography with a simultaneous rocking motion of the x-ray tube and radiographic film. This preferentially imaged a single plane in the body. The technique suffered from blurring and ghost images from adjacent layers. Imaging of transverse planes parallel to the feet was not possible because it would require x-rays to penetrate through the entire length of the body! In the 1970s, these limitations would be overcome with the advent of *digital computed* tomography (CT).

If we consider the human body as a "black box", particles or waves can be applied externally while detectors record the number of transmitted, reflected, scattered, echoed, or emitted rays, as shown in Fig. 1.

In transmission imaging, a small fraction of the incident stream of rays is measured by an exit detector on the far side of the patient. Alternatively, rays or waves that are deflected or echoed in the body can reach a smartly placed detector, as in ultrasonography and Compton-scatter imaging. In nuclear

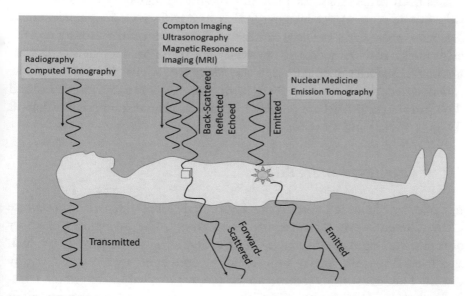

Fig. 1 Medical imaging with transmitted, scattered, reflected, or emitted radiation

medicine, a radioactive tracer is first injected into the patient and the radioactivity then lodges in specific organs or tumours of interest. Emitted particles such as gamma (γ) rays are detected around the patient. For emitted or transmitted rays, a mathematical operation known as image reconstruction, dating back to fundamental work in mathematics by Lorentz and Johann Radon (1905, 1917), can further be applied to a set of detected views or "projections". The output is an image of the patient's interior absorbed radioactivity or tissue density distribution.

Different types of radiation can reveal not only the physical architecture of the human body, but also map the physiological status of different organs and their metabolic activity. Modern techniques can track the diffusion of important biomolecules, the migration of tumour cells, and measure blood flow rates across tissue regions. Following on the heels of earlier development in astronomy and microscopy, computed tomography was first applied to nuclear medicine. Emerging rays are detected by a large external gamma camera. The most common radioactive substance is technetium-99m. It decays to its natural state by emitting low energy γ-rays that leave the targeted organ and can reach the gamma camera.

Imaging with harmless radiation using sound or radio waves has also been developed, much like sonar and radar. The best examples include ultrasonography and magnetic resonance imaging (MRI). Externally applied ultrasound waves of inaudible high frequency can produce 2D or 3D images by detecting echoes from tissue interfaces, including pre-born infants in the womb. Similarly, magnetic resonance imaging (MRI) uses radio wave frequencies above the "ham" radio band to image soft tissue. When placed in a strong magnetic field, nuclei such as hydrogen of water-like tissue produce radio waves of specific frequencies that pinpoint their location. Coils that surround the patient detect the nuclear signals to form images. More details of ultrasound (US) imaging and MRI are provided in various chapters, especially Chap. 8 (US), 11 (MRI) and 19 (US).

1.2   Catching All the Rays

In nuclear medicine, a gamma camera is rotated around the patient to capture many different views of the trapped radioactivity. This opens the possibility of reconstructing the radioactivity distribution within a targeted organ such as a liver. Abnormal uptake seen as hot or cold spots in the image signify disease. The standard 3D reconstruction process is called single-photon emission computed tomography (SPECT). Positron emission tomography (PET) is similar but it uses a narrow ring detector to image only a single transverse slice of the patient. Consecutive slices are then imaged by moving the patient into the donut hole of the scanner. PET imaging is specialized and requires positron-emitting isotopes. Positrons are particles with the same mass as an electron but with a numerically equal *positive* charge. When a positron is emitted by the nucleus, it slows down in tissue and subsequently is annihilated by pairing with an electron. Two γ-rays are promptly ejected and detected by a pair of detectors on opposite sides of the ring detector system. The γ-rays have a signature characteristic energy (511 keV), enabling clean radiation detection while rejecting stray scattered rays with lower energy. PET relies on image reconstruction concepts very similar to those used in SPECT, but restricted to a fan-beam (thin in one direction, divergent in the other), imaging one slice at a time. A set of acquired slices can be digitally stitched together to render the radioactive volume. Radioactive fluorine-18 is the most commonly used positron emitter. Positron-emitting isotopes are normally produced in a nearby nuclear or cyclotron facility. The isotope must be shipped and used promptly because it has a half-life of only 110 min. For this reason, a dedicated compact cyclotron is sometimes installed within a hospital. Radioactive sugar (i.e., fluorodeoxyglucose, $F^{18}DG$) is consumed in metabolically active regions of the brain. In functional PET imaging, the regional uptake of radioactivity is monitored while the patient is performing an assigned task such as tapping fingers or listening to music. Parts of the brain that are stimulated will "light up", unless there is disease or a cognitive impairment. Similar functional studies can also be performed with specialized MRI techniques.

2 The First Clinical CT Scanner

Allan Cormack made advances in the mathematics for numerically reconstructing tomographic images. He then built a small-scale demonstration system that produced accurate attenuation maps of simple test objects known as "phantoms". The head phantom consisted of an aluminum shell representing the human skull with interior aluminum disks embedded in plastic to mimic soft tissue tumours. In 1963, Cormack and an undergraduate student recorded γ-ray transmission profiles over a *2-day period* using a low activity cobalt-60 source. The resultant publication sparked limited attention, but a proton research group adopted his reconstruction technique to produce tantalizing images of human organs. Details on image reconstruction methods are described in Chap. 10.

Godfrey Hounsfield's first table-top CT model was similar and also used a radioactive γ-source (i.e., americium-241). The first prototype scans of plastic specimens took 9 days per slice before the source was replaced with an intense beam of x-rays, reducing data acquisition time to 9 h. In 1970, image reconstruction took another 2.5 h (per slice) on a large mainframe computer, plus another 2 h for image display using input data recorded on paper tape!

In the mid-1970s, computed tomography (CT) developed very rapidly. The procedure was also called CAT (computerized *axial* tomography) in the early days because image slices were oriented perpendicularly to the long axis of the body. This imaging detected very small differences in density between normal tissue and tumours. In French, a CT scan procedure was rightfully described by the word *tomodensitométrie* (TDM). The affordability of minicomputers from competing companies such as Data General (DG) and Digital Equipment Corporation (DEC) would keep the price of a scanner within reach of a hospital budget. Image reconstruction algorithms that had been hitherto too slow for practical diagnostic radiology could finally be used clinically. Minicomputers were the enabling technology.

In 1979, the Nobel Prize in Medicine and Physiology was co-awarded to engineer Godfrey Hounsfield and physicist Allan Cormack for their complementary contributions. In 1981, I had the pleasure of personally meeting Cormack, who was an invited guest speaker at the Cross Cancer Institute in Edmonton, Alberta where I worked as a junior medical physicist (1979–1988). He was friendly, pleasant, approachable, and unpretentious - a role model. Following his lecture presentation, he allowed me to keep his overhead transparencies now reproduced for the first time in Fig. 2. In his lecture, Cormack graciously acknowledged previous authors who had established the mathematical foundations of image reconstruction, dating back to 1905. He then explained the well-known exponential equation for transmission of x-rays passing through small cubes of heterogeneous tissue volumes (voxels) lying along a ray-line path. With enough angular projection samples (g_L values in Fig. 2), the distribution of local attenuation coefficients (μ) could be calculated throughout a tissue slice. A CT scanner therefore collects a large set of transmitted rays that criss-cross a slice of tissue in the patient. Computer processing of the measurements produces a detailed map of these coefficients at every point in the slice. The values of these coefficients can be used to form a grey-tone (or colour) image on a computer screen, applying a "paint by number" scheme for each picture element (pixel).

The first *clinical* scanner was manufactured specifically for brain imaging in the central labs of Electric & Musical Industries (EMI) in the United Kingdom under the leadership of Godfrey Hounsfield. The first patient was

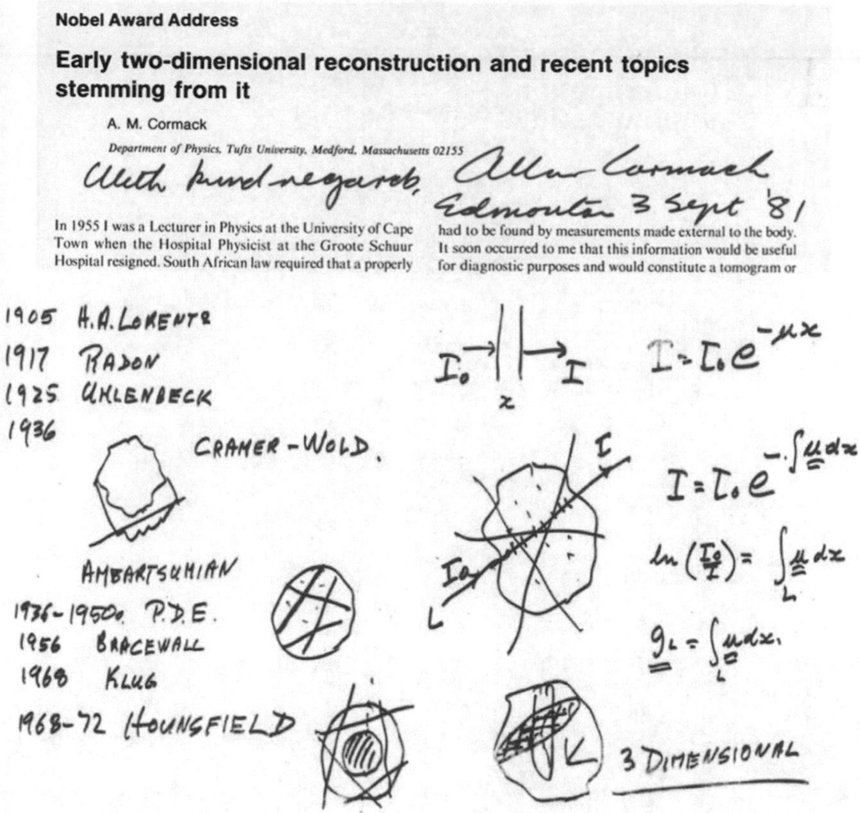

Fig. 2 Autograph by Nobel laureate Allan Cormack on a reprint of his Nobel address published in the journal *Medical Physics*, and a copy of his transparencies from a lecture delivered at the Cross Cancer Institute, Edmonton, Alberta (1981)

scanned in 1971 at the Atkinson Morley Hospital in London, England, near Wimbledon. The doctors suspected a brain tumour but fortunately for the patient, a low-density cyst was identified by CT. The new machine immediately showed its promise as a new method for distinguishing soft tissues. By 1975, 160 EMI head scanners had been installed in the United States, each at a price of $400,000.

2.1 The Beatles Connection

EMI was also the company that produced the first records of the Beatles under the label Parlophone in the UK (Fig. 3a) and Capitol Records in the US. It has been conjectured that record sales had helped to fund the development of the first scanner. However, a review of financial statements from

Fig. 3 **A** Sleeve to the first-pressed Beatles album (1963). Note the highlighted company name on the bottom right. This autographed item was valued at over US$35,000 (2012). Courtesy of Heritage Auctions (Dallas, Texas). **B** The Beatles on the Ed Sullivan Show (February 9, 1964). Photo courtesy of Express Newspapers/Getty Images

the company and research grants of the British government failed to identify a *direct* cash-flow link. This is likely a "tall tale" of music and medical
history that will continue to entertain guests at conference banquets and on
the internet.

The first long play (LP) Beatle records were coincidentally released on the
same day as the assassination of US President John F. Kennedy. The initial
reaction to the record was understandably subdued and disappointing. The
Beatles were unaware of the worldwide fame that would erupt after their
first appearance on the Ed Sullivan Show (Fig. 3b). A few years earlier, they
had signed a contract with EMI for a paltry royalty of one British farthing,
approximately 1/1000 of a British pound, per single 45 rpm record sold. By
1970, record royalties for the Fab Four resulted in only $13,500 per Beatle.
This translates to $94,500 per Beatle in today's currency. However, their even-
tual revenue sky-rocketed with a steady release of new songs, movies, and
publication of their compositions. In 2019, revenue was estimated at $17.5M
US dollars per Beatle or their heirs, 55 years after their American debut!

2.2   Whole Body CT Scanning

The first whole body x-ray CT scanner was surprisingly *not* developed by
a mega-company such as General Electric, Philips or Siemens. Dr. Robert
Ledley, a dentist-physicist, was funded by grants from the US National
Institutes of Heath (NIH) and Georgetown University Foundation. He was
familiar with digital image processing techniques and previous work by
Cormack at Tufts University. The ACTA (automated computerized transverse axial) scanner was assembled with the industry collaboration of Pfizer
Medical Systems. At this time, many of our readers will recognize the name
of this company as a supplier of COVID vaccines. The original scanner is on
display at the US National Museum of American History. The first scanners
used a limited number of detectors and required repeated movements of the
x-ray tube and detector array. This hardware would move across the patient's
width, followed by a small rotation, until a full 360° data acquisition was
completed. Eventually, faster scanners would use a full ring array of detectors
and only the x-ray tube had to be rotated, typically in a few seconds. (See the
description of four generations of CT scanners in Terry Peters' Chap. 10.)

Four-dimensional (4D) image acquisition, synchronized with the patient's
breathing, produced frames for surreal movies—a simulated voyage through
a patient's internal anatomy in 3D (Fig. 4). Diagnostic applications of x-ray
CT were followed by broader applications to dentistry, image-guided surgery
and radiation therapy.

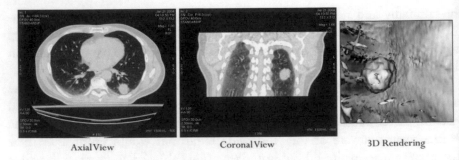

Axial View   Coronal View   3D Rendering

Fig. 4 X-ray CT Images showing a lung tumour in transverse, coronal (frontal) and pseudo-3D views. Courtesy of Drs. S. Gaede, London Regional Cancer Program and T. Lee, Lawson Health Research Institute (London, Canada, circa 2004)

Later technological developments yielded best-of-both-worlds images, combining anatomical and functional imaging information in hybrid scanners (e.g., a PET-CT, SPECT-CT or PET-MRI). The first two combinations use tissue density from CT images as the anatomical backdrop to radioactivity maps from SPECT or PET.

2.3 Reducing Radiation Exposure to the Patient

X-rays are called "ionizing" because they have enough energy to knock electrons out of atomic or molecular orbits, while sound or radio waves have a lower energy and are called "non-ionizing". The diagnostic quality of images produced by x- or γ-rays is affected by the level of radiation exposure to the patient. Use of a greater number of imaging rays reduces the salt-and-pepper noise that interferes with tissue contrast and diagnosis. However, an excessive radiation dose also enhances the risk of developing a cancer later in life or a future abnormality in very young children. A recent report in the USA reveals an average annual effective dose of 2.3 mSv per capita due to medical imaging procedures, comparable to the annual natural background radiation on earth that we are all exposed to every year (3 mSv). The mSv unit stands for millisievert, an indirect measure of risk for partial body exposure. Most of the medical radiation dose given to people is attributed to CT and nuclear medicine scans. This is not surprising because of the number of views required per 3D image and widespread medical application to the population. There is mounting pressure to "image gently" especially in younger patients. This clearly calls for a risk–benefit analysis by the doctor and patient. The immediate risk of a false or delayed diagnosis generally far outweighs the risk of developing a cancer or abnormality much later in life. In medical imaging departments, medical physicists calibrate radiation equipment, and evaluate or develop new techniques for reducing radiation dose to the patient.

2.4   What Do Medical Physicists Do?

The above topic reminds me of how many times I have been asked to explain "what is a physicist like you doing in a hospital?" to family and friends! Here is the recap why and how physicists play a role in a hospital setting. Physicists possess a strong background in atomic, nuclear, experimental and computational physics, and have elected to work in a health care setting. Often working from backstage, they support medical doctors in "doing no harm" when radiation procedures are involved. Accurate radiation dosimetry performed by medical and health physicists plays a pivotal role in quantifying radiation risk to a population from medical or industrial exposures, respectively. (Medical physicists are generally involved with patient-related activities while health physicists are primarily involved with radiation protection of the general population.) The physicists calibrate all types of hardware devices and test clinical software. They assume radiation safety responsibility for staff and members of the public while in a hospital facility. When more intense high-energy radiation is applied to treat cancer patients, their responsibility extends beyond the hospital walls where radiation levels to the public must be even further reduced to quasi-background levels. Some procedures known as brachytherapy ("brachy" is Greek for "short", referring to treating a tumour from a short distance) also require placement of radioactive sources inside the body, such as in prostate cancer treatments. Physicists and their assistants ensure that all radioactive materials are accounted for and verify that the patient will pose a minimal radiation hazard to members of the family. Measured radiation levels must comply with all government regulations.

Medical physicists are often appointed to a local university and have teaching and research duties, concurrent with providing clinical services. They provide on-the-job training to graduate students, residents, medical radiation technologists—the future generation of specialists who will be working in health care facilities. At present there are nearly 30,000 medical physicists who are members of the International Organization of Medical Physics (https://www.iomp.org/). The present world population is approximately 8 Billion people. Hence the odds of meeting a medical physicist for a world traveller is quite low: 3.84 per million—in the same order of magnitude as the odds of being struck by lightning in a year. The probability of meeting a medical physicist is up to five times greater in developed countries but considerably lower in developing nations. An organization called *Medical Physics for World Benefit* (https://www.mpwb.org/) provides advice and training in medical physics, especially to low- and middle-income countries.

3   My Graduate Studies and Early Career

After completing my undergraduate degree in Physics in 1971 at Loyola College (now Concordia University) in Montreal, I traveled westward along a relatively new super-highway 401 with a friend and fellow student of mathematics. We stopped at many Ontario universities along the way, in search of a graduate program. The final stop was in London, Ontario at the University of Western Ontario. I had arranged to meet with medical physicists off campus at the "cancer clinic" of Victoria Hospital. As I learned later, this was near the historic site where the world's first treatment of a cancer patient with cobalt-60 radiation had taken place, some 20 years earlier in 1951. At the cancer facility, I was welcomed by a tall very impressive individual— Dr. J. C. F. MacDonald. I was given a quick tour of the treatment units that included a humongous circular treatment machine made in Switzerland called the *Asklepitron-35* betatron. The machine derived its name from *Asklepios,* a skillful doctor who was eventually revered as Greco-Roman god of medicine. The "35" model refers to the circulating electron beam energy of 35 MeV—a very high energy for radiotherapy applications with either electron or x-ray beams. I met several friendly staff members and graduate students who influenced my decision and career path, particularly Dr. Donald Dawson and Mr. Ken Shortt. Before departure, I was interviewed by Dr. MacDonald in his office. A cancer patient happened to wander by the open office door, appearing somewhat confused. Dr. MacDonald immediately stood up and redirected this patient to his next appointment with compassion. ***At that exact moment, my career path was set***—I wanted to become a clinical physicist helping cancer patients in their journey of radiation therapy. I soon enrolled in the Physics Department at Western and joined the clinical physics team in the research labs of the cancer centre. I completed my master's degree working with magnetic fields on the massive betatron. Dr. MacDonald then recommended that I embark on a Ph.D. program at the University of Toronto.

In Toronto, graduate students were placed in labs of the Ontario Cancer Institute (OCI) housed in the same building as the Princess Margaret Hospital (PMH). I was assigned a 3D imaging project and surrounded by other strongly motivated students, the likes of Aaron Fenster (see Chap. 19) and Martin Yaffe (see Chap. 18) who worked on CT scanner developments. This was a life-changing time, performing research in the presence of

Canadian pioneers of radiotherapy physics—Drs. Harold Johns and "Jack" Cunningham. They were the Lennon-McCartney team of medical physics with a worldwide following and influence.

3.1 Transmission Dosimetry

This section describes the highlights and challenging moments of my Ph.D. program (1973–1977). There were strong connections of my projects to the concurrent developments of CT scanners. My thesis supervisor was Dr. Michael Bronskill, a nuclear medicine physicist at PMH. In my role as a rookie graduate student, I was assigned a clinical task to earn my stipend that covered daily living expenses. Graduate students often receive a stipend for being a teaching assistant or by providing support in some research project or clinically related activity. For my first journal publication, I worked in the evenings on a radiotherapy machine known as the X-otron—a unique cobalt-60 machine developed by the PMH-OCI physics group (see Chap. 2 by J. Van Dyk). This machine was well ahead of its time with an integrated x-ray imaging system for precisely aiming radiation beams at a tumour. In addition, a "transit dosimeter" was incorporated in the beam stopper on the exit side of the patient. It monitored the γ-rays transmitted through the patient while the beam rotated around the patient. The data were fed into an operational amplifier that automatically computed and displayed the average Tissue-Air Ratio (TAR) for individual patients to help determine the dose received in the presence of tissues with different densities. TARs are important quantities that are used for determining length of time the beam needs to be turned "on" to give a specified tumour dose to the patient. Figure 5 shows the transmission profiles for a pelvic treatment as the beam rotated around the patient. The CT principle of measuring transmitted rays for different beam angles was "staring at us right in the face". We failed to recognize that adding lateral motion of the treatment table might have provided enough projection data to create a full internal image—a near-miss of a Nobel Prize?

3.2 Compton-Scatter Imaging

My Ph.D. thesis research project was equally exciting from the start. A prototype tomographic scanner named the *Gammatome* (Fig. 6) had been manufactured by Atomic Energy of Canada Limited (AECL), a company known internationally for supplying high-intensity radioactive cobalt-60 sources for cancer radiotherapy. Arthur Compton had discovered that high

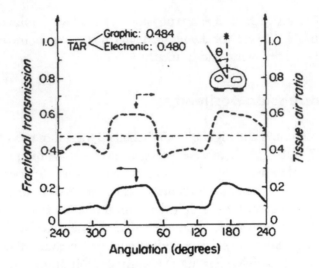

Fig. 5 Transmission (bottom) and TAR (top) profiles as the beam pivots by an angle θ through a full rotation. The γ-ray transmission decreases as patient thickness increases and when the beam passes through dense bones. Reproduced with permission from J. Battista, P. Leung, W. Taylor. Med. Phys. 3, 335 (1976)

energy photons can be scattered by electrons and exit with a lower energy (1923). The Gammatome detected a large cone of Compton-scattered rays from a small voxel of tissue exposed to a narrow incident pencil beam of cobalt-60 radiation.

Small tissue volumes (voxels) were scanned with the pencil beam and scattered photons emerging from these voxels were detected by four large scintillation detectors underneath the table top. A scintillation crystal gives off light when hit by a photon. Behind the crystal is a detector which measures the amount of light given off by the crystal, hence, a scintillation detector. By programming the motion of the tabletop, the Compton scanner had the potential to create images in *any orientation directly*, including flat or even curvilinear slices. Dr. Robert Clarke of Carleton University (Ottawa, Canada) had already demonstrated some excellent soft tissue images of human kidneys.

My first experiments were conducted in the Nuclear Medicine Department of the University Hospital in London, Ontario (circa 1974). Gerry Van Dyk was the coordinating physicist working for AECL who provided technical support. He is, coincidentally, the brother of Jacob Van Dyk, editor of this book! We were prepared to gather images of a small head-like cylinder when it was realized embarrassingly that the test object (phantom) had been left behind in the lobby of the Princess Margaret Hospital (200 km from London, Ontario)! An urgent telephone call was placed to the Physics labs in

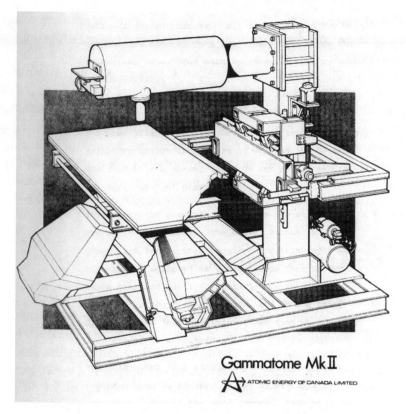

Fig. 6 Prototype of a whole-body Compton scanner. Courtesy of Atomic Energy of Canada (Internal Report CPSR-407, 1974)

Toronto and fellow student, Aaron Fenster, volunteered to rescue the situation by delivering it to us a few hours later. Data sets were successfully acquired and transferred to paper tape. (Punched paper tape was a standard medium for data storage in those days.) Mr. Lee Santon of Abraxas Engineering (company name derived from the album of guitar legend Santana) developed the program to read these tapes. He also modified a standard Sony Trinitron colour television to interface with a PDP-11 computer for image display at the Toronto physics lab.

The race to develop whole-body CT scanners had also been launched. Ledley's rapid progress to a whole-body CT scanner was a concern for me as it would diminish the motivation for developing a Compton scanner and potentially negate some content in my Ph.D. thesis. If the 3D whole body imaging puzzle was already solved by CT, what advantage would be gained by Compton imaging? In the end, Compton scanning with a cobalt-60 source yielded more accurate maps of tissue density, which was especially useful for radiotherapy dose computations. Tissue density is key to accurate computation of radiation dose distributions in cancer patients treated with comparable megavoltage rays.

It is historically interesting that Allan Cormack was also motivated to develop CT for the same application to radiotherapy. He was a part-time hospital physicist at the Groote Schuur Hospital in Cape Town, South Africa. He observed how radiation treatment plans were generated by overlaying transparencies of isodose charts for each radiation beam and summing the numerical dose values to produce the composite dose pattern. These charts assumed an *all-water patient* and his mission became clear to develop a way to account for realistic inhomogeneous tissues of different density in individual patients.

3.3   Tales of the RANDO Phantom

Soon after completion of my Ph.D. degree, I continued to work at OCI-PMH as a post-doctoral fellow—one that would provide early training in clinical radiotherapy physics. I became involved in the installation, calibration, and support of a new whole-body CT scanner in the Radiology Department at PMH. Before the installation of the Picker Synerview scanner, I conducted a series of CT scan experiments at the Hospital for Sick Children on an Ohio-Nuclear CT scanner with Marc Sontag who was about to complete his Ph.D. program. We were allocated evening hours access and performed scans with the intent of using the tissue density data sets for dose computations. In one of these night sessions, we used a human-like mannequin called the RANDO "phantom" (Fig. 7). This pseudo-patient consists of a set of adjoining plastic slices that can be disassembled for internal radiation dosimetry purposes. The electron density of its artificial organs and human bones had been accurately verified with cobalt-60 scans of individual slices on the Gammatome scanner. In one of the sessions, we transported the heavy upper torso on a wheelchair across the hospital emergency department *en route* to the CT scanner suite. The waiting patients must have been wondering what this was all about! We soon realized that being a medical physicist in a hospital requires utmost sensitivity and respect for patients at all times.

In a separate later incident, I had a travel adventure with the RANDO phantom and Aaron Fenster, also a "post-doc" at this time. We were dispatched to the CT factory of the Picker X-Ray company in Cleveland, Ohio where the future PMH scanner was being assembled. We travelled by

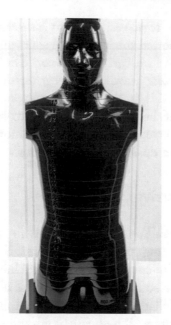

Fig. 7   Male Alderson RANDO Phantom. Courtesy of Radiology Support Devices Inc. (https://rsdphantoms.com/)

car from Toronto to Cleveland but when we arrived, we were met by pick-eters as the unionized plant workers had declared a strike (Picker picketers!). With our Canadian charm from the Great White North, the workers allowed us to drive past the picket line safely to meet our engineering hosts. We saw the scanner assembly line and met software engineers, including a very quiet individual (A.V. Lakshminarayanan) in a more secretive area of the plant. He was well known for his publication with G.N. Ramachandran on an image reconstruction algorithm that became the method of choice on commercial CT scanners because of its speed gain over earlier Fourier methods. (Fourier analysis is the study of the way that any mathematical functions may be repre-sented by sums of simpler trigonometric wave functions. Fourier methods are often used for analyzing wave patterns in digital images. See Terry Peters Chap. 10 for more details.) With the "Ram-Lak" algorithm, the interlacing of data acquisition and image reconstruction made images gradually appear on the operator's computer screen; thus, the radiologist could review the images sooner after the scanning of the patient was completed.

For this site visit, we also brought along the RANDO phantom to measure radiation doses within it. We imaged this pseudo-human on a scanner in the factory. At the end of a long day, we headed back to Toronto, with a hard stop at the US-Canadian border. The customs officer examined our dosimetry equipment; the phantom torso easily caught his keen eye. How could it not? We dismantled it and showed the individual slices and explained our mission to the agent. Within each slice, there is an array of holes that can be filled with capsules containing a *white* powder that registers local radiation exposures. The inspector examined this powder with suspicion. As we further explained our radiation science, he hesitantly allowed us to return home to Canada. This type of border interruption is not uncommon for medical physicists carrying strange radiation dosimetry equipment across international borders.

3.4   CT Scanner Arrives at PMH

A Picker CT scanner was eventually delivered for installation on the fifth floor of the PMH building at 500 Sherbourne Street in Toronto, before the hospital moved to its current location on University Avenue. A crane interrupted traffic as the scanner was swung like a pendulum until riggers could pull this million-dollar piece of hospital equipment into a large window opening at the right moment. The heart of all onlookers at street level must have stopped in synchrony with mine during this critical step. Once inside, the unit was pushed along a corridor and headed smoothly toward the scanner suite, but not without first being obstructed by a hospital firewall door. The installation crew had either incorrectly measured the outer shipping package dimensions or misunderstood the delivery pathway! Embarrassing moments like this can occur when large pieces of modern imaging or therapy equipment are being hauled into older hospitals with outdated blueprints. Equipment is often passed through a rooftop or special improvised gateways.

3.5   Tissue Densitometry for Radiotherapy Dose Computations

CT scanning proved to be highly quantitative compared with all other imaging techniques prior to the 1970s. Linear x-ray attenuation coefficients (μ) describe the fractional attenuation of x-rays removed by absorption or scattering per unit distance. For example, a μ-value of 0.20 cm^{-1} simply means that a small tissue element attenuates inbound x-rays by 20% per cm of transit thickness through the element. This coefficient depends on the type of tissue and the energy of the x-rays. To simplify the interpretation of tissue attenuation, a CT number scale was devised with a tribute to Hounsfield. This was found to be more radiologist-friendly than a set of μ-values, also with the advantage of being less dependent on the x-ray beam energy used on different scanners:

$$N_{CT} = 1000 \left\{ \frac{\mu - \mu_{H_2O}}{\mu_{H_2O}} \right\} \tag{1}$$

where N_{CT} is the CT number in Hounsfield units, μ is the linear attenuation coefficient for tissue as computed by the reconstruction algorithm, and μ_{H_2O} is the coefficient for liquid water as a reference solution. In Hounsfield units (HU), the values are more easily interpreted: in water, $N_{CT} = 0 HU$, in vacuum, $N_{CT} = -1000 HU$, and when attenuating at twice the rate of water (e.g., bone), $N_{CT} = +1000 HU$. Muscle, fat, lung, and dense bone have typical CT numbers of 30, -50, -700, and >500 HU, respectively. A paint-by-number scheme is applied to display the grey-tone image over a range of CT Numbers.

A scanner operating with a 120 kV$_p$ x-ray tube measures x-ray attenuation which depends mainly on local electron density (electrons per cm^3) that is further related linearly to gravimetric density (g per cm^3). There is a milder dependence of CT Numbers on tissue composition observed for calcified tissue, bone, or foreign objects with higher atomic number. During my Ph.D. studies, the correlation of CT Numbers with electron density was investigated *in vivo* for the first time. Figure 8 shows our lab results obtained by scanning patients on a Compton scanner and also on the CT scanner at the Hospital for Sick Children.

In our clinical study of 20 patients, linearity with electron density was excellent for soft tissues up to a CT Number of +100 and then the slope was accentuated by atomic number effects in bone. The bilinear hockey-stick form of the data in Fig. 8 has been used in density lookup tables of radiation treatment planning systems for many decades. Two points (CT Number $\approx$ 50, see question mark in Fig. 8) did not match theoretical expectation. Unexpected discrepancies like this are always intriguing for a physicist. A subsequent interview with the patient revealed that she had silicon breast implants. Silicon has an atomic number (Z = 14) that is considerably higher than for typical elements such as carbon and oxygen found in soft tissues.

Radiation therapy aims to deliver a high uniform dose to the diseased volume while minimizing the dose to surrounding healthy tissues and organs. A medical colleague once summarized this physical strategy succinctly as "hitting the tumour while missing the rest of the patient". Quantitative CT scanning opened the avenue for much improved accuracy in predicting *in vivo* dose distributions for individual cancer patients treated by x-rays, protons, and heavier ions like carbon. In the 1980s, computerized treatment

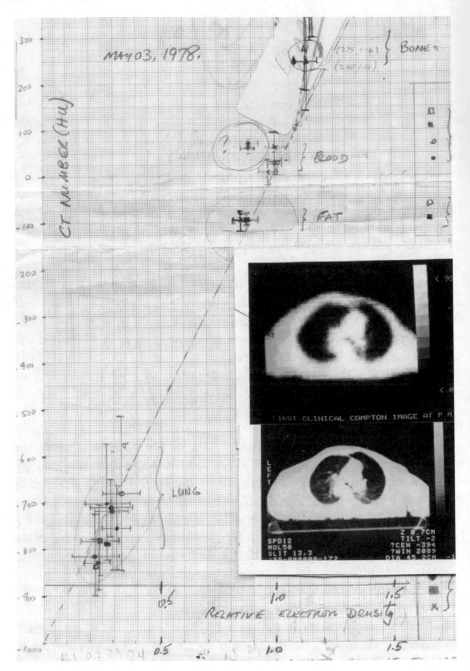

Fig. 8 Correlation of CT Numbers and normalized electron density. The top image was the world's first transverse image obtained by Compton scanning in our lab (circa 1977). The lower image is the matching CT image of the same patient from an Ohio-Nuclear scanner at the Hospital for Sick Children in Toronto

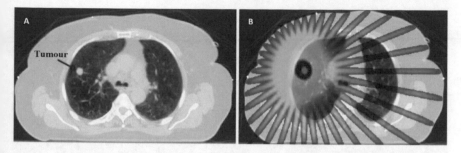

Fig. 9 a CT scan through lung regions with a small tumour. b Dose map for stereo-tactic treatment with 36 intersecting 18 MV x-ray pencil beams. Courtesy of Dr. Brandon Disher, from his Ph.D. program at Western University (2007–2013)

planning evolved rapidly with 3D graphics and consideration of scattering effects across voxels in tissue. High-precision measurements of small dose perturbations caused by ring-like air cavities were conducted at PMH (see Chaps. 2 and 7 by Van Dyk and Wong for more details). The Equivalent TAR calculation method developed by Marc Sontag as a Ph.D. student under the supervision of Dr. Cunningham became the leading "2.5D" algorithm. 3D methods based on convolution concepts followed and accounted for not only the dose dispersed by scattered x-rays but also by secondary electrons (see Chap. 5 by Arthur Boyer for more details).

In stereotactic body radiotherapy (SBRT), an isolated small tumour can be treated to a very intense dose using multiple pencil beams of high-energy x-rays intersecting at the tumour. Figure 9 shows an SBRT dose pattern overlaid onto a CT image of lung. The doses were modelled by Monte Carlo simula-tions regarded as the gold standard for dose accuracy. This method simulates the random absorption and scattering of millions of photons and electrons in tissue, one particle at a time.

4   Non-traditional CT Images

This chapter has emphasized the medical applications of 3D imaging, but there are many other applications: pre-clinical studies of rodents with micro-scanners, large animal imaging for veterinary medicine, and non-destructive imaging of inanimate objects. Some illustrative examples are provided below. The veterinary demand for 3D imaging has led to the development of a specialized scanner. Talented racehorses have significant monetary value in high stakes competitions such as the Kentucky Derby. The Equina™ scanner is shown in Fig. 10. The mildly sedated horse is being posi-

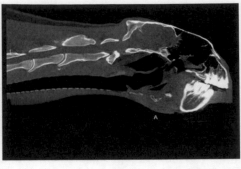

Fig. 10 CT scanning of a horse head with resultant side-view (known as sagittal) slice image of a normal neck region. Courtesy of Asto CT Inc. (https://www.astoct. com/). Accessed April 11, 2021

tioned for a head scan. A corresponding reformatted CT image is shown on the right. In this instance, the CT gantry was oriented vertically. It can also be rotated into a horizontal configuration, rising from the floor to encircle and image the lower limbs of the horse.

In an interesting turn of events at the Melbourne Cup (2019), a major stakes race with prize money of AU$8M, one of the horses named Marmelo was "scratched" before the race. This is a horse-racing term that signifies disqualification of an eligible horse before the running of a race. The veterinarian's decision was based on a CT diagnosis of an incomplete bone fracture. Such injuries pose a significant risk to all the horses and jockeys if the bone fractures during the stress of a competitive race. This decision generated considerable objections from the horse trainer and owners. The racing fans were also disappointed because this horse had a stellar reputation with a good chance of winning.

Images of inanimate specimens have sparked attention in other specialties: reverse engineering, archeology, anthropology, paleontology, art restoration, forensics, and border security. The advantage of x-ray or γ-ray imaging is the non-destructive nature of the imaging process. The specimens are intact after being traversed by the imaging rays. The Field Museum in Chicago, for example, has a dedicated CT scanner to archive its collection and reconstruct specimens in 3D in the event of unexpected damage. Figure 11 shows a collage of interesting non-medical images. The guitar image (a) is an example of reverse engineering. The internal bracing pattern of a guitar is optimized (e.g., Taylor Guitars) to achieve a better balance of tone and sustained volume of a plucked string or chord. The Egyptian mummy (b) was borrowed from the Royal Alberta Museum in Edmonton, Alberta and was

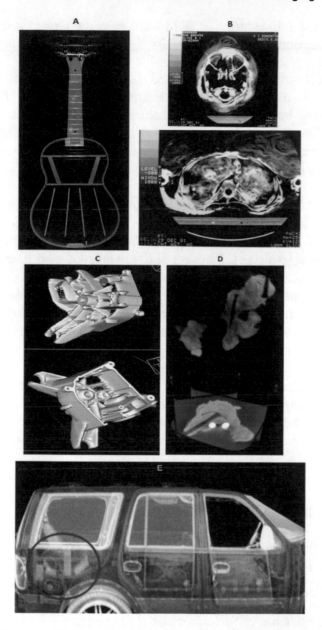

Fig. 11 **a** CT images of a Yamaha Model CGS102A guitar, **b** Egyptian mummy head and lung, **c** automobile carburetor, **d** concretions from a seventeenth century *Elizabeth & Mary* shipwreck, **e** dual energy radiography and Compton backscatter scan of a van used to smuggle organic packages (in red circle). Image credits: **a** David Ergun, Asto CT Inc., **c** Dr. "Rock" Mackie, Tomotherapy Inc. (author of Chap. 21), **d** Dr. John Schreiner, Cancer Centre of Southeastern Ontario, Queens University, **e** Courtesy of AS&E (rapiscan-ase.com). Accessed April 19, 2021

scanned by the author on a GE 8800 scanner at the Cross Cancer Institute in 1981. The contents of the lung cavity (lower image) show the embalming with linen wrappings.

Similar CT studies were conducted in 2005 of King Tutankhamun's mummy to determine his cause of death at age 19 (1324 BC). It was concluded that he probably died of natural causes rather than a suspected homicide. The automobile carburetor (c) has largely been replaced by fuel injectors in the modern car since 1990. It was imaged on a megavoltage CT scanner of a tomotherapy machine developed at the University of Wisconsin (see Chap. 21). In panel d, images were obtained of specimens from the ship *Elizabeth & Mary* sunken in the St. Lawrence seaway at Anse aux Bouleaux, Quebec. The ship was part of a large American-Anglo fleet of over 34 ships sailing from Boston *en route* to siege Quebec City controlled by the French (1690). The megavoltage CT images of the items recovered from the sunken ship were produced at the Cancer Centre of Southeastern Ontario on a modified cobalt-60 therapy machine. The specimens were investigated while kept in their anti-corrosion bath, and without risking damage to delicate internal structures. Lead shot (bright spots) and voids remaining after metal spikes had dissolved in water (black spaces) are visible in the recovered historical specimens. A most interesting contemporary large-scale application of mobile Compton imaging is the use of *backscattering* to image transport vehicles (e). Coupled with large radiographic transmission scanners, the scattered radiation helps identify camouflaged packages of drugs, contraband and radioactive goods. It also helps to spot and reduce human trafficking in trucks and trains.

5   Summary and Conclusion

In this chapter we explored how 3D images are formed by applying internal or external x- or γ-radiation and reconstructing internal features of the human body. This achievement has been recognized by several Nobel Prizes to scientists and engineers who developed practical CT systems. The evolution of these methods has vastly improved the diagnostic power for the medical community and assisted patients around the world with more accurate diagnosis and timely optimal treatment planning.

During my career, I was influenced by mentors and students who forced me to sharpen my thinking. Together we have contributed to medical imaging and radiation oncology developments. Life has its stochastic moments—random events that can affect our live path and spawn scientific advances. The benefits of "thinking outside the box" with others cannot be overlooked in multidisciplinary research with a common goal, enhancing cross-fertilization of fields especially in the fuzzy overlapping zones of all sciences, medicine, and engineering. The discovery of the helical structure of DNA with x-ray crystallography by Watson, Crick, and Franklin (1953), and development of CT imaging by Cormack and Hounsfield (1971) and of MRI by Lauterbur and Mansfield (1977) are a few convincing examples. There are deep secrets of nature yet to be discovered that will expand our understanding of the internal workings of our bodies, our brain, possibly our minds.

Acknowledgements I especially thank family members and friends for their tolerance of my persistent analogies and absent-mindedness at times. My colleagues, students, and musical bandmates have enriched my scientific career through dual exercising of my left and right brain.

Dr. Battista is Professor Emeritus at Western University in London, Canada. He obtained his Ph.D. degree in Medical Biophysics at the University of Toronto and acquired early clinical physics experience at Princess Margaret Hospital. Jerry then worked at the Cross Cancer Institute in Edmonton, Alberta His research team developed one of the first 3D radiation treatment planning systems, introducing convolution-based dose computation algorithms. He has published over 130 journal articles, several book chapters, and recently edited a book on contemporary dose computations.

Jerry is a certified medical physicist and he directed physics research of the London Regional Cancer Program while serving as Chair of Medical Biophysics at Western. He was the provincial coordinator of the medical physics residency training program in Ontario from its inception. He is advisor to provincial and national agencies on medical physics staffing levels required for radiation oncology.

Jerry is an award-winning educator at Western and known for his enthusiastic style in presentations and publications. He uses vivid analogies of physics concepts to appeal to a abroad range of audiences. He co-developed a small-scale optical CT scanner for education, used at 50 universities, and produced YouTube videos for interactive teaching of MRI principles using a guitar analogy.

Awards

2017. Gold Medal from the Canadian Organization of Medical Physicists for lifetime achievements.

2016. Kirkby Award from the Canadian Association of Physicists for service to professional physics in Canada.

2014. Western's Pleva Award for excellence in teaching and educational innovations.

2012. Fellow of the Canadian Organization of Medical Physicists and American Association of Physicists in Medicine, in recognition of significant contributions to the field of medical physics.

Multiple years. 30 awards to his students for excellence in publications and conference presentations.

17

Medical Physics in Five Easy Job Descriptions

Tomas Kron

1   So, What Are You Doing Anyway?

Thank you for asking. It is a good question which many of us medical physicists get asked from time to time. I will try to answer it by interpreting my career over the last 30 years in the light of five fictive job descriptions.

Going back, I did not know what medical physics was before I was 30 and I still have difficulties defining it now after more than 30 years 'on the job'. This reflects probably one of the most interesting features of a hugely rewarding career: there is variety, uncertainty, responsibility and ongoing change. Medical Physics in 2020 is not the same as when I started in 1990 and as we are a small profession, every member has an opportunity to contribute to the change.

A lot of this goes unnoticed: medical physicists are often in the background—and many like it this way: if everything goes as per plan they have done their job well. Medical Physicists are suddenly visible when things go wrong or when change is needed. As change comes in many shapes and things can go wrong in so many different ways defining this aspect of the role of medical physicists can be tricky. However, there are several defining features and I will try to illustrate some of the characteristics of medical

T. Kron (✉)
Peter MacCallum Cancer Centre, Melbourne, Australia
e-mail: tomas.kron@petermac.org

© The Author(s), under exclusive license to Springer Nature
Switzerland AG 2022
J. Van Dyk (ed.), *True Tales of Medical Physics*,
https://doi.org/10.1007/978-3-030-91724-1_17

physics through my own story. This is obviously subjective and anecdotal but I hope a few things are generalizable and reading about them is hopefully just as enjoyable as the career itself.

1.1   How Everything Else Builds on the Early Jobs

While having no idea about medical physics, my first jobs prepared me well for what lay ahead:

- At age 14 many kids in 'Rheinhessen', the county in the middle of Germany where I was born, would spend a week or two in autumn picking grapes. Vineyards are often steep and the weather changeable. On particularly cold days, the vintner would come through the rows with a bottle of wine to 'warm up' the troops and I still enjoy a nice dry white (not too cold if I may add). He also scolded us for throwing away the shriveled and moldy looking grapes: they are really helping fermentation. This would have been the first science lesson that stuck with me. Science is not necessarily following the 'obvious' path but nevertheless leads to tangible results. In my second year in the vineyards a friend and I worked with a group of Portuguese women who obviously knew what they were doing and finished twice as many rows than we managed to complete. We were sacked on day two, another valuable lesson for life (even if I cannot really recall what it taught us).
- A few years later I got a job as basketball coach of a junior team of 10–12 year old boys. Despite the fact that it was a small group I had real difficulties getting their attention let alone explaining the finer points of jump shots. Things changed when I followed the advice of a more experienced colleague and spent the first half hour with the team just running up and down the gym without an obvious aim (a piece of chocolate may have been involved). After completely and utterly exhausting everyone including myself, the team was ready, and still huffing and puffing, everyone was listening. I took from this that one must not ignore physical needs and listen to those who have relevant experience. Getting older, my half hour bicycle commute to work everyday replaces the running up and down the gym.
- I also got a job at the local TV station carrying cables and putting out props. Here I learned that things are not as they seem, from blue screen to lip-synching and furniture without a back. It took me years to accept that a 'virtual' environment could be stimulating and useful as a model. In any case, our family did not own a TV for many years and we brought our kids

up largely without it. A medical physics colleague realized our son's need and gave us his old TV so we could watch 'Hockey Night in Canada'. I realized that looking out for others and taking decisive action is an integral part of medical physics.

In practice, it was tutoring of high school and university students, which gave me a first glimpse of physics and medicine in combination. I had to teach physics to undergraduate medical students and there was not much appreciation for each other on both sides. Blood pressure and mathematics seem to have little in common; maybe except for the fact that the flow rate decreases with the power of four as a function of the diameter of the blood vessel as per Poiseuille's law. Power of four is a lot and explains why blood vessel narrowing is such a big deal in cardiology. I did not think much about it at the time but this lack of mutual appreciation in the presence of vitally important overlap of concepts is clearly one of the key challenges and opportunities for medical physicists.

1.2   Events in Order of Importance: Marriage, Study, Ph.D., Migration and the First 'Real' Job

I commenced my studies at the University of Mainz in Germany and was initially enrolled in Philosophy as the major. This sounded like a good way to make the world a better place and physics as a minor served well to provide the tools. As it happened in third semester both Philosophy and Physics offered a seminar on 'Space'. Philosophy started with the ancient Greek concepts and ended somewhere around Leibniz, a German philosopher and mathematician (1646–1716). The physics sessions on the other hand took Newton (a contemporary of Leibniz, both of them inventing calculus around the same time) as a starting point and very quickly moved to Einstein and relativity. As a young man, enlightenment was lost on me and I felt that the Philosophy train was stuck at the station while the Physics express passed by. For me it was physics from then on. However, a broad interpretation of what physics is has stayed with me and I was probably the only student giving a seminar talk on the science behind divining (which turned out to be not compelling).

After my physics degree, I made one of my better choices in life: I moved to Munich to work for 18 months in radiation protection at a research centre, the 'Gesellschaft fuer Strahlen- und Umweltforschung' (Research Centre for Radiation and Environmental Sciences or GSF for short). This was part of my civil service as I had been a conscientious objector to the military service which was compulsory at the time in Germany. Having a degree in physics gave me the choice of radiation protection as civil service to society. It was largely a 9–5 job (very different to my studies) and gave me ample of opportunity to explore life. One of the hangovers of this time is my ongoing love for opera from Purcell to Wagner and a bit beyond.

It had been a breath of fresh air to be out of university. However, working in radiation protection made me aware that I only had commenced a journey: physics of radiation is half of the story. Radiation starts with a radiation source and radiation transport but the effect is mediated by biology. As I had my last biology lesson in high school I felt underprepared and ready for more learning. I enrolled in a Ph.D. program at the Johann Wolfgang Goethe University in Frankfurt at the Institute for Biophysics investigating the "Metabolism of Tellurium in Man". Tellurium is a trace element of limited interest in medicine but tellurium and its isotopes constitute one of the most important radioactive contamination in the first days after a nuclear disaster and knowledge of how it is ingested and distributed in humans is of interest. Therefore, the study also introduced me to common methods used in nuclear medicine such as compartment modelling. Most of my friends did not think this topic was very exciting. That is until the nuclear disaster in Chernobyl rocked the world in 1986 about half way through my Ph.D. and many charts of radioactive clouds included tellurium—an unexpected and brief bout of attention, something physicists seem to enjoy from time to time.

I met my wife, Robyn, when doing my Ph.D. in Frankfurt and after my completion, we moved to her home country of Australia to get married in 1989. Robyn comes from the Western Suburbs of Sydney and her parents helped us to get settled. Finding a job was not as straight forward as we had hoped and I did some part time postdoc work and a bit of consultancy. My research work at the University of New South Wales was on Magnetic Resonance Imaging (MRI), a marvellous technique to obtain the most amazing images of almost everything containing water. However, part time postdoc work did not quite pay the rent and I applied for jobs in teaching, asbestos removal and finally in radiotherapy physics. To be honest, I did not have much of an idea as to what this was but at least I had by now local references in the form of my postdoc mentor, Jim Pope, and the father of a friend who was a prominent nuclear medicine physician. During my interview, an elderly gentleman came in for a few minutes, asked a polite question and left again. I got the job. As I learned later, the elderly gentleman was Professor Rod Withers, the head of the department at Prince of Wales Hospital in Sydney and one of the greatest radiation oncologists and scientists. Apparently he had stated that 'he wanted the one with the Ph.D.'—and that was it. The

Fig. 1 Professor Rod Withers to whom I owe my career in radiotherapy physics

four years I had been laboring over my thesis on 'tellurium in man' had been well spent (Fig. 1).

I still have many fond memories of my first job in radiotherapy physics and my job description gives some insights as to what medical physicists do:

- To perform quality assurance activities to make sure that potentially dangerous radiation therapy equipment is fit for purpose
- To provide an in vivo dosimetry service for patients undergoing treatment
- To support the brachytherapy service
- To do whatever else is needed to keep the show on the road from a technical perspective.

I will expanded on these topics in the next section; in brief, medical physicists provide necessary technical services to support patient care. Three words sum this up nicely: it is all about safety, quality and innovation.

2   The 1990s: A Time of Innovation

The 1990s were a perfect time to come into the field of radiation medicine: technology and new techniques were everywhere, and I was fortunate to be part of the first stereotactic radiosurgery treatments in Australia. I also was responsible for the in vivo dosimetry service, which entails measurement of radiation dose directly on the patient during the radiotherapy delivery. It was most gratifying to see patients, talk to them and be part of their care team. This involvement in patient care was also a key aspect of the brachytherapy service which involves the implantation of radioactive sources directly into the tumour of the patient. The implant was done by clinicians but the preparation and most of the handling of the radioactive sources is the job of the physicist.

2.1   On the East Coast of Australia

During my time at Prince of Wales Hospital I maintained my connection with my university mentor and continued some work on MRI in his lab out of hours. I also got my first competitive grant to buy a microwave oven to make radiation sensitive gels. Two years later my wife and I moved to Wollongong, to be in a smaller place and start a family. Robyn had studied languages and was fluent in German. She had been teaching English to migrants and later enrolled in a Ph.D. project in linguistics at the University of Wollongong.

We had said that children would not change our lives—I never was more wrong even considering some of the blunders I have made throughout my career. Our two children Edmund and Alessandra were born in regional Australia (Wollongong and Newcastle) and both appreciate science but studied arts related subjects, which is perfect from my perspective. My father had not been a scientist but was always supportive and curious about it. I always thought it a good thing that we could discuss the world from different perspectives without treading on each others professional toes.

My boss in Wollongong was Peter Metcalfe who taught me 'radiation' by imagining how a photon or electron moved through matter, step by step, one choice of interaction at a time. This is something I still do now and then as it provides 'physical' insight into how particles would see the world (and it gets you out of the chair and meandering through the room). Peter also introduced me to Monte Carlo calculations and the concept of modelling reality, something that greatly enhanced my appreciation of dosimetry.

I got involved in thermoluminescence dosimetry (TLD) and pondered the concept of the 'ideal' dose measurement tool shown in Fig. 2. TLD comes pretty close in my humble opinion, being small, of one material only, tissue equivalent and without the need to attach cables. Solid state dosimetry of which TLD is one method also brought me into contact with many colleagues who had a great influence on me. Of particular note is the late John Cameron who was Professor for Medical Physics in Wisconsin and one of the first

Fig. 2 The ideal dosimeter, which is highly sensitive but so small that it cannot be seen in the figure and of a material that blends in perfectly with its surrounds

academic medical physicists. He encouraged me to stick with TLD and not fear radiation. His quote "everything worth doing is worth doing badly" sums up my approach to medical physics nicely: it is important to patient care but given the complexity of the human condition bound to be imperfect science.

Peter Metcalfe also allowed me to participate in one of his passions, "The Book" (Fig. 3). For those who are curious there are still a few copies on sale at Amazon (P. Metcalfe, T. Kron and P. Hoban. *The physics of radiotherapy x-rays from linear accelerators.* Medical Physics Publishing, 1997). Peter's vision of a book with relatively narrow scope was astonishing and without his enthusiasm I would have never thought it possible to actually write something longer than a few pages. The book also gave me one of the most humbling experiences in my career when Jack Cunningham (who co-wrote THE defining book on medical radiation, H. E. Johns and J. R. Cunningham. *The physics of radiology.* Charles Thomas 1969) bought our book and asked me to sign it.

Finally, Peter taught me to not take administration too seriously and armed with this reassuring knowledge, I took the position as Chief Physicist at the Newcastle Mater Hospital about two hours' drive North of Sydney. My new job description was:

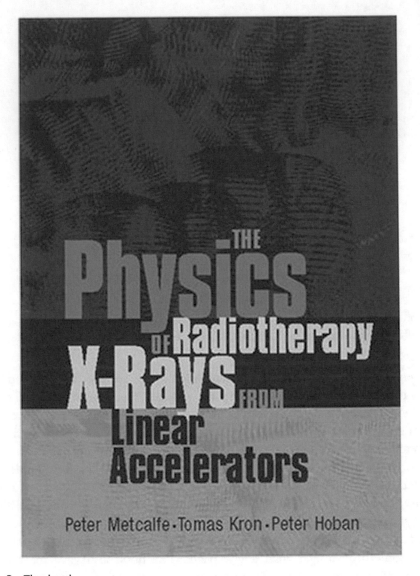

Fig. 3 The book

- To perform and organize quality assurance activities to make sure that potentially dangerous radiation therapy equipment is fit for purpose
- To act as radiation safety officer for the hospital taming the very dangers of radiation
- To ensure that our physics and radiation engineering team were appropriately dressed

- To do whatever else is needed to keep the show on the road from a technical perspective.

Newcastle in Australia takes its name from Newcastle in the UK and it is indeed a medium size city built on industry and coal, but unlike Newcastle in the UK it has some of the best beaches in the world which feature prominently in local life. When moving to Newcastle I did not know anything about this, nor did I realize that the Newcastle Mater Hospital was home to two of the largest cancer clinical trials groups in Australia. One of them, the Trans Tasman Radiation Oncology Group (TROG), is entirely dedicated to radiation oncology trials and Jim Denham, one of the founding fathers of TROG, was clinical director of radiation oncology at the 'Mater'.

2.2 Clinical Trials and Medical Physics

TROG was still in its infancy and supported by the enthusiasm of the clinicians and their trust fund. There was a fondness of clinical evidence, radiobiology and good red wine, something available in abundance in Newcastle which is located close to the Hunter Valley, a famous wine growing region. TROG organized annual meetings that featured famous overseas speakers, likeminded clinicians from New Zealand and Australia, a choir gathering around the piano bar until early hours in the morning and an adventure afternoon: Hiking a volcano anyone? Quad biking? Rafting? No previous experience required. My kids enjoyed some of the annual meetings as well. Most of all, it was a welcoming group which was very happy to include physicists and entertain their ideas about quality assurance.

My role as the first 'TROG physicist' was to read protocols, help with the radiobiological modelling for some trials and, most importantly, organize a quality assurance (QA) program modelled on emerging activities in North America (the Radiological Physics Center (RPC) in Houston, Texas, US), UK and Europe. This culminated in a dosimetric intercomparison using an anthropomorphic phantom which can be used to mimic the patient treatment journey from start to finish.

The phantom travelled with us all over Australia and New Zealand as can be seen for a visit in Queensland in Fig. 4. We tested the ability of centres to deliver a head and neck and a prostate cancer treatment. It was a wonderful experience to meet colleagues in many different radiotherapy clinics, observe that there are many ways to approach treatment delivery and, most importantly, learn that every centre genuinely tried to do the best for patients.

Fig. 4 Our anthropomorphic phantom relaxing at the Goldcoast in Queensland during an audit for TROG

The notion about the importance of clinical evidence, the role of QA to improve data quality, the need of involving patients and listening to their experience, and the necessity to publish results stem from my involvement with TROG. In addition to this, I learned that medical physicists are part of the clinical workforce and that they could ask meaningful clinical questions.

My colleagues at the Newcastle Mater Hospital were most supportive for various endeavors. This included hosting the annual Engineering and Physical Sciences Conference (EPSM) for Australia and New Zealand in Newcastle (see Fig. 5), engaging in clinical trials and chairing the state branch of our professional organisation, the Australasian College of Physical Scientists and Engineers in Medicine (ACPSEM). I was also allowed to engage with the University of Newcastle where I supervised my first Ph.D. student, Patricia Ostwald, and we managed to set-up a Medical Physics Master's degree program. I realised that teaching—at least for me—does not come just naturally. It is a lot of work and preparation, but also tremendously rewarding in particular when seemingly simple questions from students open both a can of worms and a new world of understanding.

3   The Early 2000s: International Departure

Over the years my role at the Newcastle Mater Hospital evolved and drifted more towards administration and management. I don't think I was ready for this and when an opportunity came up to join Jake Van Dyk's group

Fig. 5 The conference convener playing a blues at the closing ceremony for the Engineering and Physical Sciences Conference (EPSM) in Newcastle in 2000

in London, Ontario in Canada, I could not resist. In the late 1990s I had presented a paper by Thomas Rock Mackie in the oncology journal club: "Tomotherapy, a new concept for the delivery of dynamic conformal radio-therapy". There was polite interest amongst the clinicians and the question was asked if something like this could ever work. In 2001 I moved to Canada to take up the position of 'Tomotherapy co-ordinator' to show that tomotherapy can work and treat patients. This was a dream come true with a simple job description:

- To do whatever is needed to get the tomotherapy show on the road at the London Regional Cancer Centre at least as far technology is concerned.

London in Canada is named after the London in the UK and for a short time in history was poised to become Canada's capital. Despite being more southerly than its bigger namesake in England, it can get colder and ice skating (particular in the guise of hockey) is a big thing. So is innovation and a lot of this is linked to medical radiation. The first ever cobalt radiotherapy treatment in the world was performed here in 1951, which was one of the most important milestones of cancer medicine and one of the few events in radiotherapy that made it onto a national postal stamp as shown in Fig. 6.

As it turned out the first generation of tomotherapy units delivered absolutely fantastic dose distributions—unfortunately, there were a few reliability issues (long since resolved) and for cancer treatment hitting the target four out of five days with exquisite accuracy is not sufficient. I learned that the reliability and safety of medical equipment is expected by society to be nearly perfect. If radiation is involved the public and media will prefer this 'safety' margin to be even larger. This has important consequences which include the need of medical physicists who spend a lot of their time ensuring equipment is fit for purpose and, needless to say, very safe.

Due to the delays in moving to clinical implementation of tomotherapy, I tried to make myself useful for other tasks in the department and over time my job description acquired a few additional points:

- To support the safe introduction of new treatment techniques such as motion management on conventional linear accelerators
- To support the teaching program for radiation oncology and medical physics residents
- To do whatever else is needed to keep the show safely on the road from technical perspective.

Tomotherapy went clinical in the end in London (3rd place in the world). One of the more important features of going clinical was that all patients treated initially were part of a clinical trial. This ensured safety and quality for the innovation by requiring:

- a clear protocol and defined endpoints,
- selection criteria for patients,
- approval of an ethics committee,
- informed consent of the patients,

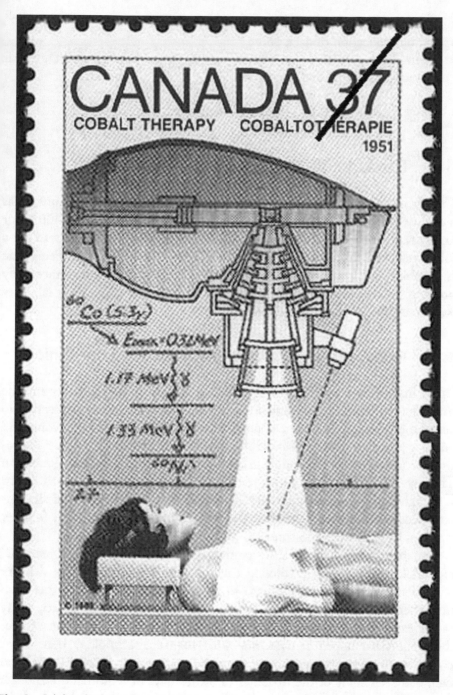

Fig. 6 Celebrating the first treatment with a cobalt-60 unit in London, Ontario. I arrived 50 years later

- quality assurance measures (including checklists),
- rigorous collection of data, and
- the need for publication so we can learn from the past and do not repeat the same thing over and over again.

Safety, quality and innovation at its best.

4   The Late 2000s: Return to Down Under

Canada is a wonderful country, but the sky of the southern hemisphere has a quality that makes people born here (like my wife) homesick. In 2005, our family returned to Australia and settled in Melbourne. I was lucky to get a job as research physicist at the Peter MacCallum Cancer Centre. Peter Mac is a public health care and research facility solely dedicated to cancer. My position was new and rather unique in the institution: while there is a very large research division my role was located within clinical services.

4.1   Technology and Techniques: Research in the Clinic

In my first years, I was floating on a wave of interest in tomotherapy and the image guidance and adaptive features inherent in the concept. However, no actual tomotherapy unit was installed in Australia for some time to come and another paradigm change was rolling in: Stereotactic Ablative Body Radiotherapy (SABR) also commonly referred to as Stereotactic Body Radiation Therapy (SBRT). The technique pushes radiotherapy to its technical limits: very high doses are delivered with high precision, small margins and all stops are pulled out in terms of image guidance and motion management. In other words, a great challenge for medical physicists.

Understandably there are also significant safety concerns and conventional radiobiology, the art of predicting the effect of a given radiation delivery on patients, does not easily extend to doses of more than 10 Gy per fraction when we were planning to give twice as much. As a matter of fact, many earlier experiences have seen significant toxicity at these high doses, something clinicians hoped to overcome with exquisite technology that allows precise focus on the target. One of the clinicians I respect hugely, Prof David Ball, had seen some of the toxicities in earlier days of radiotherapy and felt that SABR should only be introduced as part of a clinical trial with all its checks, balances, multidisciplinary input and most importantly the involvement of patients through representatives and informed consent. This led to the TROG 09.02 trial 'Conformal Hypofrationated Image guided "Stereo-tactic" radiotherapy of Early stage Lung cancer' (CHISEL). One of the take home messages for me was that a good acronym is an important feature of a successful trial—it helps clinicians to remember it and consider patients for participation while in a busy clinic. The trial was a lot of work and a great success in the end proving the efficacy of SABR in the context of early stage lung cancer and helping to establish SABR technology in Australia.

Have I told you what my job description was?

- To introduce new technology and techniques into radiotherapy at Peter MacCallum Cancer Centre
- To support clinical trials
- To teach postgraduate students and radiation therapy professionals about research skills and support their projects
- To try and attract funding to support our operations.

In the context of the fourth point I would like to acknowledge the Eric and Elizabeth Gross Foundation. I was introduced to Elizabeth Gross through David Ball and the foundation created in her name has supported our medical physics work over many years. The generosity of the Gross family and now their trustees has been a tremendous help in achieving some of the research aims we were setting out to achieve. I was particularly excited that they funded a Ph.D. scholarship—there is no better way to make science and scientists relevant not only today but also in the future.

Otherwise, the job description was a physicist's dream and, as all dreams, it was bound to end sometime. In 2014, our head of the department, Jim Cramb, retired and after a year of toing and froing, I was offered the position as Director of Physical Sciences at Peter Mac.

4.2   Interlude: How Come I Forgot to Become a Competent Programmer?

Before I continue with the timeline and job descriptions, I would like to draw attention to an interesting conundrum: nowadays the most important tool of the medical physics profession is the computer and programming ability sharpens this tool. Even if it does not make its way into job descriptions it is something that makes medical physicists even more useful and effective at their job.

I am from an era when computing at university involved Fortran, punch cards and patience. I always used computers but never mastered them.

Problem solving for me is still more related to a pen, a napkin or a white board than a computer, key board and mouse. However, I realize this short-coming, which makes me dependent on others—usually for better not for worse. It also brings back a feeling that accompanied most parts of my career—that of being an impostor by advising, teaching and commenting on rather complex issues, most of which had no perfect answer. At least from my current understanding this is one of the defining characteristics of medical physics practice: we try to provide exact answers to questions which are too complex to have one.

This takes me to a brief mention of my view that medical physicists do not make enough fuss of the fact that they are the guardians of many plat-form technologies in medicine. We often think of platform technologies as genetics, proteomics and molecular oncology. However, a lot of them are related to computing, something which becomes evident in fields such as Monte Carlo calculations, optimization and of course artificial intelligence. I was fortunate to be part of the early uses of another platform technology, 3D printing, into radiotherapy. At least in my case, this highlights also that a Ph.D. is not a one way street: I learn at least as much from the two Ph.D. students working on projects related to 3D printing in our department as they learn from me.

In the end, programming was never part of my job description, but I very much appreciate what a good programmer can do and would consider writing it in future position descriptions.

5   International Engagement Over the Years

The laws of physics are the same all over the world. Unfortunately, this cannot be said about Medical Physics practice. There are significant impediments due to access to education, resources and professional recognition.

5.1   IAEA and AusAid

Whilst working in Newcastle I got involved through the International Atomic Energy Agency (IAEA) in training of a colleague from Vietnam, Nguyen Xuan Cu from Hanoi, in 1995. It had been a very rewarding experience even if our ability to deliver comprehensive training in radiation oncology physics was probably not as comprehensive as we had hoped. One of the most impor-tant realizations from the 'training' was that my colleague was as senior as I was, had more experience in treatment planning and just lacked access to some of the resources we took for granted.

A year later I was invited by the IAEA to visit Vietnam and review radiation safety practices in public radiotherapy centres. I visited Ho Chi Minh City, Hue, Danang and Hanoi and caught up with my colleague. I remember how busy the department was and how radiotherapy even in a resource constrained environment can provide significant benefits for patients. I also had the opportunity to learn about brachytherapy treatments, some of which were still performed using radium, a very long lived radioactive material. This provided a fascinating window to history as the radioactive sources used for treatment had originally come from France and the source certificate was issued and signed by M. Curie (Fig. 7). It was inspirational to see her impact on patient care more than 60 years after her death.

Another important realization in my travels has been that the creativity amongst colleagues all over the world is one of the most important assets in medicine. Figure 8 shows a cobalt-60 radiotherapy unit in Papua New Guinea, where I was with a mission supported by AusAid, the Australian Overseas Aid organization in 2000. The mission consisted of a radiation oncologist, a medical oncologist, a radiation therapist (RTT) and me as a medical physicist. These team activities always proved to be the most effective way of understanding and supporting local practice. The IAEA QUATRO missions are an excellent example for this with QUATRO standing for Quality Assurance Team for Radiation Oncology.

The cobalt unit in Fig. 8 was more than 20 years old but a clever engineer had mounted a diagnostic x-ray tube at the back, making the unit also useful as a simulator which has been used for many years in radiotherapy treatment planning. More importantly, it also allowed taking good quality images of patients prior to treatment—something the world adopted widely many years later as Image Guided Radiation Therapy (IGRT).

5.2   Certification and Professional Recognition

Having had the opportunity to travel I quickly realized that most medical physicists all over the world know what they are doing. I assume that being grounded in a basic science probably has this effect. However, not all medical physicists appear to have the opportunity to contribute as much to patient care as they could, be it for lack of access or recognition. As such, I was very excited in 2012 to get an opportunity to work with a relatively new group, the International Medical Physics Certification Board (IMPCB). The board, which was established through the vision of Raymond Wu, has two functions:

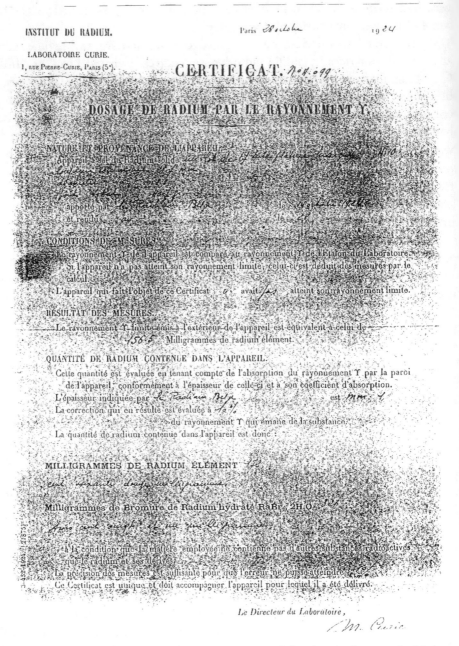

Fig. 7 Source certificate for radium sources used for brachytherapy in Vietnam signed by M. Curie

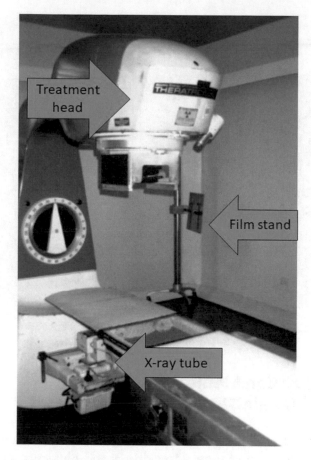

Fig. 8 Image guidance in Papua New Guinea (2000)

- Accreditation of existing medical physics certification boards with the aim to ensure all boards meet the same high quality of assessments
- Certify medical physicists from countries where no such board is available.

All this has the objective to improve medical physics practice all around the world by ensuring a suitable standard of knowledge and practice independent of location or place of practice and—possibly more subtle—giving medical physicists more opportunity to provide input in decision making by raising their profile. The importance of this activity has recently been recognized by the International Atomic Energy Agency (IAEA) in a publication on "Guidelines for the Certification of Clinically Qualified Medical Physicists". I am sure that all the colleagues in Fig. 9 would approve.

Fig. 9 Celebration on the occasion of the IMPCB accreditation of the Korean Medical Physics Certification Board in the National Assembly, Seoul, November 2015

6   2010s: Can Medical Physics and Administration Co-exist?

I was 56 when I took the position of director of our department of Physical Sciences consisting of some 45 engineers and physicists working across five campuses. At that age I felt that my ability to invent cutting edge stuff was diminishing and it seemed reasonable to take a position where I may be able to support younger colleagues. It is difficult to say if this actually works in practice—in any case there is a lot more desk work. My involvement in research is still on-going, albeit in a more hands-off fashion and usually out of hours.

My job descriptions is:

- To oversee medical physics and biomedical engineering services
- To co-ordinate technical support for radiation oncology
- To motivate staff
- To do whatever else is needed to keep the show safely and efficiently on the road from a technical and administrative perspective.

My actual day to day activities are a bit more prosaic and involve routinely 20 meetings and 1000 emails per week.

However, there are also several perks of the job and every day is different as it would be for all my colleagues. Without doubt working with physics and engineering colleagues is the most rewarding part of the job and I would find it very difficult to work alone in a small department.

Other highlights of my job in no particular order are:

- Having been able to reappoint a colleague to my original research position (after one year of haggling). Nick Hardcastle proved to be the perfect fit for the job where he has been involved to introduce several exciting new treatment approaches such as cardiac SABR and CPAP motion management... (but this is not my story anymore).
- Being involved in major projects such as the movement of our hospital to a new site into the spectacular building shown in Fig. 10. In addition to learning how to work on building sites I was involved in design and logistics of the move. We were able to secure additional space for Physical Sciences with two 'laboratories' (dosimetry and imaging). We can now justifiably say in conference talks, "I am from the 'so and so' lab..." Most satisfyingly though, the new building is liked by patients who appreciate being in an environment that looks nothing like a hospital.
- Writing mission statements, organizational charts and job descriptions. It is probably a sign of age that I enjoy putting together documents that I always thought of as 'garnish' in my earlier years. I have learned that these documents are important and the mission of our department: 'Safety, Quality and Innovation' sounds better to me every day...

7 So, What Am I Doing Anyway?

My apologies for taking so long to answer what sounded like a simple question. Based on the discussions above there is no 'standard' medical physics job and the field is ever changing. As the use of technology in medicine expands so does the role of and demands on the medical physicist who is the professional at the interface between technology and the patient. From my perspective, Medical Physicists need to manage this interface with three important considerations in mind:

Fig. 10 The new Peter MacCallum Cancer Centre in Melbourne

- *Safety*: This is obviously the first thing on medical physicists' agenda and also the most likely reason why society pays us. It is easy to make an argument that someone needs to ensure that complex technology is safe, in particular if in direct contact with vulnerable patients.

- *Quality*: This is at the core of medical physicists' practice: equipment is not just fit for purpose but performs as best as possible. This includes education, training and documentation to ensure that the equipment is not just in optimal condition but also used in the optimal way.
- *Innovation*: For me this is another key component that makes sure, safety and quality will be even better tomorrow. In practice, this is probably why many of us are excited to get out of bed in the morning.

At the core of this essay have been job descriptions. I am now in the fortunate position to draft some of them for people who do all the things I have described and more. They are—like the jobs—always in flux; however, I am sure that medical physicists always add value to the clinic. At this point in time I would summarise their job as:

- To provide services for patients undergoing diagnosis and/or treatment
- To perform and organize safety and quality assurance activities to make sure that therapy and diagnostic equipment is not only fit for purpose but also optimal for use
- To introduce new technology and techniques
- To do whatever else is needed to keep the show on the road.

Tick…

Tomas Kron was born and educated in Germany. After his Ph.D. he migrated to Australia in 1989 where he commenced his career in radiotherapy physics. From 2001 to 2005 he moved to Canada where he worked at the London Regional Cancer Centre on the commissioning of one of the first tomotherapy units. In 2005, Tomas became principal research physicist at Peter MacCallum Cancer Centre in Melbourne, Australia where he now is Director of Physical Sciences. He holds academic appointments at Wollongong, RMIT and Melbourne Universities and has been elected Fellow of the Australasian College of Physical Scientists and Engineers (ACPSEM), the Canadian Organization of Medical Physics (COMP), the International Organization for Medical Physics (IOMP) and the International Union of Physical and Engineering Sciences in Medicine, (IUPESM). Tomas has an interest in education of medical physicists, dosimetry of ionizing radiation, image guidance and clinical trials demonstrated in more than 300 papers in refereed journals and 90 invited conference presentations.

In his spare time Tomas enjoys family, opera, playing guitar and riding his red bicycle.

Awards

2020. Life time membership award, Trans Tasman Radiation Oncology Group (TROG).

2018. Inaugural "Prof. Kiyonari Inamura Memorial AFOMP Orator" at the Asia-Oceania Federation of Organizations for Medical Physics in Kuala Lumpur November 11–14, 2018.

2014. Order of Australia Medal (OAM) for services to medicine, research and education

2013. Distinguished Services Award of the Australasian College of Physical Scientists and Engineers in Medicine (ACPSEM).

2013. One of 50 outstanding medical physicists featured at the International Conference of Medical Physics (ICMP) in Brighton, UK, to celebrate the 50th anniversary of the International Organization for Medical Physics (IOMP).

18

Ever Bored? Not Likely!

Martin Yaffe

1   Another Prairie Boy

It was on a very cold day with the promise of spring barely in the air when the phone rang. "When are you coming?". "Who's calling? I asked". "Harold Johns", the voice replied gruffly, sounding perhaps surprised that I needed to ask this question. "So, when are you coming?". For a moment, I was speechless.

It was 1973 and I was living in my hometown, Winnipeg, Canada, newly married and working toward the end of a master's degree in physics applied to medicine. My supervisor, Dr. Arthur Holloway, worked in a cancer clinic and had a background in clinical physics, but his interests had turned to virology. His skills would have been in great demand these days. But he had given me a project to work on in a new area, medical imaging. This was an area in which nobody in the lab had any experience. The lab, in the basement of the Winnipeg General Hospital, was a mixture of scientists supporting the radiation therapy facility in the cancer clinic and those working with Holloway's virology research program. I was pretty much on my own. There were no fellow students in the lab, just a few younger employees who took me under

M. Yaffe (✉)
Sunnybrook Research Institute and University of Toronto, Toronto, ON, Canada
e-mail: martin.yaffe@sri.utoronto.ca

© The Author(s), under exclusive license to Springer Nature
Switzerland AG 2022
J. Van Dyk (ed.), *True Tales of Medical Physics*,
https://doi.org/10.1007/978-3-030-91724-1_18

their wings, so I sought out and found kindred spirits—people interested in imaging and computers, wherever they existed in the rambling old hospital.

Winnipeg is in the prairies where land was readily available, so rather than being compact and high, the hospital consisted of a dozen or so separate buildings of varying height, connected together underground by tunnels to shield workers and patients from the long, frigid winters. It was a lot of fun exploring these tunnels on my way to find a coffee, take classes or meet with some of my new colleagues.

At the time, there were only a couple of courses specific to my new discipline. Dr. Holloway suggested that I make up a program of courses from different departments. I was delighted! As an undergraduate, I had looked forward to studying a wide range of subjects—getting an education. But after first year Science, I was given clear instructions that I needed to focus. No more, philosophy, chemistry, economics, languages, etc. Physics, math, more math and more physics. I had really not been prepared for such specialization. But now, along with my radiation physics and radiobiology courses, I was free to take a course in anatomy in the medical school (we dissected a cadaver), and physiology for engineers, to study human vision.

I had great colleagues in Winnipeg. I had come from an undergraduate honours physics program where there were a lot of bright serious people—as well as a few very bright ones who were also serious but loved to have fun. Alan Wright was one of them. A very sharp guy who learned about computers early and as a student, spent part of his time at the hospital figuring out how to make digital nuclear medicine images. He had a keen and amazingly creative mind. He combined an easygoing nature with a powerhouse ability to get things done. He really taught me how you can work hard with full attention and then party hard.

2   The Spark

It was as 4th year undergrads that a group of us were asked late one afternoon by the department chair if we would be willing to take the visiting lecturer out to dinner. Well the bill was being paid, so there was little chance that hungry (and thirsty) students would say no. But of course, first we had to listen to some dull talk. Well, that "dull" talk changed my life. The speaker was Dr. Harold Johns, who was doing a cross-Canada tour to stimulate students to consider entering the emerging field of medical physics.

Most of our lectures were delivered as pre-recorded television which a couple of hundred of us watched in a darkened theatre, doing our best to stay awake. But this was very different. Here was a guy who was a bit over 50 years old and literally bounced into the room, exuding energy. He would walk up to a student, poke him in the chest and ask a question like—"what do you know about pi mesons?" "Do you think you could treat cancer with them?", or "what do you think happens in the first billionth of a second after radiation hits a cell?". He went on to describe with great exuberance the research that he was doing with his students and colleagues in his department at the Princess Margaret Hospital in Toronto. His enthusiasm was contagious, and the students, especially me, were literally luminous with excitement. When the lecture ended, the audience, rather than trickling out of the room to their commute home, stayed in the theatre for half an hour peppering Johns with questions.

Although I usually travelled by city bus, I had managed to borrow my father's pride and joy—a huge mint condition, 1968 Dodge, resplendent with bronze metallic finish, reclining seats and a vinyl roof. By today's standards it was a cruise ship of a car, and pretty much handled like a ship. Johns, his suitcases, and Alan Wright, along with my other miscreant physics friends, Peter Taylor and Jim Rae, piled into the car and took Johns to what to us was a luxurious restaurant (it had a liquor licence), The Old Spaghetti Factory. Johns was noticeably impressed that a bearded young punk like me would be driving such an establishment car.

Well we had a wonderful time. Johns offered to buy us a drink and to his surprise, we all ordered whiskies, straight. As he sipped his beer, he alternated between regaling us with stories of the excitement going on in medical physics research and interrogating us as to our experience and perceptions as students. By the time I dropped him off at his hotel, I knew what I wanted to do "when I grew up". And that's what led to my being in the master's program studying physics applied to medicine when two years later Johns called me on the phone.

By then I knew who Johns was—inventor of cobalt radiation therapy, author of a textbook on medical physics, translated into many languages, Officer of The Order of Canada, and an international superstar in his field. As I was contemplating what to do next after my master's I was also deeply immersed in what was happening in the world at the time. The Vietnam Conflict was still blazing, with many people on both sides dying. The political turbulence in the U.S. spread quickly to students in Canada. I had just got married and the influence of feminism was entering my life. Much like today, it was a time of great turmoil. And inspired by the Dustin Hoffman character in the film The Graduate, I had bought a little red Austin Healey sportscar (which I still drive today). There was no way that I could imagine myself working in the sphere of somebody like Harold Johns and, although I applied to some university Ph.D. programs, I couldn't even conceive of being accepted to work with him. So I didn't apply.

When his call came, I was shocked. I was also surprised that when I mentioned his offer to my wife she immediately said, "sure, let's do it!". So, the sports car went into storage and off we went to Toronto.

3   Life in the Big City

The Johns lab was very different from what I had experienced in Winnipeg. As a rookie in the lab, I had to earn my stripes. At first, Johns challenged everything I did. This was long before we did our writing on computers. It was all typed or written out. He liked to do things longhand, simply using a tube of Liquisilk glue to join pieces of ruled paper together to form long, scroll-like documents. If there was an error, he would use a ruler to tear the sheet neatly, remove an offending paragraph and glue in additional paper for a correction. Eventually his assistant, Stellis, would type these up.

Johns, had an abrupt nature, not hesitating to poke you emphatically when he felt that you needed to think harder. Often, he would read your carefully written document, hold it up between you and slowly rip it to shreds before filing it in the trash bin. If it was late in the day, he would tell you to meet him the next morning at eight to re-write.

What I learned in those sessions was how to sharply focus your thinking on a problem and how to persist until you had a workable solution. Dr. Johns had two more senior students at the time—Aaron Fenster (see Chap. 19) and Don Plewes. These two and the lab technician, Dan Ostler more or less adopted me and provided mentoring to prepare me for my sessions with Johns. Also, Johns used to invite me into his office when he was working on a paper with Aaron or Don and let me watch. While he was more respectful toward them, it was not uncommon for him to fix one of them with his laser-like glare which he held on them for what seemed like minutes and then say something like: "Plews" (he never pronounced the "e" that made it rhyme with Lewis; instead he made it rhyme with "news"), "If you sent that to a journal, they'd crap all over you". Or, as he slowly ripped up a piece of writing that Aaron had proudly submitted, he'd say at a similar slow pace, "well (rrrrip) Fenster (rrrrip), your (rrrrip) writing (rrrrip) is improving." So, rather than feel discriminated against, I simply realized that the standards were high, and I'd have to present my best game at all times.

4   Flashback and Punch Cards

Back to Winnipeg for a moment. For people today, it is hard to imagine how primitive computers were in the early 70s. But even then, their usefulness in research was unquestionable. In Winnipeg, as a summer student, my supervisor had been given some software for gamma ray spectroscopy analysis by a colleague in the U.S. and I was given the assignment of making it work for my supervisor's project. I didn't realize it at the time, but it provided me with a great opportunity. My supervisor had spectra (graphs of the number of measured gamma rays versus their energy on the horizontal axis). Each spectrum looked like a series of mountain peaks, but the limitations of the instruments available made the peaks broad. To improve their resolution, one could perform a mathematical operation on them called a deconvolution. It's sort of like bringing a lens into focus. The math was hard to do on the computers available at the time, so the trick was to transform the data by performing a Fourier transform. Then by simply multiplying the transformed data by a set of correction factors and retransforming (called an inverse transform) data back into what we referred to as "the real world", the spectral peaks would be crisp and more useful. An example of Fourier analysis is instead of expressing a passage of music as a time sequence of notes, sorting it into the different frequencies and listing the distribution of how many bass, midrange and high notes were in the piece. (See Terry Peters' Chap. 10 for a further description of Fourier transforms.)

The software, called a "fast Fourier transform", was written on a set of punch cards—cardboard file cards punched with holes representing the data in ASCII (binary coded letters and numbers) form—one line of code on a card. The program consisted of two heavy boxes of cards, which every time I had to run it, I carried from the physics building where I edited the code on a keypunch machine to the engineering building where the shared mainframe computer ran as a 24/7 service. This generally occurred in the small hours of the morning in the − 20 °C Winnipeg winter and I remember slipping on an icy sidewalk and dumping several hundred cards out of the box onto the snow. I quickly learned to draw a diagonal line across the tops of the cards with a felt marker to make it easier to recover from this sort of problem.

The benefit in Dr. Johns' lab was that I was the guy who knew about Fourier transforms and this was an extremely valuable tool as the lab started to work on developing imaging systems and examining their performance. In this case, an image, which was a spatial map of brightnesses, could instead be broken down by Fourier analysis into a distribution of "spatial frequencies" where low frequencies represented coarse structures in an image and high represented fine detail. Fourier transforms were also at the heart of the mathematical reconstruction of three-dimensional or cross sectional images for the development of computed tomography and magnetic resonance imaging which were in their earliest phases at that time.

5   Missionary Zeal

Dr. Johns had been born in Chengdu, China to Canadian church missionary parents and he had spent his early years there, roaming about small communities in the mountains of Szechuan province with his father, a no-nonsense disciplinarian who believed strongly in devotion to duty and hard work. He learned to be focussed and driven to succeed at whatever was his mission. When the family eventually returned to Canada, he brought that to his graduate work in physics and later to the University of Saskatchewan where he built a strong medical physics research group, concentrating on developing and refining radiation therapy. There he developed the first (or possibly the second—there is some debate as his unit and a competitor, built by Eldorado Mining and Refining Ltd., were used to treat patients within a week or two of each other) cobalt-60 radiation treatment system and carried out pioneering work on radiation dosimetry and treatment planning. Dr. Johns and his work in Saskatchewan were actually mentioned in the film "First Man", about the astronaut Neil Armstrong whose daughter had suffered from a brain tumour. Later, he began work on "The Physics of Radiology", a textbook which he referred to jokingly (I think) as "The Bible".[1] This book truly became a guide to those working in radiation oncology all over the world and was published in multiple languages. While the book and its various editions consumed many of his evenings after a hard day at the lab, Johns reverse bragged that he earned about two cents per hour on his textbook writing efforts.

When I joined the lab in Toronto, Johns had many former prairie boys from Saskatchewan and thereabouts in the labs at the Princess Margaret Hospital. These included Jack Cunningham, who became a co-author of "The Bible", Gordon Whitmore, Bob Bruce and James Till. Each of these scientists branched out into his own area of cancer research and became a star in his own right. The one "prairie girl", Sylvia Fedoruk, who had worked with Johns in Saskatchewan stayed there, working for many years as a radiation therapy and nuclear medicine physicist and eventually taking on roles as the Provost of The University of Saskatchewan and later the Lieutenant

---

[1] Original title: The Physics of Radiation Therapy. This was generalized as HEJ's interest broadened to diagnostic applications.

Governor of that province, representing the Queen of Canada. No slouches, these Johns alumni/ae. But I can only imagine what a challenge was faced by a woman in this 1950s macho world. And Sylvia more than met that challenge.

But, a pattern was established in Saskatchewan and later brought to Toronto where the lab life didn't end at 6 PM, but carried into a vigorous social environment, bringing together the scientists' spouses and families. In the prairies, this was a sort of necessity to deal with the long, frigid winters, but in Toronto became the glue which transformed a job into a community. When I entered the scene, being part of Dr. Johns' world was an all-encompassing experience. Everything he did, he did with great gusto and competitiveness. This included curling, squash and waterskiing among other activities. He was also an avid golfer, but this was one activity that, possibly due to cost, seemed to exclude student participation.

But almost anything became a competitive exercise. When occasionally he offered to buy his "boys" (at that time all of his graduate students and technicians were male) a coffee, it involved a race down eight flights of stairs from his lab to the hospital cafeteria, where we spent 20 min being grilled on our science or else regaled with anecdotes from his childhood experiences in China. And then a competitive race back up the eight flights. And although he was in his 50s, he was always at the front of the pack. And then there was all the arm wrestling, practical jokes, and other combative competitions that he imposed on his students.

Johns had three daughters, two who had already married medical physicists and, I believe, he had always desperately missed having a son. This may explain the boisterous behavior.

While by today's standards it seems hopelessly male and would not fare well, I think that, much like military training, there was a purpose to it—like it or not. And often I didn't like it so much. It built camaraderie and bonded the students together as a group—if only to complain about Johns' latest outrageous shenanigan. The experience created colleagues in whose scientific judgement and advice I could place great trust for the rest of my career and some wonderful lifelong friendships.

6   Good Timing

I arrived at an exciting time in the Johns lab. His scientific focus was just beginning to shift from radiation therapy and radiobiology to medical imaging so that he was almost as new to the area as I was. This was a great opportunity for me as I had done my master's degree in this area and had done a lot of reading and thinking about imaging. Dr. Johns was interested in a new approach to forming images from x-rays, called electro-radiography or ionography.

To make an x-ray image or radiograph, first you have to stop the x-rays transmitted through a body part with some sort of detector while preserving the spatial pattern of the x-rays. Then you have to convert the absorbed energy into a signal that could be recorded. Traditionally, this had been done using photographic film, which contained a thick silver bromide emulsion that stopped some of the x-rays. When chemically developed, the silver bromide was converted to silver and the amount of silver at each location in the film was proportional to the number of x-rays hitting the film. When placed on a bright fluorescent-lit illuminator pane (lightbox), the processed film looked dark in areas where many x-rays passed through the body and whiter where their number had been reduced by passing through bones, foreign objects or other dense structures in the body. To improve the efficiency of x-ray absorption and use less silver (which is costly) some systems used an additional "screen", a layer of dense phosphor material pressed tightly against the film. The screen converts the absorbed x-ray energy to light for which the film is much more sensitive than it is to x-rays.

In ionography, the film was replaced by a high pressure, high atomic number inert gas like xenon or krypton which absorbed the x-rays and became ionized. The number of ions produced in a given location is proportional to the x-rays transmitted by the patient. In this case, high electric fields set up across the volume of gas caused the ions to drift toward collection electrodes without spreading sideways (this would blur the images) and then the charge would be converted into a visible image.

This was an amazingly exciting and demanding project in that it involved the need to understand x-ray physics, the properties of gases and complex theory around electric field properties and motion of charged particles in gases. While Don Plewes worked out the electric theory, others in the lab built apparatus to actually make electroradiographs. Making x-rays required voltages of up to 120,000 V and our homemade equipment was somewhat primitive. I remember sitting at my desk working on something when there was a sizzling noise a few feet above me as our technician, Dan, energized one of the high voltage lines that was strung from the lab ceiling to make an image exposure. No doubt we were violating several building codes by doing this experiment. I don't even mention the vessels that we built to contain these highly pressurized gases—essentially these were potential bombs.

Despite being a junior member of the group, I got to sit in on all of the scientific discussions surrounding the work in the lab and had the benefit of seeing how scientists look at and analyze information, how to be a critical thinker and how to design equipment. This was of amazing value. Dr. Johns was a firm believer that if you can't buy a suitable piece of equipment, you can make it. For this reason, he made sure that the research division in the old Princess Margaret hospital on Sherbourne Street was fitted with well-equipped electronic and mechanical shops. Each had a creative and highly skilled manager and these two inspiring people, Jim Webb (mechanical) and Bill Taylor (electronics) were key participants at many of our meetings. These two gentlemen, while respectful of Dr. Johns' great energy, drive and creative spirit, were able to listen to his ideas and tactfully steer him from unworkable, idealistic dreams (and potentially lethally explosive designs for pressure vessels) to inventions that could actually do the things he was trying to accomplish. As students, it was wonderful watching this display of diplomatic re-engineering and this was a process that unfolded many times during my graduate years.

Buoyed by the excitement and promise of ionography, Dr. Johns steered me toward what became my Ph.D. project, the design and construction of an x-ray detector array for CT, based on high-pressure xenon ionography. In this case, though, the output of the detector channels (400 in all, arranged on an arc) would be connected to 400 separate amplifiers. The signal from each of these would have to be digitized to produce the data samples needed to reconstruct a CT image.

Thus began the real excitement! Johns was a builder and always felt that you could do a better job at something if you built it yourself. He said that what he loved was the sight of "good Canadian boys with grease under their fingernails" Yes, at that time, they were mainly boys, but fortunately, things have changed since. So, I began to learn about how ions moved under electric fields in inert gases, pressure vessels, and low-noise electronics. And I had the unusual opportunity to meet many of the fascinating and knowledgeable people that Johns seemed to be connected to through his vast personal network.

One evening as I was getting ready to leave the lab for my subway ride home, Johns said—"Meet me at the airport at 5:30 tomorrow". "AM?", I asked. He looked at me like I was an idiot. "Where are we going?" "I'll see you tomorrow. Don't be late."

The next morning, my eyes barely open, we boarded a small plane headed to a small airport on Long Island, New York. I love flying and I enjoy small planes, but I barely got a chance to experience the flight or look out the window, because Johns kept me busy discussing equations, running ideas past me, drawing designs for new detectors on napkins. This barely stopped when, as we approached our destination, the captain announced that there was a problem—the landing gear wouldn't deploy. They were going to try to do this manually. We moved to other seats nearby as a crew member carrying a wrench-like tool started opening a hatch in the floor of the plane and began working with the mechanism, an intense look on his face. Johns, unfazed, continued discussing a particular concern that he had with constructing the electrode shape. I tried to remain cool.

A few minutes later the captain announced that he "thought" that everything was OK, but they would do a fly-by past the airport tower to let them have a look and then also cover the landing area with fire retardant foam. As we landed, with a dozen firetrucks with flashing red and amber lights surrounding the runway, Johns was still sketching with a pencil. That night, after spending a day at a small start-up company talking to an engineer who designed high pressure gas radiation detectors for NASA, we flew back to Toronto without incident. I think of that as a Harold Johns sort of day.

7   Life in the Lab

Because of his reputation, as Johns' students, Don, Aaron and I got to meet many of the world's leading scientists and leaders in the medical imaging world. One morning I was summoned into his office. "Yaffe, meet Godfrey Hounsfield". Hounsfield, who had worked as an engineer for the British giant, EMI (producer of the Beatles music), had, as a side project, developed the first workable CT scanner. We spent an incredible day discussing ideas for how to improve CT images. He later became Sir Godfrey and received the Nobel Prize in Medicine. A few years later at a conference, Johns beckoned me over to him and introduced me to Allan Cormack, the co-recipient of the Nobel Prize with Hounsfield. Working in another, distant country, South Africa, he had developed the mathematics that would be integral to the reconstruction of the CT x-ray data to form images. Interestingly, many of these outstanding scientists like Cormack, were multitalented, in his case, as an artist. Years after our meeting, his daughter, Jean, an accomplished biostatistician in her own rite, and now one of my colleagues, working at Brown University, sent me beautiful drawings that Cormack had done.

Coming in to work always held surprises as Dr. Johns was a magnet for international visitors. There was no pomp and ceremony though—straight to the science with minimal distraction. If food was needed, there was no question of going to a restaurant, or even, in many cases just to the hospital cafeteria. Once when Dr. Karl Reiss, the head of research for the industry giant, Siemens, visited us, his lunch entertainment consisted of our going down to a tiny room with vending machines from which he was offered a (probably lukewarm) hotdog, accompanied by a plastic packet of mustard and a Pepsi. Even as graduate students we were a bit shocked, but Reiss, a consummate gentleman, said not a word.

After much struggle, I was able to build a benchtop CT scanner. I had had to drive out to a nearby industrial plant to program the digital milling machine (these systems had only recently become available) to build the large gas pressure vessel. I had hand-soldered each of the 800 electrodes and built the electronic circuits by hand, working with a younger student engineer, Ralf Brooks, to design the digitizer. Ralf later went into the aerospace industry to help design the CanadArm for the Space Shuttle. Years later, when he was the director of a factory that produced ultrahigh resolution charge-coupled devices (CCDs), we worked together again to adapt these devices for use in digital mammography.

Working with Ralf, I played the part of a semi-supervisor of his research thesis and I discovered the difference in the way that physicists and engineers look at life, at least work life. Engineers learn how to design things very well and to meet specifications and deadlines. Physicists tend to be somewhat less disciplined, adopt an attitude of "I've built this good enough to work, so I'm happy", but are very curious about how and why something works (or doesn't work). Maybe I'm overgeneralizing, but maybe not. I'm impressed by engineers and their skill and find a mixed team of scientists and engineers very effective in moving ideas forward, especially if these are devices or software intended for clinical use.

The CT system and its evaluation was the basis of my Ph.D. thesis, which I had to defend at two separate oral examinations, the "departmental" and "The Senate". The names are really misleading. The departmental is by far the most demanding. No questions are out of bounds. At one point, an examiner prefaced a question with—"I can't possibly pass you on this examination because…". With a doomed feeling, I did my best to defend my work and my position while he sadly shook his head from side to side. I knew that I was sunk, but I figured that I might as well keep trying—nothing more to lose at this point. At the end, the vote by the examiners was unanimously favourable. I asked the professor who had given me the difficult time how he could have approved, given what he had said. He replied, "I was just trying to see how you would react". Hmm….

8   Another Big Break!

Dr. Johns was readying himself for retirement. His Parkinson's disease which was just beginning to progress when I joined the team, had become much worse. Don Plewes had graduated and had taken a position in the U.S. and I was amazed when Dr. Johns offered both Aaron Fenster and myself junior academic positions, as lecturers. This was the first small step toward becoming a professor. The pay was not high, but he was going to let us take over his lab! It is difficult for a scientist to get off the ground in a career and here was an opportunity to move into the momentum of an established laboratory, surrounded by people who I knew I could work well with. This was truly incredible!

Dr. Johns set us up in a tiny shared office, a former hospital sunroom, side by side with a column between us that we continually were bumping into every time we got up from our desks. But it was the start of a career. We were so fortunate that Dr. Johns had a well-equipped lab, lots of space and a good budget.

Unlike today when budgets are highly constrained and there are countless rules with which we spend a lot of time complying, it was a time of what seemed like unlimited possibility—and plenty of surprises.

One late afternoon, Dr. Johns, who by that time had ceremoniously told me that it was time to call him Harold, called us into his office. "Boys, we have a surplus in our equipment budget. I want you to give me a list of equipment that you need for the lab by tomorrow morning. It has to be ordered in the next two days or we lose the funds." "How much do you need to spend?", I asked. "$100,000. No, better make it $110,000." Well, it was Christmas in July. We set to work, going through catalogs from suppliers. We were like kids in a candy store, able to get instruments that we never dreamed of having available to us. We had no idea when or if a windfall like this would occur again, so we tried to think long-term. I didn't much like the idea of being pressured to make hasty decisions, but basically it was—take it or leave it. So, we took it. The money was spent. And I feel good that 40 years later some of that equipment is still operating usefully in my lab.

Another time, Harold called us in and asked, "Do you have passports? If you don't, you'd better get them fast." He then dispatched us on a trip to visit some of the major medical imaging research labs in Europe, Philips in Eindhoven in the Netherlands, and in Hamburg, Germany, and then Siemens in Erlangen, Germany. At each place we were welcomed and, because of Harold's reputation, were treated like rock stars. The experience was amazing. We were wined and dined and later, when the older, more staid officials went home to bed, the younger scientists and engineers took us out to the seedier nightspot areas—and some were pretty seedy indeed! We learned a lot about how research was done in industry. Philips, a giant corporation was active in many other areas than medicine and at one point we were invited into a lab to see the prototype of what later became the DVD player. A few years later, I was in a similar sort of lab near Stanford University in California and, protected by a glass display case, I saw the world's first computer mouse.

While the trip was incredibly useful, it ended sadly for me. Late one night near the end of the visit to the Siemens labs, I received a call in my hotel room that my father had died and the next morning I began the exhausting trip—taxi to Nuremburg followed by connecting flights to Frankfurt, London, Toronto and finally Winnipeg for the funeral. It was a long day. Passing through these multiple countries and airports was just a blur as I thought about the man who had wanted to be an engineer, but had been blocked by poverty and the need to get any job to help support his family, the great economic Depression of the 1930s and World War II. He had never complained about the fact that he had been a bright student and had never been able to have the career he desired—friends of his had told me about that in confidence. He just wanted to be sure that his three sons had the chance to pursue their goals in life. And we did.

9   Life as a Medical Physicist

Medical physics was a lot different from my picture of what a physicist was, Newton, Einstein, Rutherford, Roentgen, Fermi, people like that. Of course, our work was based on theirs and we used what they had learned every day in what we did. But medical physics is applied physics—we get to use the principles of physics, not just to learn more about physics, but to solve immediate real-world problems of health, and that's what appealed to me most, because at heart, I have a bit of an engineer in me. In my job, Dr. Johns arranged that I would spend part of the time in the lab, and I began working to obtain a real graduate appointment, so that I could supervise graduate students. Meanwhile, I began helping one of Johns' colleagues, Dr. Ken Taylor, in teaching medical residents in Radiology the physics they needed to know to do their work, and more important from their perspective, to pass their professional examinations. For most of them, the medicine part was relatively easy, but the physics was a killer. So, they were appreciative when a young scientist was willing and patient in helping them with this. My goal was not just to help them pass the exam, but to actually understand the material, because I felt that this was so fundamental to a career in which you produced images using …. well there's no other way to put it, using physics. Images were produced using x-rays, ultrasound, gamma rays (magnetic resonance imaging (MRI) was still in the future) and, in my mind, the only way to do a good job, and perhaps innovate and improve, was to really understand how these forms of energy could most effectively be turned into medical pictures.

When I wasn't in the lab or teaching residents, I was earning my keep, working as a hospital physicist in the Radiology Department at St. Michael's Hospital. Although I had been in Toronto for 5 years, I had had little reason to wander into other hospitals beside the Princess Margaret before this. Beginning my career as a scientist, I had reason to spend time working or at meetings at other Toronto hospitals and I learned about the rich and fascinating variation in culture and style among them. PMH, as it was known, was a relatively small hospital with a very British tradition. On Christmas Day, the Medical Director, accompanied by two kitchen staff, would wheel a cart through the halls and serve Christmas dinner to the patients and those staff who had to work that day. As a Jewish guy, it was not unusual for me to come in on Christian holidays and one year I had the delight of experiencing this ritual. It was truly a magical sort of place and I suspect that events like this (and the annual faculty dinners where everybody ended the evening with Port and cigars—yes in a cancer hospital) are now ancient history.

St. Michael's was quite different. As one can guess from its name it was a Catholic hospital, run by nuns; its Director was the intimidating Sister Mary. At least she was intimidating to all the staff, who seemed to both fear and respect her. The hospital also had a distinctly Irish flavour and the accent and names, especially among the support staff rang strongly of the Emerald Isle. In contrast, when I entered Mt Sinai Hospital on University Avenue for the first time, it harkened to the Bar Mitzvah's in my youth in Winnipeg. Everybody there seemed to be related to somebody else there or at least knew his (or my) uncle. In a way, both hospitals felt a bit like family businesses.

I learned a lot at St. Michael's. As a graduate student, I learned about the physics of imaging. At St. Mike's I began to understand how things worked in the real world, more from a medical perspective. I saw how patients were affected by the imaging procedures. Were the images good enough to detect and diagnose their problem? How much radiation did they receive from some of these procedures? X-rays are ionizing radiation, meaning that the individual rays (which physicists refer to as quanta) carry enough energy to ionize atoms in cells and potentially cause mutations. If enough of these occur and with bad luck, these can cause cancer, birth defects and other problems. I started to learn about the relationship between the amount of radiation that we are exposed to and the probability of these effects. I also became an expert at measuring radiation exposures.

The radiologists (Drs. Bruce Bird, Norm Patt, Ron McCallum, Bill Weiser) at St. Mike's and Lynne Mongeau, the technical manager for Imaging, were very kind and tolerant of a young academic suddenly immersed in the clinical world. I think that they found me slightly amusing—I knew a lot about the technology and almost nothing about medicine. They talked to me about the procedures and what they were trying to see in the images. I attended radiology rounds (weekly review meetings of patient cases) where residents displayed images and discussed their observations, working toward a diagnosis, while the more senior radiologists peppered them with questions, both to guide them and to test their skills. And I, in turn, introduced quality control procedures to try to ensure that the imaging was consistent and of high quality and that the amount of radiation used to make the images was ALARA—as low as reasonably achievable.

10   HARPing Around

It was radiation dose that brought me to St. Mike's in the first place. A few years earlier Dr. Ken Taylor had done a survey of the amount of radiation used for particular radiology procedures. Not only were the doses high in some cases, but they could vary enormously—in some cases, a factor of 20!—even for the same procedure across hospitals or between different imaging rooms in the same department. Harold, who was a strong believer in scientific publication helped Ken Taylor organize a manuscript for a journal. Meanwhile Harold and Ken called together leaders in the Ontario radiology community to warn them that when the publication came out it was bound to be picked up by the media and there would be a furor. They recommended that the radiologists develop a plan to correct the problems of dose variation and high doses and to do this quickly.

Nothing happened. And, as predicted when the publication appeared it was picked up by the national newspaper in an article by Barbara Yaffe (no relation). And, all hell broke loose. A government commission was formed to review the problem. This led to the provincial Healing Arts Radiation Protection (HARP) Act and regulations, requirements and standards. As a radiation expert, I was asked to be on two of the panels and eventually became their Chair. Thus began my dubious relationship with government. I had the opportunity to help guide the development of the radiation safety regulations in the Province of Ontario and I wrote the Handbook that described the details of complying with these regulations. Having been associated with this work led to invitations to advise on federal radiation policies in Canada and to lead international panels and task groups, with the American Association of Physicists in Medicine (AAPM), the American College of Radiology, the International Commission on Radiation Units and Measurements (ICRU) and the International Atomic Energy Agency (IAEA). In my career I often discovered that one seemingly small event leads to others, some of which attain real significance.

I also began to appreciate the business side. Medical imaging is heavily equipment oriented and if you want to understand the technology you have to interact with the people who know about it and ideally those who design it. I did a bit of this on the visit to the industrial labs in Europe, but being on top of this requires ongoing monitoring of developments not only in academic research labs, but also in industry. Working in hospitals, medical physicists usually only get to meet and talk with the commercial people in companies. Their focus is on selling equipment and technical discussions tend to be enthusiastic, but very superficial. Harold had taught me how to connect with the scientists in the labs, who at heart were much more interested in the science and technology itself, had a deep understanding and enjoyed talking about their work with scientists like myself.

I have been lucky in having been able to strike a good working balance. On the one hand I have managed to keep a healthy distance from the influence of the commercial people in industry, while conveying to them that it is good to have scientists like me interested in their products. On the other, I have had mutually productive research collaborations with scientists and engineers in industry labs, and in this way have been able to work with absolutely leading edge ideas and technology, something that is often not possible in a pure academic environment.

After a while, I was able to get an appointment in the School of Graduate Studies. This meant that I could supervise graduate students, and this has been one of the most rewarding aspects of my professional life. Once again, fate, in the form of Harold Johns, played an important role. Almost at the same time, two very bright students applied to our department, Department of Medical Biophysics. They happened to be Paul Johns, Harold's nephew and Ian Cunningham, Jack Cunningham's (Harold's long-time book-writing collaborator) son. Both of these guys easily got into the department on their own merits, but I'm sure that Harold had a hand in seeing that Paul was assigned to my supervision and Ian to Aaron Fenster. I never met Paul's father, but he certainly had the Johns spirit in him—keen mind, highly disciplined and a walking definition of the word, "stubborn".

What a great way to start! Paul had excellent ideas, worked like a dog. It was my job to keep him moving in the right direction. Our first project together was on measurements of x-ray absorption properties of breast tissue. Paul immediately caught on to the idea that medical physics research is highly multidisciplinary. He made a connection to a pathologist at another hospital who was able to supply us with well-characterized tissue samples. He designed and built an apparatus that would give us the precision that we needed to do the measurements. And then he made the measurements with painstaking care. In the course of doing the experiments, he discovered an error in what was a standard measurement technique in nuclear spectroscopy, came up with a sophisticated mathematical correction, contacted the author of the definitive textbook in this area with the correction and published it in a Swedish journal. In doing so, he interacted with the editor of the journal who was interested in our work and also happened to be a Nobel laureate. Not bad for a rookie grad student! The publication of the tissue measurements remains one of our most cited papers. Paul then became curious about how the scattering of x-rays in the breast caused loss of quality in medical images and carried out similarly meticulous studies. After finishing his Ph.D., he went on to an illustrious career in Ottawa where he eventually became a university Department Chair, and interestingly, continued to study the effects and possible new uses of scattered radiation.

11   Going Digital

Bob Nishikawa, my second student began by analyzing in great detail the statistics, or "noise", as we call it, in x-ray imaging. Around this time, the idea that earlier detection of breast cancer by doing regular x-ray mammogram examinations of women who had no symptoms of cancer, would save lives gained attention. A strong interest developed, particularly in the U.S. in improving these detection methods. Faina Shtern, a Russian-American medical doctor had an influential position at The National Cancer Institute in the US. The Cold War and the war in Vietnam had both ended and the U.S. military had a brief period where things were a bit quiet. Dr. Shtern decided to organize a conference which was titled, "From Missiles To Mammograms" right in the Capitol. In fact, the poster for the meeting had a stylized image of the dome of the Capitol Building looking either like a breast or a missile. Because of my work, I was invited to speak on how we could further improve the performance of mammography. Faina was a petite person with the energy of an explosion and the determination of a military general. In fact, she had invited scientists and engineers from all the major U.S. military research labs to see how their swords could be beaten into ploughshares to fight breast cancer. One of the statements that was proudly made (which was unfortunately a gross oversimplification) was that "if we can detect a missile in space, a thousand miles away, we should be able to find a tiny tumour in the breast right in front of us".

When it was my turn to speak, I completely deviated from my invited assignment and said that mammography as it was being done was pretty much a mature technology and I couldn't really see that tweaking its perfor-mance would buy us much. Instead, I suggested that we could probably have more impact by going digital—developing an electronic detector to record the transmitted x-rays through the breast and use computer image processing and display to enhance the detection of cancer. I suspect that this direction was already in the cards, but certainly, at this meeting it seemed to shift the focus into a new, higher gear. When I returned to Toronto, I told Bob that we needed to start designing a digital mammography system and he jumped at the opportunity. The earlier work that he and Paul and I had done plus my experience with detectors for CT prepared us well for this daunting task.

Starting as students together and continuing throughout my career, Don Plewes has been responsible for putting some of his creative ideas into my head. I think he has so many ideas that he knows he can't pursue them all and he knows that once he's planted something with me, I'll likely do it. He's also a smooth talker. I'm sure he had something to do with Dr. Shtern's having invited me to present in Washington. After that meeting, he also was key in forming what was initially called The National Digital Mammography Devel-opment Group (NDMDG). Because he and I worked in Toronto, Canada we quickly became the International DMDG. The idea was to accelerate the introduction to the clinic of what we knew would become an important new diagnostic tool by working collaboratively among multiple academic, clin-ical and industry groups. This kind of collaboration was unusual, much like what we see in the case of war and national emergencies like COVID. But this was no emergency—it was fun! We were invited to visit many of the military research labs and those of private industry that often did contract work for the military. We had the chance to work with technological devices that were developed at unbelievably high cost—the sort of cost that only the U.S. military could deal with and adapt them to make digital mammography a reality.

It was a rarified atmosphere. We interacted with military generals and CEOs of start-up companies. It was sort of like a big, fun game, but one with a serious purpose. We got to meet the clinical icons of breast imaging and the brave women who had suffered from breast cancer and who were determined to enable the kind of research needed to improve the situation for those who would come after them. These "advocates" made it their business to understand the science just about as well as we did, and they kept us honest and focused in our work. I loved the idea of taking the toys away from the military and using them to cure, rather than kill people. And I think many of them did too.

Two of the amazing people I met were Dr. Etta Pisano, a young breast radiologist and Dr. Kullervo Hynynen, a brilliant medical physicist from Finland, who was doing medical imaging research at Harvard. Dr. Pisano became the de facto clinical leader of the IDMDG, and we have worked closely together since the 1990s, while Kullervo was eventually lured to Toronto to become my boss. He is a world leader in image guided therapy, where he uses ultrasound energy to cook or ablate diseased tissue under the precise guidance of ultrasonic or MRI imaging.

The development of digital mammography began in my lab and those of others. We came up with ideas, and tried to predict the performance of a proposed imaging system using computer models and we built lab prototype systems to the extent of our capabilities. But with the formation of the IDMDG we began working in the big leagues. Our two main industry collaborators were the gigantic GE and the infinitely smaller Fischer Imaging. While we maintain an active collaboration with GE researchers to this day, at the time we worked more closely with Fischer who had adopted a design that was much like the one we had developed in my lab. Fischer was run by its charismatic CEO, Morgan Nields and his chief scientist Mike Tesic. Morgan was a highflier—a highly competitive downhill racing skier, who could easily show up in his office wearing a cast. He was a visionary who was constantly bouncing ideas off us. Mike was a beefy guy who came from a Serbian background and had boundless energy. He had a handshake like a vise and would travel pretty much anywhere to enjoy his two favourite things, high quality sushi and playing tennis The two were a great team—one constantly juggling a million ideas and plans and the other focussed like a laser on getting the job done.

I remember when four or five of us from Toronto would fly to Denver and walk across a field of prairie dog hills between our hotel in an outer suburb to the factory, the little rodents popping out of their burrows checking us out as we tried not to twist an ankle in one of their holes. Working with a company was exciting. We would work from 8 AM almost until dawn the next day. Morgan, the CEO would be there working right beside us. If a precision part or bit of electronics was needed, Morgan would get on the phone. Then we would drive 20 miles to a nearby town where an engineer would have just made it for us and then rush it back to the plant to attach it to the system and test it. When the system was working, we set up a clinic right there in the factory, bringing in professionals- a mammography technologist and a couple of radiologists to evaluate our results. Possibly the first cancer seen on digital mammography was imaged on a volunteer in that factory.

Through the IDMDG we finally had a working digital mammography system (much of the heavy lifting was done by our industry collaborators). We were now ready to test the system in a clinical trial and it was Etta who put together the Digital Mammography Imaging Screening Trial (DMIST) to do this. This took us years to get funded, but in the end showed that for many women, digital mammography was more sensitive in finding breast cancer than film mammography. And now, most mammography is done on digital equipment. Our work had impact!

12   The Hot Tub

One of the delights of being a research scientist in my field (at least pre-COVID) was attending conferences. There is a real international community of scientists. Overall, they are friendly, fun-loving, and really interesting group of people. And over a career, you meet the same people over and over again, but in different cities, countries, and situations. Imagine getting off a plane in Beijing, Girona or Manchester after an extremely long flight, checking into your hotel and entering the elevator, only to meet a couple of good friends from New Zealand or Mexico. This would happen frequently enough that it became familiar and you would barely register surprise after a while. It was great to sit at the bar, walk on the beach, stroll on The Great Wall, or go for a hike in a country that was new to both of you with these international colleague-friends in between conference sessions. Often those discussions in these exotic settings would lead to good new ideas for my research or collaborations on projects between our labs. This is part of the richness of being a scientist.

My students were accomplished in many different areas. At one conference in Gifu, Japan I was walking one evening with two of my former students, searching for a particular restaurant. In preparation for the trip, I had learned a fair amount of conversational Japanese, but my reading skills were zero. James and Andrew, both Caucasians, had spouses who were of Chinese background and had sent their kids to learn the language. Andrew had developed skills in reading text in Kanji, the Chinese characters which are also one of the symbol sets used in Japan. So, he could pronounce the signs that we read, but not know their meaning, while I could recognize the names when he said them. In that way, as a team, we found our way to the restaurant. We used the same approach in interpreting the menu and had ourselves a real feast.

But for many years, probably my favourite spot was the hot tub at our conference hotel in Newport Beach, California. To leave the Canadian winter in February was in itself a treat, but this understated hotel, next to a very pricy shopping mall, had outdoor hot tubs and a group of us, mainly from colder climes, would congregate in the hottest, the so-called "pro tub", around 10 PM. We would chat about just about everything, but there were always vigorous debates around whatever was the latest scientific issue at the time. "Did xxxxx, when he presented his idea about yyyyy know what he was talking about?" We always arrived on a Saturday evening as the conference started on Sunday and walking through the hotel there was always a wedding or other fancy event taking place. The participants were dressed to the hilt, some whose outfits were on the skimpy side. In the lobby, this provided a real cultural contrast from the scientists at the meeting where males and females were always simply and unimpressively attired, usually in shorts or jeans, except for a few older Europeans working in industry, who wore jackets and ties or suits.

After a few years the hotel became too expensive for the conference and it moved to less enticing quarter in the suburbs of San Diego.

13 The Big Move

Another, more significant move for me took place in 1990, when Mark Henkelman, a colleague at Princess Margaret and an international leader in research on magnetic resonance imaging, suggested that he and seven other imaging scientists move into new research space at Sunnybrook Hospital. Princess Margaret, located in downtown Toronto, was simply running out of space and had been discussing relocating for several years, but was having difficulty making a decision. While we had excellent connections with colleagues there and we had reservations in leaving, we wanted to grow and the opportunity to form a new imaging group was irresistible. Mark moved to Sunnybrook first as Vice-President of Research and after six months, he gave us the "go-ahead" wave and the rest of us joined him there.

14   The World of Students

I have supervised many graduate students. When I began, they were only a couple of years younger than I, but somehow over the years they kept getting younger? This is one of the most enjoyable parts of being a scientist. I found that each student presented me with a combination of a major challenge (How am I going to get this person to think about this problem the right way?) and a great joy (These students are so smart and creative. How do I keep up with them?) Frankly, working with young, enthusiastic people makes one feel younger themselves and contributes to a great spirit. Figure 1 is a picture with some of my students and the technical group in around 2001. I no longer supervise graduate students, but we have undergraduate students and even high school students working in the lab regularly. My lab is filled with people between the ages of 18 and well over 60 with a rich mix of gender, ethnicity, race, and personality type. It is not a boring place, especially when (pre COVID) we would bring in food to share at our weekly Wednesday group meeting.

I have been fortunate to have been involved in some great projects, some of which, like digital mammography, have had international impact. Another

Fig. 1   Some of my students and technical group ~ 2001

such project was motivated by my colleague Norman Boyd, an outstanding epidemiologist. He was interested in knowing whether we could get an idea of the composition of the breast by measuring something called "breast density" by analyzing mammograms. He believed that density was a way to predict risk of breast cancer. Together with my students Curtis Caldwell and Jeff Byng, we looked for methods by which we could characterize the amount and patterns of density represented on mammograms. Curtis was interested in the mathematics of fractals to analyze the texture of the images while Jeff developed a quantitative computer tool, called CUMULUS, for measuring the area of tissue that appeared dense on the mammogram. Curtis' work gained attention in the fractal community and he was invited to a conference in France, where he met Benoit Mandelbrot, widely considered to be the "father" of fractal geometry. Meanwhile, the CUMULUS algorithm became a gold standard for breast density research and over the years we sent copies of the software to many labs around the world. It also opened fruitful collaborations with researchers interested in the connection between breast density and possible causal factors like genetics, diet, and exercise. Density measurement is still being used in research into the causes of cancer and possible ways of reducing risk.

15   New Directions

While my early work had focused on designing x-ray detectors and building machines, my lab gradually shifted attention into developing methods for measuring the performance of these detectors and the quality of medical images. In addition to designing sophisticated analytical methods we also simplified what we learned in that work to create tools for quality control (QC)—monitoring to test if clinical imaging systems were operating properly on a day to day basis. We created QC systems that were used in the DMIST study and for the Ontario Breast Screening Program. More recently we have spread the tools and techniques more widely through my work with committees in the International Atomic Energy Agency and elsewhere.

After cancer has been detected, tissue is removed from the patient, first in a biopsy using a hollow needle to perform the actual diagnosis of cancer and then in the cancer surgery itself. Much can be learned from that tissue under the microscope that will predict if the cancer is aggressive or not and also how it can best be subsequently treated. One of our breast surgeons came to me and my student Gina Clarke and asked if imaging could be used to improve this process of diagnosis and characterization. Eventually, after much work by Gina, who had come to me from the Upper Atmospheric physics program at another university, this led to the creation of The Biomarker Imaging Research Lab (BIRL) where we specialize in applying the techniques of medical imaging to the field of pathology. Suddenly I had to think a lot more about molecular biology, chemistry, and surgery. This is one of the fun things about medical physics—it is so multidisciplinary. I began talking to new (for me at least) types of researchers and clinicians and learning new language and ideas. BIRL has led to some collaborations with some outstanding scientists that may well change the way that cancer is diagnosed and treated in the future. Figure 2 shows some of my imaging colleagues at Toronto Sunnybrook in ~ 2006.

The lab works to "photograph" cancers on a molecular basis, depicting in ultra-high resolution images the grouping or arrangement of individual key molecules related to cancer. These arrangements can then be correlated to information on how aggressive or "sleepy" a cancer will be and what is the best approach to treatment—using a tough enough treatment to kill the cancer cells without unduly harming normal surrounding tissues.

As I enter my 70s, I see some of my students retiring, while some scientists older than I are still very much fired up about their work. I see a need to step

Fig. 2 My imaging colleagues at Toronto Sunnybrook ~ 2006 Happily, we're much more diverse today!

back at least a bit to make room for future generations of scientists. At the same time there are a few aspects of the work that I have been doing that would benefit from some closure, and I plan to continue working in these areas.

16   Exotic Places, Enlightening Discussions

I can't emphasize enough how much fun it is to be a scientist in my field. It makes me hesitant to, how should I phrase it?—"shut down my computer". I mentioned going off to some part of the world, running into a colleague, having a glass of wine, a bottle of beer or a cup of coffee together in some unforgettable spot. And you never know what ideas will emerge. Figure 3 shows some of our students at the SPIE conference in San Diego, USA in 2007—mixing work and play in February.

Fig. 3 Our students at the SPIE Conference in San Diego in 2007—mixing work and play in February. (SPIE = formerly the Society of Photographic Instrumentation Engineers, later the Society of Photo-optical Instrumentation engineers. It's a society that organizes conferences and exhibitions on topics such as optics, photonics and imaging engineering.)

Just for fun, here are some of the places and some of the people with whom I've done this:

The Antarctic Peninsula—Dr. Stuart Foster—world leader in ultrasound research.

A wood-panelled club in Stockholm with pictures of kings lining the walls—Kai Sigbahn, Nobel laureate in nuclear physics.

A high-tech amusement park in Yokohama—Kunio Doi, pioneering medical physicist.

Wellington, NZ. where Lord of The Rings was filmed—Ralph Highnam, co-founder and CEO of Volpara Health Technologies.

Figuera, Catalonia (Spain) Salvador Dali Museum—Morgan Nields, Mike Tesic CEO and chief scientist of Fischer Imaging.

Beach on Coronado Island, California—John Boone, outstanding medical physicist and present editor of the journal Medical Physics.

Volcano, Hawaii—Dr. Etta Pisano—leader in breast radiology.

Waterfalls in Iceland (during Midnight Sun)—Dr. Jennifer Harvey—Chair of Radiology, University of Rochester.

Three Gorges area, China, on a mountaintop, surrounded by wild monkeys—Aaron Fenster, Terry Peters—brilliant Canadian medical imaging researchers (see Chaps. 19 and 10, respectively).

Steam Engine Museum in Manchester—Sir Michael Brady, Professor at Oxford, co-founder of Volpara.

Coffee House in Haight-Ashbury (Summer of Love) district, San Francisco—Dr. Jack Cuzick—epidemiologist who developed widely-used risk model for breast cancer.

St. Gely (Rhone District of France)—Donald Plewes—multi-talented medical physicist.

Somewhere in rural Cambodia—Dr. Roberta Jong—breast radiologist and long-time colleague.

Lille, France—Matts Danielsohn, Swedish medical physicist and entrepreneur.

Art Gallery in Sao Paulo, Brazil—Dr. Pamela Ohashi, immunotherapy researcher.

Mexico City during Day of The Dead—Dr. Maria Ester Brandan—Chilean Medical Physicist.

Great Wall of China—Dr. Robert Smith—outstanding breast cancer epidemiologist.

City hall in Bremen Germany—Prof Heinz Otto Peitgen—fractal research pioneer.

Dive bar in San Antonio Texas—Paul Johns, my first grad student.

Dive bar in Atlanta Georgia—Melissa Hill, my last grad student.

(and that's just a sample).

17 Final? Chapter

I have always been interested in cancer screening—imaging people who don't outwardly display symptoms of disease on a routine basis (say each year) with the goal of finding and treating small early cancers and in this way both reducing cancer deaths and also allowing patients to avoid the harsh therapies required to treat advanced disease.

About 10 years ago I started working with computer models to predict the most effective way to do breast cancer screening. What detection methods should be used and how frequently? At what age should people begin to get examinations and at what age can they stop? The best way to answer such questions is to do large clinical trials where participants are randomized to receive different screening regimens or possibly no screening. Some trials were done in the past, but these require tens of thousands of participants (or more), are extremely expensive, and take a decade or more to get results. It is simply not practical to answer all research questions in this way. Computer modeling lets us build on relationships that we already understand from past studies, alter the conditions, and predict outcomes. For example, is it better to start screening at age 40 or age 50? Should we vary the interval for different ages? I have found the use of models very helpful and wherever possible try to validate them against real world observations.

But the models, as well as much of the other data that I have seen, often predict screening strategies very different from the policies that our health care systems have been advised to implement and I am convinced that, with breast cancer, we can do much better at saving lives and sparing women from enduring treatment for advanced disease.

Scientists usually draw a line between what they learn in their research and what happens in the world of actual health care delivery. As I approach the end of my career I am increasingly convinced that as scientists we have an obligation to engage the public and those who make health care policies with the results of our work so that it can have impact and do good. Most recently, I have been working with other scientists, physicians specializing in breast cancer and women who have had to deal personally with the disease to inform decision makers. As scientists in medicine we have the role of improving the understanding of disease, developing better ways to detect, treat or ideally prevent it, but also to figure out how to effectively translate our knowledge into improved health.

Martin Yaffe, C.M., FRSC. Ph.D. is a medical physicist who leads a 23-member research lab at the Sunnybrook Research Institute in Toronto, Canada where he is Senior Scientist in Physical Sciences, Professor of Medical Biophysics at The University of Toronto and holds the Tory Family Chair in Cancer Research. He is co-Director of the Imaging Research Program at The Ontario Institute for Cancer Research. His multiple research interests focus around approaches for better detection, diagnosis and management of breast cancer. He helped pioneer the development of digital mammography, created quantitative techniques for studying the role of breast density in cancer and more recently established a lab for digital pathology, applying imaging science, radiomics and machine learning in studying biomarkers to guide diagnosis and treatment of cancer. His lab has developed methods for contrast-enhanced mammography and for quality control in mammography and tomosynthesis. He has led Task Groups on breast imaging for the AAPM. ACR, ICRU and the IAEA and is currently Canadian Study Chair of the large TMIST study of breast tomosynthesis. He has had a strong, longstanding interest in breast cancer screening and currently is using microsimulation models to optimize screening regimens.

Awards

2021. Named as Fellow of the Royal Society of Canada (RSC), especially for his exceptional work in imaging as it applies to earlier diagnosis of breast cancer. Recognition by the RSC is considered to be one of the highest honours an individual can achieve in the Arts, Social Sciences and Sciences, as well as in Canadian public life.

2018. Doctor of Science Honora Causa from The University of Manitoba, his alma mater.

2016. For publications with his graduate student (now Professor) Paul Johns, The journal, Physics in Medicine and Biology, identified their 1987 paper, "x-ray characterisation of normal and neoplastic breast tissues" as one of the 25 most important publications in the past 60 years.

2015. Invested as Member of The Order of Canada, the country's highest civilian honour.

2014. Honorary Fellow of The Society of Breast Imaging.

Part VI

Medical Physics: More than Commercial Developments

"Try not to become a man of success. Rather become a man of value."

Albert Einstein, theoretical physicist, and recipient of the Nobel Prize in Physics "for his services to theoretical physics, and especially for his discovery of the law of the photoelectric effect."

19

Lessons Learned on an Opportunistic Career Path

Aaron Fenster

1   Preparation is Necessary for the Experiences that Create Growth

Perhaps it is egotistical to think that one's path in life can give insight into others' paths, nonetheless, the value of this chapter may be a little entertaining but with a little insight thrown in. Of course, insight into one's career trajectory is only understood when examining it looking backward, and finally understanding the trajectory and hinge points in one's career. Being near the end of my academic career and looking backward, my career path in medical physics appears straight and unwavering, i.e., undergraduate program in math, physics, and chemistry to a professor with a medical physics research lab. But a closer examination reveals hinge points, opportunities grabbed, and lessons learned that I share. Superficially, my path is as follows.

A. Fenster (✉)

Robarts Research Institute, Western University, London, ON, Canada

e-mail: afenster@robarts.ca

J. Van Dyk (ed.), *True Tales of Medical Physics*,
https://doi.org/10.1007/978-3-030-91724-1_19

Undergraduate degree—Math, Physics, and Chemistry.

Master's degree (photochemistry)—Medical Biophysics, University of Toronto.

Ph.D. degree (x-ray imaging) —Medical Biophysics, University of Toronto.

Post-doctoral Fellowship (CT imaging) —Medical Biophysics, University of Toronto, and the Ontario Cancer Institute.

Assistant Professor—Medical Biophysics, Radiology, University of Toronto, and the Ontario Cancer Institute.

Associate Professor—Medical Biophysics and Radiology, University of Toronto.

Professor—Medical Imaging, Medical Biophysics, and Robarts Research Institute, Western University, London, Ontario.

The following sections describe the career path and the opportunities at key points along that path that shaped my career. While these may not necessarily be entertaining, the lessons learned may be instructive. Note that in some places, the science and details of the involved technologies get a little deep for the non-medical physicist, you will see that a quick read through these sections allows the stories to continue.

2   In the Middle of Every Difficulty Lies Opportunity (A. Einstein)

After completing my undergraduate degree in 1971, I was told that with a degree in physics, chemistry, and mathematics I could get any job I wanted, as I am in the baby boomer generation with a belief that the world will be shaped by us. With too many options, I could not make up my mind about what career to pursue. After graduation, on a summer day, while walking down a Toronto street, I met a classmate who told me that he will be starting his master's thesis work in the Department of Medical Biophysics of the University of Toronto, which was located at the Ontario Cancer Institute (OCI) and the Princess Margaret Hospital (PMH). This sounded cool, and with nothing else to do, I walked over to the PMH in my torn blue jeans, surplus army jacket (given to me by my USA draft dodger friend), beard, and long hair (remnant of the hippie era), and was shown to Professor Harold Johns' office, as he was the Chair of the Department.

Being then in the "love" culture, I thought that we will interview each other about the possibility of entering the M.Sc. program. Instead, Professor Johns called in two other younger professors and spent one hour haranguing me on how I dressed and my long hair and beard, which signified a counter-culture not appropriate for the graduate department located inside a hospital. I stood my ground and argued with these three intimidating professors, and then left.

Walking out of the building I asked myself should I have been insulted by the treatment of someone who wanted to be a graduate student, should I want to pursue working in such a "straight" department? But, the possibility to work on a research project that may have an impact on people's lives convinced me that I wanted to work at PMH. I was not deterred by the "square" professors, cut my hair (not my beard), put on clean pants, and went back. It was already late in the summer and all graduate positions were filled. When I met Professor Johns again, he took me into his lab and showed me a desk and said I can join his lab and take this desk. However, there was still another student at that desk who was packing up his belongings. Later I found out from the other students that the student was being asked to leave the Department and that I came at a time when an opening in Professor Johns' lab was created by another student being asked to leave—a coincidence or another student's misfortune was my opportunity.

Lessons Learned

Don't let a bad experience turn you away from a good opportunity.

Be receptive to chance opportunities as they may change your life.

3   Don't Cling to a Project Just Because You Spent a Lot of Time on It

My M.Sc. thesis was on the use of ultraviolet (UV) flash photolysis for studying the interaction of UV with DNA nucleotides. This work was part of Professor Johns' research, which was focused on studying UV and gamma irradiation products of pyrimidines in water solutions. The lab's research during that time included studies of two classes of photoproducts, the dimer, and the hydrate. Members of the lab showed that the hydrate arises from reactions of the singlet state, whereas dimers in solution arise through the reaction between a molecule in the triplet state, with one in a ground state. My specific master's thesis project was on investigations of the temperature on quenching of the pyrimidine triplet states.

Towards the end of my M.Sc. thesis work, Professor Johns came back from a sabbatical leave in the U.K. where he spent time with Professor J. W. Boag at the Institute of Cancer Research, Sutton (UK) learning about x-ray-based electrostatic imaging. With this background, he decided to stop his research on the effects of UV and gamma irradiation on the initiation of cancer and planned to start one of the two labs in Canada focused on medical imaging (the other was at the Montreal Neurological Institute). Since I was interested in continuing my studies in a Ph.D. program, the choice was to continue with another professor in the department as there were many working on various biological aspects of cancer, effects of radiation on initiation of cancer, and the use of radiation to treat cancer or to change directions completely and continue with Professor Johns on a project focused on medical imaging using electrostatic approaches. Professor Johns' expertise in this area was only based on a 6-month sabbatical leave and no other students were working on a topic even remotely close to this new area.

The safest Ph.D. work would be to join one of the other labs and continue to investigate the effects of radiation on initiating cancer with a topic focused on radiobiology. But, Professor Johns' two mottos listed below convinced me to continue with him on this new area of research.

> Physics conquers all.
> If you can't see it, you can't hit it, and if you can't hit it, you can't cure it.

He was firmly convinced in his vision that by applying physics, a new and improved x-ray imaging method could be achieved. His high level of confidence was infectious and energizing. A new graduate student joined Professor Johns' lab (Don Plewes) and the lab began to consider the development of a new imaging method. At that time, xeroradiography was being used as a mammographic imaging approach. This approach had a unique property of edge enhancement, which was produced by the method used to develop the latent electrostatic image on the xeroradiography plate. The image development method was based on the powder cloud approach, which involved blowing a fine blue powder over the plate. The charged powder particles were attracted to the opposite charge on the plate and rendered the image visible. However, the fringe field at regions of sharp charge gradients caused the powder to pile up on one side on the charge edge and be absent on the other side producing enhanced image edges.

With Professor Johns' expertise in ionization chambers used for radiotherapy dosimetry, an experience he developed during his sabbatical at Professor Boag's lab, and the two students' (Plewes and me) innovative spirit, we developed a new x-ray imaging method—ionography. The method was based on high pressure (typically 10 atm) xenon ionization chamber, in which the electrodes were separated by 1 cm and were curved with a radius of curvature focused at the location of the x-ray tube's focal spot. A Mylar sheet was held against one face of the inner electrode. When the xenon chamber was connected to a 10 kV power supply and an x-ray exposure was used to irradiate an object above the xenon chamber, the electric field in the ionization chamber caused the charge in the chamber to drift to the electrodes and be collected on the Mylar sheet generating a latent electrostatic image. After the exposure, the pressure in the chamber was relieved, the chamber opened, and the Mylar sheet was removed. The latent electrostatic image was then developed by the powder cloud method, generating a visible image with edge enhancement as in xeroradiography.

It was clear that we were onto a new x-ray imaging method, which we published in the first year, first issue, and the first page of the new journal Medical Physics (Fenster A., Plewes D., Johns H. E. Efficiency and Resolution of Ionography in Diagnostic Radiology. Medical Physics 1: 1, 1974).

Don Plewes and I continued to pursue the ionographic imaging method and published a series of papers on the subject describing various innovations and improved understanding of the edge enhancement effect. Our development of this imaging method resulted in much scientific interest, including commercial interest, until digital radiography was developed and shown to be a superior method for x-ray imaging. After Don Plewes and I completed our Ph.D. degrees, Professor Johns stopped his lab's research on this topic and began to search for a different imaging approach.

Some may think that a project or Ph.D. work of an innovative technology that becomes obsolete so rapidly signifies failure. However, throughout the years that we worked on the project, we continued to be challenged to invent various aspects of the method including the use of a liquid instead of gas, reading the electrostatic image outside of the imaging chamber without opening it, and methods to control the edge enhancement effect. The experience of focusing on innovations and realizing them by building a complete imaging system transformed the way I looked for research projects and shaped my research career to this day.

Lessons Learned

Pursue innovative ideas without hesitation, but be prepared to change direction if a different approach is evident.

or

Don't hold on to a project after its time to end it, as new opportunities will appear.

4 Opportunity is Missed by People Because It is Dressed in Overalls and Looks like Work (T. Edison)

It was 1976. I received my Ph.D. and Professor Johns offered me a post-doctoral position to work on the next hot medical imaging development—computerized tomography (CT). Don Plewes left for a position in Rochester, N. Y. and Martin Yaffe (see Dr. Yaffe's chapter for his side of the events, Chap. 18) had already joined our lab and was completing his Ph.D. Since our lab believed we can do anything by applying physics principles and we were not afraid to take risks, we began to design a CT scanner. Our approach was to develop a third-generation CT scanner (see Terry Peters' Chap. 10 describing four generations of CT scanners). In commercial third-generation CT scanners, the x-ray tube and detectors are mounted on a support, which rotated around the patient while collecting measurements of the transmitted x-rays through the patients. However, we began to construct a proof-of-principle table-top system, in which the detector system and x-ray tube were stationary, but the object rotated. Since Professor Johns' expertise was based on our ionography experience with xenon ionization detectors, we began to design a third-generation detector system by mounting small tantalum plates in a radial pattern pointing at the x-ray tube's focal spot. The tantalum plates were placed in a pressurized chamber filled with high-pressure xenon gas. The plates were connected to a power supply so that any ionization products drifted to the electrodes and were measured. Martin Yaffe began to develop the detector system and I began to develop the CT reconstruction software.

Searching for any published articles on reconstruction methods, I came across a thesis from New Zealand by Terry Peters (author of Chap. 10 in this book and current lab mate). In his thesis, he described a CT reconstruction method suitable for third-generation CT scanners, which was based on his supervisor's (Professor Richard Bates) work. After studying Terry's thesis and trying to understand complex mathematics, I came across Dr. Ramachandran's paper on CT reconstruction, which was based on the central-slice theorem, but rearranged so that the CT image can be generated in two steps: convolution and back-projection. This approach was much easier to implement, which I did on a Digital Equipment Corporation PDP 11–55 computer with 64 K memory. It was very exciting to build a CT scanner and solve all the problems with issues of third-generation scanners, of which ring artifacts were the most severe. Although we learned a great deal about building imaging systems, which included hardware and software subsystems and published many papers, the work came to an end when Professor Johns retired, and General Electric (GE) developed their commercial third-generation CT scanner, which we could not outcompete. Nonetheless, again we gained significant knowledge and confidence that we can try to compete with industry.

Lessons Learned

If you keep fixing the old, you may miss the next big thing.

Knowledge gained while competing with industry may be useful to industry.

5   Everything that Touches Your Life is an Opportunity if You Discover Its Potential Value (Wallace D. Wattles)

In 1978, Martin Yaffe and I joined the Department of Radiology at the University of Toronto and the Ontario Cancer Institute (OCI) as faculty. Our graduate appointments were in the Department of Medical Biophysics and we began to recruit graduate students to work on our projects. Although my lab continued to work on various innovations related to CT imaging, I was also looking for other projects as I perceived that CT would become a mature technology. After attending a seminar at the OCI about the limitations of external beam radiation therapy, which described the need for monitoring of the patient's position and movement while the treatment beam is on, it became clear that it may be possible to use the beam exiting the patient to create an image, i.e., portal imaging. Together with a graduate student (Peter Munro), we developed a system that made use of a fluorescent screen in the beam path and a 45° mirror to reflect the fluorescent image to a camera positioned outside the beam path.

After demonstrating the method in the radiation oncology department of the Princess Margaret Hospital and becoming aware that other researchers were working on various portal imaging systems, I approached the leaders at OCI about patenting our method and commercializing it. Unfortunately, I received no interest in any aspects of commercialization, as the response was that commercialization will contaminate the pursuit of knowledge. However, it was clear that achieving a wide impact in one's research beyond regular publications would be through the translation of the lab's innovations to the private sector that will make the technology widely available. Thus, at a meeting of the American Association of Physicists in Medicine (AAPM) in 1991, I met a CEO of a start-up company (S & S Inficon) and at dinner, I explained our technology and invited the company to develop a product. Having no patents and being naïve about commercialization, but interested in making the technology widely available, I transferred all our knowledge and software to the company. This company developed the product and began selling it as a portal imaging option within a year. A short time later, the Radiation Oncology Department at PMH purchased two portal imaging systems, demonstrating that we lost an opportunity to participate in further development of the product and also realizing some financial return on research. This experience stayed with me and my students as it demonstrated the potential of commercializing technology and making it widely available through the private sector. As well, by collaborating with the licensing company, there could be an opportunity to partner with the company and continue to develop the technology. The knowledge gained about how not to lose an opportunity to commercialize innovation from the lab and make them widely available was a hinge point in my career path.

Lessons Learned
Impact can also be realized through private sector translation.

6   We Cannot Become What We Want by Remaining What We Are

Some people change jobs often because of dissatisfaction in their current position or desire to increase their salaries, and some individuals stay in their position for many years because they are comfortable in the situation or they have built a "jail" around themselves that they cannot leave. Many who reach their mid-life around their 40th birthday and are very comfortable in the environment, which consists of long-term friends and colleagues, begin to question whether they want to continue in the current path or to seek a change that will challenge them and energize their lives. Of course, the resolution of this crisis takes many forms. On my 40th birthday, I was presented with an opportunity by a radiologist colleague (Dr. Barry Hobbs) who was offered the Chair-Chief of Radiology position in London, Canada at the University of Western Ontario. He asked me to move with him to London and start a medical imaging research program in the department in partnership with the staff radiologists. My wife just had a baby, we purchased a new house, I had developed an active research program with graduate students, and we had many good friends in Toronto. Nonetheless, I told Dr. Hobbs that I would visit the university in London and will let him know whether I had any interest to move.

After visiting London and investigating this opportunity, it was clear that I was faced with another hinge point: staying where I was and continue my well-funded research or move to London and restart my research and build a new imaging research program. It was clear that moving to London presented an invigorating opportunity to start a research facility and shape it in the way I thought would bring success. The opportunity in London was to build the research program in a new research institute—the Robarts Research Institute. Since there was no imaging research in the Institute, the imaging program would have to be built from "scratch". Based on my experience in Toronto, I considered the opportunity and believed that the elements needed to bring success are much more than physical but also cultural as they relate to the lab.

Physical Elements

Sufficient lab space that would not be in danger of eroding

A critical number of scientists at inception

Start-up funds to buy necessary equipment

Access to graduate students

Cultural Elements

A collegial culture among investigators

Collaborative culture seeking partnerships with clinical and basic science departments

> A culture of sharing resources among investigators (e.g., money, students).

Obtaining the physical resources required negotiations with many leaders in the London community—university, hospitals, and philanthropy. Sufficient space to allow growth was found at the Robarts Research Institute, which is located on the university campus but attached to the university hospital. This location seemed ideal for the following reasons:

(1) it avoided being located in a hospital with constant space pressure from the hospital with a primary clinical care mission,
(2) the proximity to the hospital location (internal link) provided easy access for scientists and physicians to travel between the hospital and the Institute (especially in Canadian winters),
(3) the location on the university campus provided easy access for graduate students to the rest of the resources at the university including facilities and class locations, and
(4) the location on the university campus also provided easy access to potential collaborators with the skills needed for our research program.

After accepting the offer in 1987 to be the Director of the new lab (the Imaging Research Laboratories) and presenting the vision for the new lab to the CEOs of the three hospitals and a philanthropic family in London resulted in funds allocated to our new lab to recruit three scientists to allow us to start our lab with a critical number of scientists. Presenting the vision to the staff members of the Radiology Department resulted in additional start-up funds from their clinical earnings allowing us to purchase equipment. GE Canada also saw the importance of our vision and made available multiple x-ray imaging systems for our research.

Although the physical elements are important in establishing a new lab, the cultural elements are equal, and at times more important, as all the acquired physical elements may be squandered if the cultural fabric of the lab is toxic. Thus, to be able to establish a trusting and collegial culture immediately, we recruited two of my ex graduate students to faculty positions at the University of Western Ontario (Dr. Brian Rutt who was a post-doc at UCLA, and Dr. Ian Cunningham who completed his Ph.D. at the University of Toronto). The third member joined us from Winnipeg (Dr. Ting Yim Lee). Our lab at the Robarts Research Institute had a traditional "jail" model for research labs, i.e., hallways with labs behind doors, which are closed, separating the institute lab members from each other.

Our vision was to have an open lab without walls, including common student areas and experimental spaces. With the open lab approach to the lab design, a collaborative spirit could be established leading to sharing research, equipment, staff, and students—some called it a kibbutz model, which is a collective settlement in Israel.

With this vision, we convinced the management of the Institute to allow us to design our space differently from the rest of the building. Furthermore, our vision and a mantra that we kept repeating were focused on interdisciplinary partnerships, collegiality, and graduate student training. Over the next decade, we established collaborations within the lab, across the university, the London hospitals, and other universities, leading to multiple large research initiatives. Over the following years, we recruited additional scientists based on our success (approximately one new scientist every two years) and expanded the lab, which now (32 years later) occupies about 35,000 sq. ft. (3250 m^2) with 18 scientists, about 100 graduate students, and many large multi-investigator and institutional grants. Currently, these initiatives have led to us acquiring state-of-the-art imaging equipment within the Imaging Research Laboratories including four magnetic resonance (MR) scanners (two 3T, 7T, and 9.4T), CT scanner, 3D angiography system, and 12 ultrasound machines, and other infrastructure, totaling $60M.

7   If Opportunity Doesn't Knock, Build a Door (Milton Berle)

The section above describes that we now have a large number of graduate students in our lab. However, the path to generating such a large number of students for a new program that started with only four faculty members was not simple. By 2000, it was clear to us that relying on attracting students only from Medical Biophysics of the University of Western Ontario would not sustain our research and would certainly not allow expansion of the program. It became clear that we needed to involve graduate students from other disciplines and that we needed to be able to partner with other university departments to be able to build interdisciplinary research programs. Since early in the development of the Imaging Research Laboratories we were quite a small lab and not well known in the university, we could not get the attention of other departments to be able to have allocated or attract the best graduate students. Thus, in 2001 (three years after establishing the Imaging Research Laboratories) in partnership with a scientist in the Faculty of Engineering (Dr. Wan K. Wan), we convinced the Dean of Engineering of the importance of Biomedical Engineering for the university and its faculty. We wrote and submitted a proposal to the Whitaker Foundation to establish a Graduate Program in Biomedical Engineering (BME) at the University of Western Ontario.

Our proposal involved a request for funding for three faculty members to be matched with three faculty members provided by the university. Getting an allocation from the university for three tenure track positions required many negotiation sessions with Engineering departments as they would need to give up positions allocated to their departments. The critical issue we faced was also their fear that the BME program would attract graduate students away from their departments. Furthermore, concern by the university provost was also that establishment of a BME program would weaken the Medical Biophysics Department by attracting students away from that department. However, the Chair of the Department of Medical Biophysics (Dr. Peter Canham) recognized that offering a graduate program to those students interested in obtaining a graduate degree in the Faculty of Engineering would increase the number of graduate students as any student interested in a higher degree would have multiple options through three faculties: Engineering, Medicine, and Science.

Getting these positions allocated to BME required intervention from the university provost, who believed in the advantages of establishing a BME program as well as the possibility of obtaining external support from the Whitaker Foundation. With the support from the central administration at the university and the Deans, we were able to get the matching positions commitment and were successful in obtaining support from the Whitaker Foundation. Our BME graduate program was launched in 2001 with me as its first Director and with three streams: medical imaging, biomechanics, and biomaterials, and began to attract graduate students.

Since our program was new, we could shape it with our vision. We resisted pressure to recruit many students and decided to shape the program based on excellence. Thus, admission into our program was set at a high level and we began to attract excellent students with a potential to be competitive for Canadian Scholarships. Our program grew and now we have about 75 graduate students in the program. Based on our success, in 2019, the BME program became the School of Biomedical Engineering with a new undergraduate program.

Lessons Learned

A new program requires a new vision

Launching a new program requires support at high levels

High standards in a new program will lead to success.

8 A Person Who Fears Making a Mistake Never Tries Anything New (A. Einstein)

As our Imaging Research Laboratories grew, the need for collective funding opportunities for core support of our programs also grew. In 1997, the Ontario government realized that universities are underperforming in terms of converting their research to economic impact. With advice from various sectors, they established a funding program (the Ontario Research and Development Challenge Fund—ORDCF) to increase the economic activity through funding of partnerships between industry and universities in the Province as a way to stimulate the translation of innovations at universities to the private sector. A secondary benefit would be to develop highly qualified trainees who will find career paths in industry and become leaders in their companies.

Since members of our Imaging Research Laboratories were already collaborating with companies, we realized that the ORDCF program was an ideal fit for our research. After contacting some individuals who were connected to the Ontario government, one of us (Dr. Brian Rutt) led a five-investigator proposal in partnership with multiple companies. The proposal was funded in 1998 with a total budget of $5M over 5 years.

Unfortunately, my former colleagues from the OCI who were now at the Sunnybrook Research Institute (SRI) in Toronto were not funded by this program, which created a significant problem in the management of the ORDCF. Discussion with our contact in the government made it clear that province-wide excellent proposals that brought multiple institutions together in partnerships with industry would be the preferred approach. With this information, I approached my former colleagues at the SRI, and described that the vision of integration of research across multiple institutions would be the best path to funding from this provincial funding initiative. After many discussions, we established a leadership team consisting of Dr. Michael Bronskill from SRI and me to lead this initiative. We hosted a retreat to which we invited medical imaging researchers from all universities in Ontario to cluster around cohesive programs with industrial partners. Five themes emerged and we began to write the proposals. We also negotiated with the government ministry that controlled the fund and proposed an envelope fund to which we could apply with proposals focused on medical imaging. These discussions led to the establishment of a $50M envelope for medical imaging proposals. Over the next 3 years, and multiple submissions of the proposals, five of our proposals were funded with a total budget of $50M matched with $50M from industry, and $50M from our universities and institutions.

These funded programs increased the collaboration among medical imaging investigators in Ontario, which led to additional joint funding opportunities. Furthermore, to enhance the collaboration among the medical imaging investigators in the province, in 2002, Dr. Bronskill and I established a 2-day annual symposium, which we called the Imaging Network of Ontario (ImNO). The ImNO has been holding the annual meeting for the past 19 years with approximately 400 people registered in the 2021 virtual event.

The ImNO meetings have been important for trainees as they provided opportunities to meet other medical imaging trainees in the province, establish collaborations, and start discussions on job opportunities. In addition, medical imaging researchers in Ontario continued to collaborate and pursue joint projects, creating a collegial and collaborative culture among medical imaging researchers in Ontario. With these large grants, our medical imaging program in London grew substantially over the next decade with the recruitment of additional scientists and a large number of trainees.

Lessons Learned
Collegial interactions with other institutions lead to rewarding research partnerships.

9   Opportunity Only Knocks on the Doors of Those Who Prepare

The work in my lab before 1994 was focused on x-ray imaging (radiography and CT). While we continued to work on the development of CT systems for imaging specimens and small animals at very high resolution, discussions with radiologists in London, emphasized that opportunities in the use of ultrasound imaging are growing due to advances in ultrasound technology. Although we had no experience in ultrasound imaging, we began to explore opportunities for innovation in furthering ultrasound applications. To understand better where our lab would have an impact on the use of ultrasound to meet a clinically unmet need, my graduate students and I attended ultrasound sessions at various conferences. At the 1991 Radiological Society of North America (RSNA) meeting, I attended an ultrasound obstetrical session where a radiologist presented the difficulties in diagnosing fetal malformations using 2D ultrasound imaging. Since our lab was quite experienced in 3D visualization through our work on CT, the potential use of 3D ultrasound came to mind. At that time, there was a company (Kretz Ultrasound) developing 3D ultrasound systems using an integrated probe with an internal motorized approach to scan the piezoelectric elements. (The piezoelectric effect is the ability of certain materials to generate an electric charge in response to applied mechanical stress.)

In our lab meeting, we discussed alternative methods to generate 3D ultrasound images. The method we agreed on was to use an external motorized hand-held fixture that would accommodate any commercially available 2D ultrasound probe. By activating the motor, the ultrasound probe would be caused to translate or rotate. As the ultrasound probe was moved, conventional 2D ultrasound images from the ultrasound machine would be acquired into a computer via a video frame grabber. Since we controlled the motion of the ultrasound probe, the positions of the acquired 2D images relative to each other were known to us. Therefore, the acquired images could then be placed in the correct relative position in three dimensions using a 3D reconstruction algorithm. Although this approach seems feasible, we needed a clinical application that was not explored by others. Since 3D ultrasound imaging was being used for fetal imaging, we searched for other applications.

The chair of the Department of Radiology (Dr. Hobbs) introduced a young radiologist (Dr. Donal Downey) to us and we began to discuss potential applications of 3D ultrasound. In one of these discussions, he related to us the difficulty he had with a prostate biopsy procedure. He explained that the use of the conventional 2D imaging approach with a trans-rectal ultrasound probe does not give him 3D context to perform the biopsy. He was concerned that in sampling the prostate with approximately 12 biopsies, he may not be spatially sampling the prostate as he planned as he had no spatial cues, and wondered whether the high rate of false-negatives was due to non-ideal sampling of the prostate. We immediately suggested that we could develop a trans-rectal 3D ultrasound system so that he would have a 3D view of the prostate, and also develop guidance software to allow him to sample the prostate properly.

Within a few months in 1992, we had developed the system and generated the first 3D ultrasound image of the prostate. The approach involved a motorized mechanism that rotated the trans-rectal probe with a side-firing ultrasound array. By rotating the probe over about 160° in about 12 s and collecting a 2D ultrasound image every 0.44°, we were able to obtain a sufficient number of 2D ultrasound images to reconstruct the prostate in 3D (see Fig. 1a). We also developed a linear scanning mechanism that translated the ultrasound probe over a 5 cm distance in about 10 s. Using this method, we generated 3D Doppler ultrasound images of the carotid arteries (Fig. 1b).

Based on my experience at the OCI on the failure to act on a commercialization opportunity, we began to file patents and showed the images to investors and companies. Based on the interest from these individuals, we spun out a company (Life Imaging Systems—LIS) to commercialize our 3D ultrasound technology. After obtaining local funding to hire three engineers, we were able to develop a range of hardware and software tools needed to further develop the 3D ultrasound systems. We licensed the hardware and software patents to LIS and raised about $4M from a venture fund to expand the team and commercialize the technology. Although my radiology colleague (Dr. Downey) and I were founders of the company, we stayed in our hospital and academic positions respectively, but were active in the company's technology development and testing.

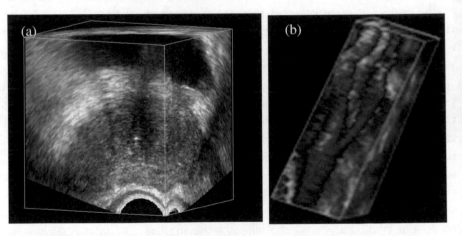

Fig. 1 **a** 3D ultrasound image of a prostate. The ultrasound probe is located in the small black circular region at the bottom of the image. The prostate is the darker concave region above the probe extending about two-thirds of the way upward in the image. **b** 3D Doppler ultrasound image of the carotid arteries. Red shows blood velocity towards the head, and blue shows blood velocity towards the heart

LIS expanded to about 20 people with engineering and marketing teams, developed collaborative partnerships with Nucletron, B&K, and other companies, and developed multiple 3D ultrasound imaging systems. Although the technology was very advanced, the model for the company was to work through other companies and agents to sell the products. Unfortunately, LIS had no control of sales, which lagged. LIS tried to raise another round of financing, but the timing was wrong, as it was during the dot.com bust. LIS could not raise any funds and had to close in 2005.

Since my lab did not assign the patents to LIS and the licensing agreement had a clause about reverting the license to my lab should LIS cease its operations, all the patents reverted to my lab. In addition, I hired the key developers at LIS back into my lab, gaining the advantage of 8 years of developments, multiple patents, and key 3D ultrasound software algorithms. We licensed and sold patents to multiple companies (Nucletron, Hitachi, B&K, Resonant Medical, Eigen, Focal Healthcare, Enable Technologies, and Varian) and continued our development and clinical translation of 3D ultrasound imaging for prostate biopsy, carotid imaging, and applications of image-guided interventions (e.g., focal liver tumour ablation, gynecological brachytherapy, prostate brachytherapy).

10 If You Do not Change Directions, You May End up Where You Are Heading (Buddha)

In the early 90s, one of my graduate students (David Holdsworth) was using a solid-state detector as an optical detector coupled to an image intensifier to generate x-ray images using a slot-scan approach. The approach made use of a low-noise digital camera based on a 512 × 96 element charge-coupled device (CCD) operating in the time-delay integration mode. By combining the digital camera with an x-ray image intensifier, radiographic images could be generated by scanning a slot beam of radiation with the benefit of greatly reducing the detection of scattered radiation. The applications we were targeting were for angiography and mammography with the advantage of scatter rejection.

This research work was being performed in an open lab module in the basement of the Robarts Research Laboratories. The basement had a total of six x-ray imaging modules in which other graduate students worked on their experimental projects. Since the modules were open, the students were able to form a collaborative and collegial culture and help each other in their research. Another student (Brian Reid, who died tragically during his graduate work) was working on developing a micro-CT imaging system under the supervision of Dr. Ian Cunningham. David saw the possibility of using his experimental setup to also generate micro-CT images. By rotating a specimen between the intensifier and the x-ray source, he collected projections that could be reconstructed into a CT image. He positioned the specimen closer to the source and using a small focal spot, a high-resolution CT image could be produced. To demonstrate the concept, he then used a small dead fish from his aquarium and generated a high-resolution CT image of the fish. This image clearly demonstrated to us that very high-resolution CT images could be produced using a cone-beam imaging approach.

Discussion with two other master's graduate students (Michael Thornton and Hao Li) led to the possibility to spin-out a company focused on developing a very high-resolution CT scanner (micro-CT). The two graduate students completed their master's degrees and together with a business partner spun out a company called Enhanced Vision Systems (EVS) with David Holdsworth and me as founders. We obtained an agreement from our institute to allow the incubation of EVS in our lab and use the resources in our lab to develop the technology. At the time, we were not sure of the market for a micro-CT scanner, but we had two major possibilities: non-destructive testing and pre-clinical imaging of mice. To help generate the pre-clinical market, I used some images generated by the team and traveled to a few leading labs in North America and showed our images and described the possibility to use serial micro-CT imaging of mice to follow progression and regression of disease models in mice as part of various experimental protocols. Although I was convinced of the potential of micro-CT in pre-clinical imaging, I got no interest as the researchers could not see the need for these types of CT images or perhaps it was my inadequate sales pitch.

EVS continued to develop the micro-CT scanner and explore its use for non-destructive testing such as ore samples and soil samples, as well as imaging of biological specimens. The micro-CT scanner developed by EVS made use of the cone-beam CT approach to generate 3D images of small specimens based on the know-how developed at our imaging laboratories. Importantly, the technology was being developed with government grants without any outside investment or dilution.

After a disagreement between the major partners of EVS, Michael Thornton bought out his business partner and focused EVS on pre-clinical imaging of mice. At the same time, our lab began to collaborate with Eli Lilly-Canada. At one of our meetings, we described the micro-CT scanner approach for imaging biological specimens. We got immediate interest, as at the same time, Eli Lilly was evaluating the effect of one of their drugs on the bones in mouse models. They had many hundreds of bone samples and were laboriously examining these bones by serial sectioning. The possibility of using high-resolution CT with a resolution of a few microns would allow them to quantify the trabecular structure of the bones and obtain valuable information on the effect of the drug. Thus, they contracted EVS to build for them a micro-CT scanner with sufficient funds in the contract to build a second one to remain in our lab (see Fig. 2). EVS was successful in building the micro-CT and was able to demonstrate the system to other researchers. With these successes, EVS began to sell these systems to other pharmaceutical companies and researchers. EVS then expanded and continued to grow with the development of a second system for CT imaging that could be used to image a complete mouse and not just specimens.

By 2002, EVS's reputation for very high-quality micro-CT images grew and was identified as best-in-class by GE as a leader in x-ray micro-CT technology. At that time, GE Medical Systems was expanding into the molecular imaging sector and saw the need for high-resolution CT imaging of specimens and small animals. After a lengthy negotiation, GE Medical Systems acquired EVS in 2002, but kept the company in London and expanded it to over 20 people with the core team that was trained in our Imaging Research Laboratories. EVS continued to collaborate with our research labs and developed additional micro-CT scanners to the benefit of both EVS and our research laboratory.

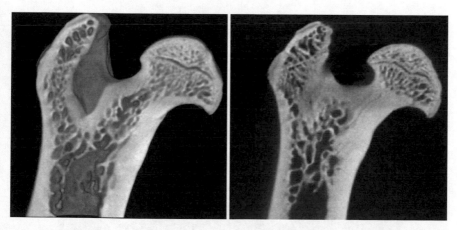

Fig. 2   3D Micro-CT images of the femur of a mouse obtained at 55 kVp. The voxel size is 0.025 mm with isotropic voxels and 20 min acquisition

11   The Opportunity of a Lifetime is Seldom so Labeled

With the patents from LIS and the experience with EVS, our lab continued to develop 3D ultrasound-based systems. Each development was triggered by a clinician who described to us an unmet clinical need that could be addressed with the application of 3D ultrasound imaging technology. Two early discussions with physicians triggered two major areas of research and grant funding.

The first was by Dr. David Spence, a leading neurologist who was focused on the prevention of stroke throughout his career and for which he received the Order of Canada. He described to us the need for a method to quantify the volume of carotid plaques for use in monitoring the progression and regression of the carotid plaques in response to medical therapy and lifestyle changes. We developed a 3D ultrasound imaging system in which the ultrasound probe from any manufacturer was placed into a motorized fixture, which translated the probe while 2D ultrasound images were acquired into a computer and reconstructed into a 3D image (see Fig. 3).

The imaging system was tested in multiple clinical trials together with another researcher (Dr. Grace Parraga) at the Robarts Research Institute and was shown to be very useful in monitoring changes in the carotid arteries. Based on the utility of our imaging approach, with the help of the Western University business development arm (WORLDiscoveries), we

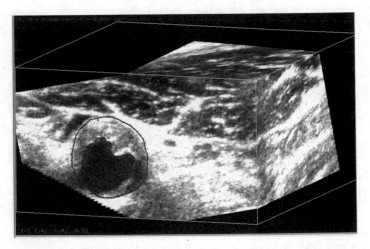

Fig. 3 3D ultrasound image of the common carotid artery with the plaque outlined in red. By segmenting the plaque in serial slices, the burden of plaque can be quantified

sought investors to spin out a company. After traveling to China and pitching the technology to a potential investment firm, we spun out a company (Enable Technologies) in London and Tianjin, China.

The second was by Dr. Cesare Romagnoli, who was an interventional radiologist and head of ultrasound imaging at the London Health Sciences Centre. After Dr. Downey left our centre, Dr. Romagnoli continued to perform prostate biopsies. With his insights, we developed an MR-3D ultrasound fusion prostate biopsy system (see Fig. 4) and licensed the technology to Eigen (a company in California). The technology was further developed by Eigen who is selling the system globally.

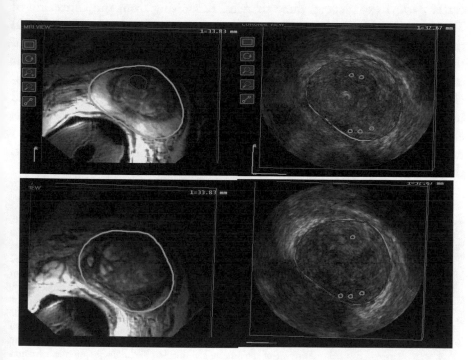

Fig. 4 Two views of a registered 3D ultrasound image (on the right) with an MR image of the prostate (on the left). The red outlined regions are suspicious areas in the prostate that were identified in the MR image and superimposed on the registered 3D ultrasound image. The small orange circles are the location of the biopsy

12   Are You Really Successful or Just Comfortable?

Throughout one's career, new opportunities present themselves as new research projects and jobs. Deciding to act on any of them is difficult and requires a leap of faith as well as leaving one's comfort zone. While staying with the same research field or location can be rewarding and lead to a successful career, tackling a new research direction or taking a new job can be energizing. By challenging oneself by change broadens one's horizons that can lead to discoveries and partnerships. Looking backward on my career path, I experienced these changes. Relocating from the University of Toronto to the University of Western Ontario opened opportunities to build the Imaging Research Laboratories, the Biomedical Engineering Program, the Centre for Imaging Technology Commercialization, and spinning out companies. Changing research directions in the following sequence: photochemistry, x-ray imaging, CT, micro-CT, ultrasound imaging, image-guided intervention, broadened by research experiences, kept my research programs always challenging and never comfortable. Whether the changes such as I experienced are suitable for you is difficult to gauge, but without a doubt, they promise a very interesting career.

Aaron Fenster is a Scientist at the Robarts Research Institute, founder, and past Director of the Imaging Research Laboratories at the Robarts Research Institute. He is a Professor and Chair of the Division of Imaging Sciences of the Department of Medical Imaging at Western University, the founder and past Director of the interdisciplinary Graduate Program in Biomedical Engineering, and past Director for the Biomedical Imaging Research Centre at Western University. In 2007, he became the Director of the Imaging Program at the Ontario Institute for Cancer Research (OICR). In 2010, he became the Founder, Acting CEO and Centre Director of the Centre for Imaging Technology Commercialization (CIMTEC)—a federally funded Centre of Excellence in Commercialization and Research. Currently, he is its CEO. Fenster's laboratory has been pioneering the development of 3D ultrasound (3DUS) imaging and image-guided interventional systems in radiology, oncology, urology, neonatology, and vascular imaging, some of which were successfully translated into clinical use and to companies. Most recently, his lab has developed 3DUS image-guided interventional systems for prostate biopsy and brachytherapy, breast brachytherapy, and MR-guided focal laser thermal prostate ablation in the bore of the MR magnet, and 3DUS focal liver tumor ablation system.

Awards

2020. Member of the Order of Ontario. The Order of Ontario is the province's highest civilian honour for an Ontarian who has shown the highest level of excellence and achievement in any field, and whose impact has left a lasting legacy in the province, in the country and around the world.

2013. Selected by the International Organization for Medical Physics (IOMP) as one out of 50 medical physicists "*who have made an outstanding contribution to the advancement of medical physics over the last 50 years.*" This recognition was given as part of IOMP's 50th anniversary.

2011. Fellow of the Canadian Academy of Health Sciences. A Fellow is considered one of the highest honours within Canada's academic community in recognition of leadership, academic performance, scientific creativity and willingness to serve.

2010. *Canadian Organization of Medical Physicists (COMP) Gold Medal*. This is the highest honour that COMP bestows on one of its members in recognition of an outstanding career as a medical physicist who has worked mainly in Canada.

2007. Premier's (Ontario) Discovery Award for Innovation and Leadership. The Premier's Discovery Awards celebrate the research excellence of Ontario's finest accomplished researchers by highlighting their individual achievements and demonstrating Ontario's attractiveness as a global research centre.

20

Medical Physics and Artificial Intelligence (AI) of Image Interpretation

Maryellen L. Giger

1 Becoming a Medical Physicist Before I Knew the Field of Medical Physics

In the late 1970s, as a mathematics and physics major at Illinois Benedictine College (IBC), I was offered opportunities that shaped my future and led me down the path to medical physics, before I even knew that medical physics was a field. IBC has gone by many names, initially St. Procopius College, then IBC, and now Benedictine University.

During my senior year, I conducted radiation research with Professor John Spokas using phantoms made of tissue-equivalent plastic. The tissue-equivalent plastic had been developed in the 1950s by Francis Rudolph Shonka in the Physical Sciences Laboratory of St. Procopius College. Shonka was an alum, and later faculty, of my alma mater as well as of my future graduate university, the University of Chicago. The tissue-equivalent plastic (designated as A-150) mimicked muscle and continues to this day to be included in Exradin Detectors from Standard Imaging for use in dosimetry and measurements of radiation.

During my summers, I worked at Fermilab (Fig. 1b, c) in the beam diagnostic group as well as in the Neutron Therapy Clinic, in which I honed my

M. L. Giger (✉)
University of Chicago, Chicago, IL, USA
e-mail: m-giger@uchicago.edu

© The Author(s), under exclusive license to Springer Nature Switzerland AG 2022
J. Van Dyk (ed.), *True Tales of Medical Physics*,
https://doi.org/10.1007/978-3-030-91724-1_20

Fig. 1 **(a)** Rose Carney at the board at IBC. **(b)** "Protons are being accelerated through Linac's nine cavities for the first time at 200 MeV. Here, at the control console, are Robert Goodwin and Mike Shea. (1970)" **(c)** "Miguel Awschalom, Lionel Cohen watch final preparation of Fermi Lab's Neutron Therapy Center" (1976)

software programming skills. I learned breadboarding/prototyping methods in order to design printed circuit boards that would be used in the detector systems monitoring the radiation along the accelerator beamline, a source that also created the neutron beams for use in the Neutron Therapy Clinic for treating cancer.

As I was completing my undergraduate double degree in physics and mathematics, a new major was introduced at my college—health science. When I added health science as a third major, I was introduced to organic chemistry and microbiology. While interesting, I could never leave the problem-solving aspects of physics, even though some of my classmates were making their path to medical school.

At IBC, my advisor was Professor Rose Carney (Fig. 1a), who taught me the rigours of mathematics and opened the doors to many opportunities. She had previously spent time on the Manhattan Project while a research assistant at the University of Chicago's Metallurgical Laboratory. Dr. Carney nominated me for a Rotary Fellowship, which I received and allowed me to attend the University of Exeter in England, where I earned my master's degree in physics. I conducted computational research on quantifying the ECGs (electrocardiograms) of subjects with the potential to assess crib death, i.e., sudden infant death syndrome. My advisor at the University of Exeter, Vernon Wynn, taught me that research was 99.9% frustration, however, the 0.1% when one has discovered, or accomplished results, made it all worthwhile.

The stars were aligning. I had found the field of medical physics and started down the path to the University of Chicago. Also, the many opportunities granted to me when I was in college led me to want to "pay it forward". And so I did, when I later became a professor and would offer summer research opportunities to high school and undergraduate students in my lab (more later).

2   Medical Physics and the Digital Era—Analog to Digital

In 1979, I entered the Medical Physics Graduate Program, which included medical physics faculty from the Departments of Radiology and of Radiation and Cellular Oncology at the University of Chicago. It was here that I learned first-hand from many of the leaders in medical physics including Charles E. Metz, Kunio Doi, Franca Kuchnir, Lawrence Lanzl, Lester Skaggs, and Charles Pelizzari, as well as from my fellow classmate, Chin-Tu Chen.

It was during my graduate years that radiology was entering the digital era, converting from screen/film analog combinations to digital detectors, which yielded images with discrete pixel sizes and quantization. *[Note that radiology was already being revolutionized by the 1972 invention of the CT (computed tomography) scanner.]* My Ph.D. research investigated basic imaging properties in digital radiography. [I conducted my Ph.D. research in the Department of Radiology Kurt Rossmann Laboratories, the director of which was Kunio Doi, my dissertation advisor.] Digital radiography includes those imaging systems that acquire two-dimensional projection data in a digital format to yield an x-ray transmission medical image, e.g., a digital chest radiograph. My research focused on the physical image quality (spatial resolution and noise) of digital radiographic images and its effects on the detectability of signals in radiographic noise. A signal-to-noise ratio (SNR), derived from statistical decision theory that included a visual response function and internal noise of the human observer's eye-brain system, was used to demonstrate the effect of various parameters of digital radiographic imaging systems on threshold contrast for detectability (Fig. 2).

Over the next decade, the "x-ray film" was replaced by the "digital image", the hanging of films on a light box or alternator was replaced with the computer monitor, and the "film room" was replaced by the Picture Archiving and Communications System (PACS). These advances were not always an improvement in image quality but were a definite improvement in efficiency. Today's trainees in radiology and medical physics may only know of films through radiology jargon.

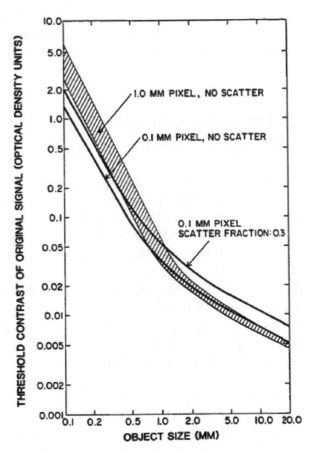

Fig. 2 Effect of pixel size and scatter on the threshold contrasts of unsharp square objects on a digitized X-Omatic Regular/XRP screen/film system (1985). This threshold contrast can be viewed as a measure of detectability with this figure showing how detectability is reduced in the digital radiographic image as the pixel size increases. https://doi.org/10.1364/JOSAA.4.000966

3 Medical Physics and Artificial Intelligence—Human Vision to Computer Vision

The benefit of a medical imaging exam in terms of its ability to yield an accurate diagnosis depends on both the quality of the imaging technology (e.g., the chest radiograph, the mammogram, the CT scan, the magnetic resonance image (MRI)) and the quality of the interpretation (e.g., the radiologist).

The interpretation of medical images by radiologists is limited by the human eye-brain system including various search and perception problems. These limitations can result in missed detections, diagnostic misclassifications, subjective—as opposed to quantitative—assessments of changes (i.e., disease progression), and errors in clinical decision making for patient management.

Understanding human perception yields precautions for radiologists as they tackle the "Where's Waldo?" situations, the satisfaction of search problem, image and structure noise, distractions, fatigue, data overload, the varying subtleties of disease states and normal findings, radiologists' prior training and experience, and the somewhat endless non-image-interpretation tasks associated with a radiology practice. Perception in radiological interpretation was, and still is, an active field within image science including quantifying human observer performance, assessing the role of physical image quality on such performance, and understanding interpretation errors. Imaging scientists Robert F. Wagner, Charles E. Metz, Harrison Barrett, and others worked to establish a unified theory of medical imaging along with mathematical expressions for image quality, detectability, and decision making.

The limitations of the eye-brain system are especially exacerbated in screening, such as in breast cancer screening programs, where most of the images being interpreted are normal (i.e., non-cancerous). Note that it has been shown that breast cancers on mammograms can be missed even though in retrospect they are visible—including both subtle and obvious cancers.

Having artificial intelligence (through computer vision and machine learning) yield reproducible quantitative "interpretations" was seen as a means to effectively and efficiently improve the image interpretation process.

In the mid-1980s, I was part of a University of Chicago team of imaging scientists and radiologists who established the field of CAD (computer-aided detection (CADe) and computer-aided diagnosis (CADx)). Kunio Doi, Heang-Ping Chan, and I submitted the seminal patent on CAD that was granted March 6, 1990 (Fig. 3). The success of CAD was rooted in this

United States Patent   4,907,156

Doi , et al.   March 6, 1990

**Please see images for: (Certificate of Correction) **

Method and system for enhancement and detection of abnormal anatomic regions in a digital image

Abstract

A method and system for detecting and displaying abnormal anatomic regions existing in a digital X-ray image, wherein a single projection digital X-ray image is processed to obtain signal-enhanced image data with a maximum signal-to-noise ratio (SNR) and is also processed to obtain signal-suppressed image data with a suppressed SNR. Then, difference image data are formed by subtraction of the signal-suppressed image data from the signal-enhanced image data to remove low-frequency structured anatomic background, which is basically the same in both the signal-suppressed and signal-enhanced image data. Once the structured background is removed, feature extraction, is performed. For the detection of lung nodules, pixel thresholding is performed, followed by circularity and/or size testing of contiguous pixels surviving thresholding. Threshold levels are varied, and the effect of varying the threshold on circularity and size is used to detect nodules. For the detection of mammographic microcalcifications, pixel thresholding and contiguous pixel area thresholding are performed. Clusters of suspected abnormalities are then detected.

Inventors: **Doi; Kunio** (Willowbrook, IL), **Chan; Heang-Ping** (Chicago, IL), **Giger; Maryellen L.** (Elmhurst, IL)
Assignee: **University of Chicago** (Chicago, IL)
Family ID: 22081174
Appl. No.: 07/068,221
Filed: June 30, 1987

Fig. 3 The first (of many) patents in computer-aided diagnosis

interdisciplinary team of faculty experts in both imaging science and radiology, which also included Heber MacMahon, Robert A. Schmidt, Carl J. Vyborny, Charles E. Metz, and Robert M. Nishikawa. The introduction of computer vision to radiology was accepted due to the presentation of this technology as an aid to radiologists (as opposed to a replacement), thus the name, computer-aided detection/diagnosis.

Initial clinical tasks addressed by CAD were the detection of abnormalities on mammography and on chest radiographs. Since routine digital radiography was still in development, initial CAD research was conducted on radiographs that were digitized, taking approximately one hour to scan one chest radiograph into a digital image and then four hours to process that digital image to yield an output of potential nodule candidates! The early components included bulky film digitizers, computers, storage devices, and displays (Fig. 4).

Development of CAD required knowledge of the physical image quality of the image acquisition system, the radiographic presentation of the disease as well as normal structures, the mathematics of computer vision, and the human behavior involved during medical image interpretation. Although the image analysis and machine learning within an AI algorithm may be complicated, the computer–human interface needed a level of simplicity in order to enable the user of the AI to be both effective and efficient. Multiple output modes were developed (Fig. 5) including (a) numerical

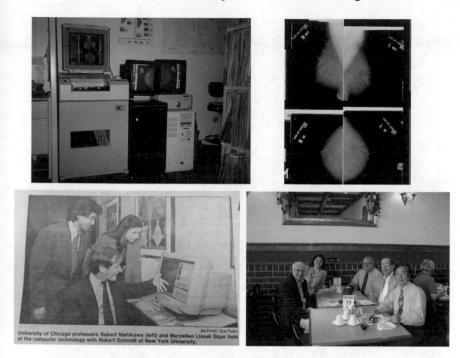

Fig. 4 (Top left) First CADe prototype system for computer-aided detection of clustered microcalcifications and mass lesions on digitized mammograms. (Top right) Example of the computer output showing a true-positive detection and a false-positive detection. (Bottom left) Excerpt from the Chicago Sun Times newspaper updating the public on the AI advancements in breast imaging at the University of Chicago. (University of Chicago, 1994); (bottom right) Casual lunch on radiography and CAD with Art Haus, myself, Charles E. Metz, Carl Vyborny, and Kunio Doi (1990s)

output in terms of the computer-estimated likelihood of malignancy, (b) picture-based output involving similar images automatically retrieved from an online atlas of images, and (c) graphical output showing the unknown case (small arrowhead) on a histogram relative to all cases in the training database. Validation of the computer vision/machine learning algorithms was, and still is, conducted on both the algorithm itself (stand-alone evaluation) and on radiologists interpreting images without and with the computer aid (Fig. 5).

Once a suspect region is detected, radiologists need to characterize the region and assess its likelihood of being an abnormality, e.g., a cancer. This characterization task could benefit from a computerized analysis of the region, in order to yield a diagnostic AI aid such as an estimate of the likelihood that a lesion in question is cancerous (Fig. 5). Note that this is not a detection task but a classification task. A major research effort in my lab has been on the classification of breast lesions as malignant or benign from computer vision-machine learning on mammograms, breast ultrasounds, and

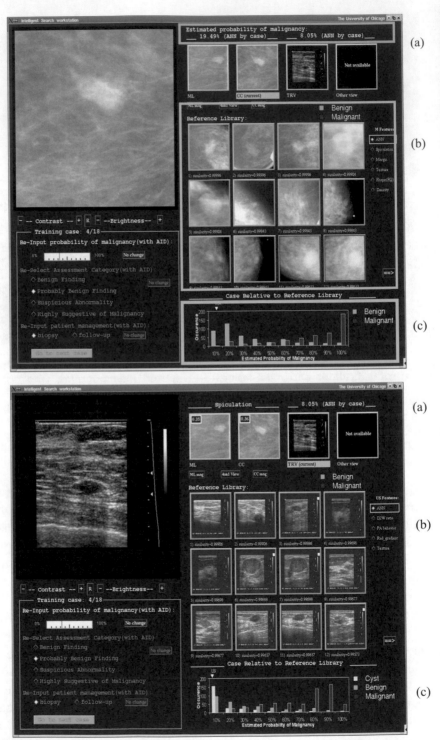

Fig. 5 (continued)

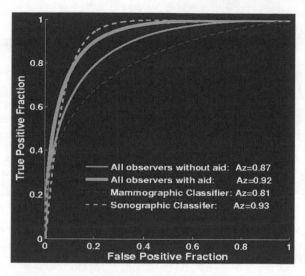

Fig. 5 (Top) Intelligent multi-modality CAD workstation (for mammography and ultrasound) in which the machine learning output was presented **(a)** numerically in terms of likelihood of malignancy, **(b)** pictorially with similar images automatically retrieved from an online atlas of images, and **(c)** graphically showing the unknown case (tiny arrow just above the histogram) relative to all cases in the atlas. (U.S. Pat. 6,901,156, May 31, 2005). (bottom) Demonstrated improvement in radiologists' performance from a reader study when the computer aid was used in their interpretation of mammograms and breast ultrasounds. This is a receiver-operator characteristic (ROC) analysis. The closer the curves are to the upper left, the better the results (Karla Horsch, 2006; https://doi.org/10.1148/radiol.2401050208)

breast MRIs. Developing computer algorithms to segment and then extract features (characteristics) of the lesion has included both human-engineered radiomic features and deep learning outputs. These computer outputs can serve as "virtual biopsies," which could yield quantitative "biomarkers", when an actual biopsy is not practical. Such a four-dimensional (3D plus time) virtual biopsy could automatically segment the tumour from the surrounding background, be non-invasive, include aspects of the complete tumour, and assess a heterogenous nature within the tumour interior, such as the variation in contrast agent uptake as demonstrated in Fig. 6.

When investigating AI in medical imaging, there is much to be appreciated; it is not just the AI algorithm/code itself but the entire workflow, including the end user (Fig. 7). Such considerations require multi-disciplinary teams including radiologists, medical physicists, computer scientists, statisticians, pathologists, and technologists. In developing AI for medical imaging, one needs to consider the clinical need/question, as well as how the AI will be used and who is using it.

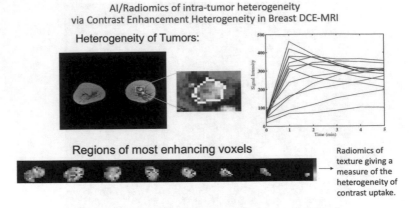

Fig. 6 Schematic example demonstrating intra-tumour heterogeneity of the contrast agent uptake within the vascular system on dynamic-contrast-enhanced breast MRI (Weijie Chen 2006)

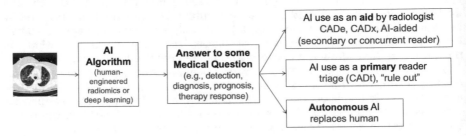

Fig. 7 Schematic diagram of the various considerations necessary in the development and successful implementation of artificial intelligence in medical image interpretation

Around 2012, the term "radiomics" arose, an extension of CAD and defined as a method to extract data mineable features from medical images for use in clinical tasks (i.e., diagnosis) and in multi-omics discovery studies (i.e., radiogenomics).

In medical imaging, the hype and hope of machine learning with deep learning has been escalating for the past decade. While some in computer science hastily predicted replacement of the radiologist, it has been clearly seen that current algorithms will instead augment and enhance radiologists' interpretation. While some focused tasks will be conducted by computers, radiologists will be required to integrate the various computer outputs in the overall interpretation of images and subsequent patient management recommendations. It will be a long time before all medical interpretations have associated AI methods, as there are many diseases and their stages, imaging-system presentations, image acquisition/scanners with different protocols and physical image quality levels, and patient populations. Besides time needed for the development of the various AI methods, the testing within the United States Food and Drug Administration (FDA) clearance/approval process and further evaluations during post-market clinical use will also take time.

It is interesting to consider what has changed and what has not changed with AI over the decades. We now have faster computers, larger datasets of images, more advanced algorithms including deep learning, a broader realization of the additional needs and means to incorporate AI into clinical practice, and the fact that more algorithms across more and more modalities and disease sites are being developed. However, much remains the same. We have the same clinical tasks of detection (finding/locating a signal in an image), diagnosis (characterizing/classifying the signal as disease or non-disease, and decision making on patient management through integrated diagnostics. We have the same concern for "garbage in, garbage out," the same need for rigorous statistical evaluations, and the same need for developing and assessing the AI algorithm while keeping in mind the physical image quality and acquisition protocol used in obtaining the image, as such variations in the image acquisition may lead to variations in AI performances. We also have the same worry of the potential for misuse of the AI device (i.e., such as off-label use), that could lead to erroneous management of the patient.

In training and testing, we have the same requirement for a sufficient number of cases to span the distribution of disease and normal (non-disease) presentations; a major challenge is the lack of diverse data. While diverse data substantially contributes to reducing bias in medical AI algorithms, it is not sufficient. Also needed are methods for assessing the diversity in data; for establishing fairness metrices; for developing algorithms with reduced bias, for testing and then mitigating biases in algorithms and evaluating their generalizability; and for understanding the impact of biased algorithms in clinical practice.

Medical physicists are clearly essential for AI to succeed, especially with their expertise in the physical basis of imaging, in quantifying and assessing effects of image quality, in metrology, and in measuring the performance level, repeatability, reproducibility, and generalizability of AI algorithms. As AI becomes more integral to and integrated in both diagnostic and therapeutic medical physics, such skill sets will be vital in the training of the next generation of medical physicists.

As is now emerging, successful AI systems will use multiple human-engineered feature-based and deep-learning-based subsystems to exploit the benefits of each (Fig. 8).

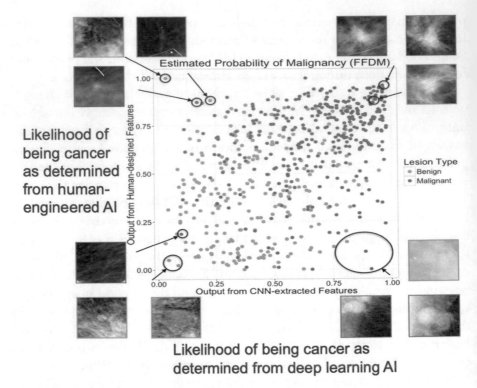

Likelihood of being cancer as determined from deep learning AI

Fig. 8 Demonstration of outputs from human-engineered CADx/radiomics based AI and deep transfer learning AI in the task of distinguishing between cancerous and non-cancerous breast lesions on dynamic contrast-enhanced MRI (DCE-MRI). (Antropova 2017; https://doi.org/10.1002/mp.12453) (CNN = convolution neural network, a class of artificial neural network, commonly applied to analyze images; FFDM = full-field digital mammography)

Additional challenges remain in AI for medical decision making. One needs to understand the clinical question/task and one needs to understand how, and by whom, the AI algorithm will be used in clinical practice. Also, clinical implementation relies on user trust in the AI output. AI methods will need "explainability" and "interpretability", i.e., techniques to help the radiologist understand why the AI system yielded a specific output.

4   Entrepreneurship and the Academician: Commercialization as a Route to Healthcare for the Public Good

In the 1990s, the company R2 Technology licensed our CADe patents from the University of Chicago. R2 incorporated (and further developed) the methods and software code into a CADe system, called the ImageChecker, which in 1998 became the first FDA-approved system for CADe in medical image interpretation, that is, as a second reader in breast cancer screening mammography. The R2 team included Bob Wang, who earlier in his career was one of the inventors of rare-earth screens as well as flat panel displays for radiography, Jimmy Roehrig, a former high-energy physicist, and Wei Zhang, a former research associate from the University of Chicago Rossman Lab, further illustrating the role of medical physics in the early days of AI of medical imaging. [An interesting side note is that the ImageChecker included early deep learning with a convolutional neural network (Fig. 9).]

The public had seemed ready to welcome the use of computers in medical image interpretation. In the early days of CADe, a radiologist colleague noted that his patients never said, "Doctor, don't bother having the computer read my images."

While the process of translating our CADx research and developments to the public had already started through patenting, commercialization started in the academic year of 2009–2010, which ultimately resulted in the creation of a University of Chicago start-up company, Quantitative Insights, that worked to translate my lab's research on computer-aided diagnosis (CADx) into the artificial intelligence product QuantX, used to aid in the interpretation of DCE-MRIs (dynamic contrast-enhanced magnetic resonance images) of breast cancer. The commercialization was initiated when a team of five University of Chicago students—a medical physics graduate student from my lab (Yading Yuan), a medical student, and three MBA students—entered my lab's research CADx prototype in the university's 2009–2010 New Venture Challenge, with me and Gillian Newstead (an expert breast radiologist) as advisors. Of the initial 111 teams, after two rounds of presenting the science and the business perspective, only nine teams made it to the final round, one being our team. In 2010, the company—Quantitative Insights (QI) was established with the goal of creating QuantX as the product (Fig. 10). QI benefited from experts in the Chicago Innovation Mentors program, from funding from the University's Innovation Fund, and access to the University's technology-transfer office, which had become the Polsky Center for Entrepreneurship and Innovation. With the new company's focused team,

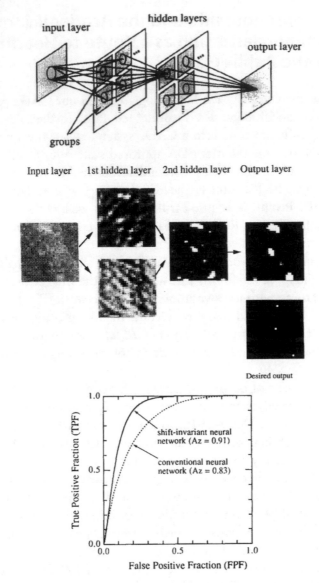

Fig. 9 The first use of deep learning—convolutional neural networks in medical imaging, demonstrated in the computerized detection of clustered microcalcifications in digital mammograms. "The advantage of the shift-invariant neural network is that the result of the network is not dependent on the locations of the clustered microcalcifications in the input layer. ... Approximately 55% of false-positive regions of interest (ROIs) were eliminated without any loss of the true-positive ROIs. The result is considerably better than that obtained in our previous study using a conventional three-layer, feed-forward neural network." (Wei Zhang 1994; https://doi.org/10.1118/1.597177)

University of Chicago Giger Lab CADx Workstation

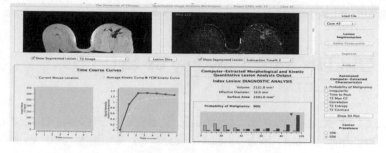

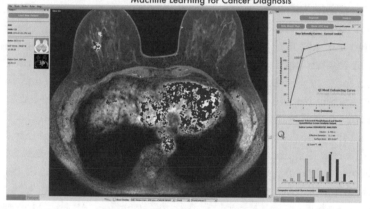

Fig. 10 (Top) Giger lab's CADx workstation for aiding in the interpretation of breast MRIs. (Bottom) The first FDA-cleared, machine-learning driven system to aid in cancer diagnosis (CADx). QuantX, commercialized through a University of Chicago start-up company, Quantitative Insights, was translated from the Giger lab via a 2009–2010 New Venture Challenge program

which included former medical physics graduate students from my lab (Robert Tomek and Michael Chinander), the prototype QuantX became a product. In July 2017, QuantX, through the de novo process, became the first FDA-cleared machine-learning system for use in cancer diagnosis, the first CADx system (FDA DEN170022). QuantX was ultimately sold in 2019 to Qlarity Imaging, a Paragon Biosciences company.

The journey from identifying the clinical problem, conceptualizing a solution within a research lab, and then translating and completing the FDA review is a long one, with many steps along the way. The trek involves investigation of how to convert the eye-brain process of image interpretation to mathematical methods and software, robustness assessment, validation, and translation and commercialization through company start-up. All stages are crucial, and with each milestone, the sense of accomplishment is great.

5   "And to Think They Pay Us to Do This" and "Gotta Dance"

The life of an academic medical physicist includes both research and education. One hopes that one's research leads to findings that will benefit humanity and that one's teaching impacts the next generation of medical physicists.

My involvement in medical physics education ranged locally from didactic teaching, advising graduate students, and serving as Director of our CAMPEP-approved Graduate Program in Medical Physics/Chair of the Committee on Medical Physics at the University of Chicago. Graduate students in my lab have gone on to careers in academia, clinical (both diagnostic and therapy physics positions), government (NIH, FDA) and industry. Nationally, while on the board/executive chain of AAPM, I co-led summits on the future of graduate education of medical physicists, leading to rigorous guidelines for strong didactic education through graduate programs, clinical training through diagnostic imaging or therapy physics residencies, and alternative pathways into medical physics through the establishment of certificate programs. Attention to and periodic review of topics for such educational programs are vital to the future of medical physics. At the University of Chicago, the oversight of our strong medical physics graduate program is now led by Samuel Armato.

In both research and education, I have learned that communication is the key to progress, and that miscommunication is usually the root cause of problems and misunderstandings. As a student, then junior faculty, and then senior faculty, I had been greatly influenced by Charles E. Metz, my mentor and a pioneer in ROC analysis (receiver operating characteristic analysis; a method of cost/benefit analysis in the evaluation of medical imaging system). He taught me communication skills—both verbal and written—in explaining the mathematics of imaging science, in explaining one's research design in a paper, and in teaching. His words of wisdom also included "And to think they pay us to do this" and "Gotta dance," which described his passion for his life's research at the University of Chicago and the importance of delving completely into the problem, finding the solution, and communicating it effectively.

Over the years, I am constantly reminded of these sayings as I proceed through my academic days as researcher, intuitive mathematician, medical physicist, innovator, educator, entrepreneur, communicator, and leader. Although my days are packed with many meetings, the highlights of my day are times spent with individual students as I attempt to pass to them the joy and methods of conducting research. I feel greatly accomplished when my student becomes my colleague.

6   The Lab Behind the Professor—Running a Research Lab and Team Science

As one's research grows through increases in grant funding, lab members, and collaborators, one relies on more than one's expertise in the field but requires insight to the dynamics and interplay between lab members. I attempt to understand each person's strength and interests, and then funnel that into their own *niche* within the lab. Figure 11 shows the members of my lab at the time of writing this chapter during the summer of 2021. While often such a photo would be taken after a lab lunch at some Chicago restaurant,

Fig. 11 Current members (summer 2021) of the Giger lab including (starting at the top left) Maryellen Giger, Karen Drukker, Hui Li, Heather Whitney, Chun-Wai Chan, Li Lan, John Papaioannou, Alexandra (Sasha) Edwards, Madeleine Durkee, Qiyuan (Isabelle) Hu, Jordan Fuhrman, Lindsay Douglas, Natalie Baughan

today we are in the midst of the COVID-19 pandemic. I had learned from my Ph.D. advisor, Kunio Doi, to bring one's lab together beyond the lab setting—through lunches, dinners, and picnics, and thus I continued the tradition with my own lab.

Running a research lab involves mentoring and sponsoring, where sponsoring involves recognizing opportunities for those in your lab, and passing such opportunities along to them so that they can grow in research, education, and leadership. Simultaneously, the lab itself grows.

Members of my lab are also involved in the many collaborations with other research groups from around the world—expanding both the lab's research and the lab members' exposure to research styles, opportunities, and networking.

Notably, many former and current members of my lab have exceled in their niche and contributed substantially to the growth of AI in the field of medical imaging. Beyond some of the many already noted in this chapter, are the contributions from Fang-Fang Yin as my first graduate student; from Zhimin Hou and Hui Li on the role of AI in breast imaging to assess breast cancer risk; from Matthew Kupinski, Karen Drukker, and Heather Whitney on rigorous methods of statistical analysis including evaluation of reproducibility and repeatability of AI in interpretation of medical images; and from Michael Chinander and Joel Wilkie on the role of computer vision of bone trabeculae on skeletal imaging for assessing osteoporosis and osteolysis, as well as Martin King, Neha Bhooshan, Andrew Jamieson, William Weiss, Christopher Haddad, Adam Sibley, Kayla Robinson, and Jennie Crosby. In addition, there are many more former graduate students, some with graduate students of their own, that are my "grand-students".

As I mentioned earlier, I benefited greatly from opportunities granted to me when I was an undergraduate student, and thus I am driven to say thank you by "paying it forward". Thus, each summer, my lab grows with typically six summer students (medical students, undergraduate students, and high school students) with the mentoring of these students shared among the senior researchers and senior graduate students in the lab. This activity of teaching the research chain not only provides research opportunities to the summer students, but allows the graduate students to grow from mentee to mentor.

Team science also includes collaborations within and outside of one's own research institution. Collaboration expedites AI research. Major collaborations with my lab have included Isabelle Bedrosian from MD Anderson on using imaging in breast cancer risk assessment, John Shepherd from

UCSF/University of Hawaii in developing quantitative 3CB (three component breast) imaging, Claudia I. Henschke and David F. Yankelevitz from Icahn School of Medicine at Mount Sinai on low-dose CT imaging, Peifang Liu and Yu Ji from Tianjin Medical University Cancer Institute and Hospital for validation of AI for breast MRI, and multiple breast radiologists and computational genomic data scientists involved in the TCGA/TCIA breast MRI mapping team. (TCGA = The Cancer Genome Atlas; TCIA = The Cancer Imaging Archive).

However, with any collaboration, it is important for all to "play nice in the sandbox," which I often stress in my talks on team science. For effective collaboration, "all I really need to know, I learned in kindergarten".

7   Leadership in the Field of Medical Physics/Medical Imaging

I was trained as a medical physicist/imaging scientist and continued to increase my knowledge over the years by routinely attending the American Association of Physicists in Medicine (AAPM) annual meeting and the SPIE Medical Imaging annual conferences, as well as the annual meetings of the Radiological Society of North America (RSNA) and then the biennial meetings of the International Workshop of Digital Mammography (IWDM). Interestingly, the IWDM had its initial meeting within a SPIE conference, became its own biennial meeting, and then transformed into the International Workshop of Breast Imaging (IWBI) to cover more breast imaging modalities. Additional meetings filled my plate including meetings of IEEE, AIMBE (American Institute of Medical and Biological Engineering), and many one-time focus meetings, such as with National Institutes of Health (NIH) and FDA workshops.

At these meetings, my learning and networking were enhanced during the coffee breaks where details of research were pondered, and future collaborations were established. Eventually, I was asked to join committees of these various organizations. My start with AAPM was enabled by Lawrence Lanzl, a therapy medical physicist from Chicago who added me as a member of the AAPM's Commission on Accreditation of Medical Physics Programs, which I later chaired. The Commission ultimately transitioned from AAPM and into the independent CAMPEP (Commission on Accreditation of Medical Physics Education Programs). I would later serve on the CAMPEP Board.

Within AAPM, I served as abstract reviewer, then as director for the annual meeting's scientific program, then as AAPM Board member, then as

AAPM Board Treasurer, and then as the 2009 AAPM President. Note that as President of an organization, one has the privilege of establishing ad hoc committees to assess new avenues for the organization. While one is thanked for being a volunteer, it is I who would thank the various associations for giving me such experiences and enabling me to grow in my leadership skills, which for the most part, were taught "on the job".

AAPM is well-known for its expertise in assessment of physical image quality and methods of rigorous evaluation of imaging systems (both hardware and software systems). In 2008, as President-elect of AAPM, I met with each of the three AAPM Councils—Science Council, Education Council, and Professional Council—to understand their needs. Science Council mentioned the need for technology assessment and, throughout the nation there was much talk on clinical effectiveness, of which technology assessment was a part. Thus, I established an ad hoc Committee on the Establishment of a Technology Assessment Institute, which ultimately became the Technology Assessment Committee (TAC), a standing committee under Science Council with Bill Hendee as chair. Notably, Bill Hendee was a role model for me of how one should volunteer, give back to one's professional organizations, become a leader, and always take the time to stop and talk with others. To this day, I still refer to his manner of interacting with others and how he was able to effectively and efficiently run a workshop. From 2014 to 2017, I served as chair of TAC. TAC is a committee under the AAPM Science Council (led then by John Boone) and is responsible for technology assessment efforts of AAPM, including assessments of imaging and therapy technologies, and other activities that enhance the research potential and quality assessment capabilities of medical physicists.

My experiences serving in professional societies have included valued opportunities for engagement with external organizations. As AAPM President, I presented AAPM comments at the Federal Coordinating Council for Comparative Effectiveness Research Listening Session on May 13, 2009. I shared that it is important to realize that technology assessment studies are a subset of comparative effectiveness studies, that the reach of comparative effectiveness includes both diagnostic and therapeutic procedures and systems, and that medical physicists play a vital role in conducting such studies [http://www.aapm.org/publicgeneral/ComparativeEffectiveness.asp]. To the FDA in 2010, I presented on technology assessment for CAD research (FDA-MIPS Workshop 2010) with a focus on having a sufficiently large dataset for independent evaluation of CAD algorithms while preserving the integrity of the testing dataset. My dream of having such a dataset continued over decades to finally reach realization as you will read in the next section on the Medical Imaging and Data Resource Center (MIDRC) with its both open commons and sequestered commons.

SPIE is a much broader organization than AAPM, although for most of my professional life, as with many of my imaging scientist colleagues, I saw SPIE only through the lens of its Medical Imaging conferences held each February. My SPIE role model was Bob Wagner (Robert F. Wagner), who excelled in the mathematics and science of medical imaging but also in his methods of interacting and communicating with others to ensure clarity and understanding, in order to push the science to the next stage.

My leadership journey through SPIE Medical Imaging followed a similar path in service as my AAPM journey, moving from committee member, to CAD conference founding chair, and then rising to the Medical Imaging Symposium chair as well as to editor-in-chief of the new SPIE Journal of Medical Imaging. However, my broader leadership in the full SPIE organization was slightly unexpected. One day, while in my office at the University of Chicago, I received a call from Maria Yzuel, a professor of physics from Barcelona, Spain, who worked in medical optics and diffraction image theory, and who was then the 2010 Immediate Past President of SPIE. She was creating the election slate and asked me to run for Board member of SPIE. I was surprised and figured they needed some nobodies on the slate, so I said yes to help out, never thinking that I would get elected. Long story short, I was elected, and then in later years, was put on the slate for the presidential chain and was elected, serving as SPIE President in 2018. With such participation, I was able to bring attention to medical imaging to the wider SPIE optics and photonics community, the discourse resulting in mutually beneficial advancement of both.

An example of such a mutually beneficial, multi-society collaboration was the start of the NCI-AAPM-SPIE Grand Challenges that resulted from multiple discussions I had with Larry Clarke (of National Cancer Institute— NCI) at SPIE and AAPM meetings. We asked Sam Armato to lead the effort starting with the LUNGx SPIE-AAPM-NCI Lung Nodule Classification Challenge of 2015, and the grand challenges continue today, including those with MIDRC.

Thus, I recommend that when asked, one say yes as it can initiate an entirely new journey with additional learning experiences, new colleagues and collaborators, and world travel demonstrating how the language of science and engineering transcends geographical boundaries and customs (Fig. 12). In addition, scientists, as leaders of such organizations as AAPM and SPIE, can only be successful due to the outstanding executive and supporting staff

Fig. 12 Various events and trips with AAPM and SPIE. (Top row) AAPM FOREM on Imaging Genomics (2014); (second row) AAPM Women's Luncheon and my family after being honoured with the AAPM William D. Coolidge Gold Medal (2015); (third row) Example of my travel as SPIE President (2018), here with Maria Yzuel, Allison Romanyshyn of SPIE, and Ignacio Moreno (far right in photo) in Barcelona to celebrate the 50th anniversary ceremony of SEDOPTICA and (right) with other leadership, including C. Dainty (Past President OSA), A. A. Azcarraga (President RSEF), M. F. Costa (President SPOF), H. Michinel (Presiddent Elect EOS and Secretary General ICO), E. Solarte (President RIAO), I. Moreno (President SEDOPTICA), Jesus Lancis (chair of RNO organizing committee), M. Giger (President SPIE). This figure also illustrates the ongoing need for more diversity in leadership roles

of the organizations. My accomplishments would not have been possible without these individuals.

8   Medical Physicists and MIDRC

The relatively new field of data science had been growing over the past few decades. Data science is well matched to medical physicists who already work in interdisciplinary fields involving medical domain experts, imaging scientists, computer scientists, statisticians, and clinical end users. In 2018, Bruce Tromberg, Director of the National Institute of Biomedical Imaging and Bioengineering (NIBIB) of the NIH, along with Krishna Kandarpa, Director of Research Sciences and Strategic Directions at NIBIB, began convening workshops on AI in Medical Imaging with various academics, professional organizations, industry, and other federal agencies. Issues being considered for the successful development and translation of machine intelligence in medical imaging included (a) the absence of large, high-quality, and diverse medical image datasets, (b) the urgency to integrate siloed databases and knowledge bases, (c) the need to develop non-redundant efficient AI tools, and (d) the importance of creating an ecosystem of stakeholders to develop effective and efficient clinically-validated AI applications with which to improve patient management and clinical outcomes. One of the goals of the workshops was to identify a "use case" on which to demonstrate and address these issues. At this time, I was also serving on the NIBIB Advisory Council and participating in their strategic planning for the coming years.

Simultaneously in 2018, as AAPM President-elect, Cynthia McCollough met with Science Council, Education Council, and Professional Council to understand their needs for her next year as President of AAPM. Data science was a major topic, and thus she called a roundtable discussion with the various medical imaging organizations. I was asked to chair the new Data Science Committee that was formed under Science Council with aims to coordinate, steer, and organize AAPM efforts in the fields of Big Data, Radiomics, Machine Learning, and Imaging Metrology and Standards.

And then COVID-19 struck, and COVID-19 became the use case, and the Medical Imaging and Data Resource Center (MIDRC) resulted.

From the start of the COVID-19 pandemic, it was clear that medical imaging would play an essential role in any comprehensive epidemiological approach to defeating this disease. Consequently, a pressing public health need arose for the aggregation, analysis, and dissemination of COVID-19 medical images and associated clinical data. In response to this need, the American Association of Physicists in Medicine (AAPM) (represented by me and Paul Kinahan), the American College of Radiology (ACR) (represented by Etta Pisano and Michael Tilkin), and the Radiological Society of North America (RSNA) (represented by Curtis Langlotz and Adam Flanders), along with 20 other academic institutions, private practices, and the FDA, established MIDRC (midrc.org), hosted at the University of Chicago through funding from the National Institute of Biomedical Imaging and Bioengineering (NIBIB). MIDRC includes a data commons (on the Gen3 platform; Robert Grossman) that supports the public access and analysis of imaging data and metadata (Fig. 13).

MIDRC leverages the existing infrastructure of its participating organizations and institutions and provides coordinated access to data. MIDRC harmonizes data management and data analysis activities at five critical stages: (1) intake and de-identification, (2) semi-automated quality assurance, cleaning, and annotation/labelling of images and associated data, (3) exploration and analysis using machine learning algorithms, (4) distributed access and interoperability through open application programming interfaces (APIs) to create a data ecosystem, and (5) rigorous methods of evaluation to expedite

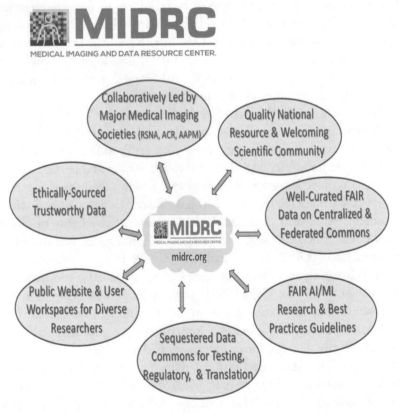

Fig. 13 Uniqueness of the Medical Imaging and Data Resource Center, which is co-led by AAPM, ACR, and RSNA, hosted at the University of Chicago, and funded by the NIBIB of the NIH

AI within the healthcare community. These data and the associated machine intelligence analyses will support essential biomedical research to address (i) improved detection, differential diagnosis and triage of individual patients, (ii) prognosis, including prediction and monitoring of response to intervention for improving patient outcomes, (iii) population-based surveillance and prediction of resurgence, and (iv) assessment of post-COVID and "long haul" patients. MIDRC will accelerate the public dissemination of answers to urgent COVID-19 questions based on research conducted by its users.

The multi-society, collaborative MIDRC data commons has two input portals within ACR and RSNA, and a public-facing portal on the open source Gen3 data platform at the University of Chicago. While ACR and RSNA bring their expertise in real-time and batch ingestion of imaging and non-imaging data, cleaning, labeling, and curating, AAPM brings its expertise in physical image quality and harmonization, and in metrology including tailored distributions, standards, and evaluation metrics. One of the many unique attributes of MIDRC is the generation and maintenance of a sequestered data commons (along with the open commons), which will allow for trustworthy independent evaluations of developed AI algorithms to yield performance metrics for specific clinical questions/tasks across specific populations, allowing for effective and efficient translation of AI through regulatory pathways and post-market performance evaluations. It is quite apparent that medical physicists are clear leaders and collaborators in the success of MIDRC, which is next expected to grow beyond COVID-19 and expedite AI development across multiple medical imaging use cases (disease sites and imaging modalities).

As COVID-19 took hold across the world, my lab immediately began research on AI methods to interrogate thoracic radiographs and CTs, using imaging studies from the University of Chicago COVID-19 datamart and from collaborators in China. Driven by my current graduate students, AI techniques have been developed for (a) distinguishing between COVID positive and COVID negative imaging studies, (b) assessing patient prognosis using surrogate markers such as whether or not the patient had been put on steroids, or whether or not the patient had been intubated or sent to an intensive care unit, (c) monitoring patient on treatment, and (d) sequestering data for independent testing. The students, along with the senior members of my lab, all immersed themselves in their research and in their participation in MIDRC.

9   Closing Remarks

As I wrote this chapter, I kept thinking of the many events in my life as a medical physicist and the multitude of outstanding individuals who established paths for me, walked with me down new paths, or who are the future path makers. I am sure I have forgotten to mention various individuals, research findings, and events. I have also thought about my family, who grew up at AAPM meetings and accompanied me on various SPIE travels. I have found it enjoyable and beneficial to introduce my family to my life's work.

I have learned many lessons, and I hope I have imparted many to others. It is important to identify your passion in medical physics whether it be research, education, and/or clinical, and then make a difference. It is also important to the field that we educate, excite, and open doors for the next generation. Pay it forward.

Maryellen L. Giger, Ph.D. is the A.N. Pritzker Distinguished Service Professor of Radiology, Committee on Medical Physics, and the College at the University of Chicago. She has been working, for decades, on computer-aided diagnosis/machine learning/deep learning in medical imaging and cancer diagnosis/management. Her AI research in breast cancer for risk assessment, diagnosis, prognosis, and therapeutic response has yielded various translated components, and she has used these "virtual biopsies" in imaging-genomics association studies. She extended her research to the analysis of COVID-19 on CT and chest radiographs and is contact principal investigator on the NIBIB-funded Medical Imaging and Data Resource Center (MIDRC) (midrc.org). She has more than 250 peer-reviewed publications (from over 500 publications), more than 40 patents, and mentored over 100 trainees (medical physics graduate students, residents, medical students, undergraduates, high school students). Giger is a former president of both AAPM and SPIE, member of NIBIB Advisory Council of NIH, and Editor-in-Chief of Journal of Medical Imaging. She is a member of the National Academy of Engineering; Fellow of AAPM, AIMBE, SPIE, SBMR, IEEE, IAMBE and COS; and was cofounder of Quantitative Insights [now Qlarity Imaging], which produced QuantX, the first FDA-cleared, machine-learning driven CADx system to aid in cancer diagnosis.

Awards

2022. SPIE Harrison H. Barrett Award in Medical Imaging. This award is presented in recognition of outstanding accomplishments in medical imaging.

2019. TIME Top 100 Inventions of 2019, for QuantX, the system invented in the Giger lab and translated through Quantitative Insights, incubated in Polsky Center, and now further commercialized through Qlarity Imaging.

2015. William D. Coolidge Gold Medal from the American Association of Physicists in Medicine. This award recognizes an AAPM member for an eminent career in medical physics—highest award given by the AAPM.

2013. Selected by the International Organization for Medical Physics (IOMP) as one out of 50 medical physicists "*who have made an outstanding contribution to the advancement of medical physics over the last 50 years.*" This recognition was given as part of IOMP's 50th anniversary.

2010. Elected to the National Academy of Engineering (NAE) of the United States National Academies.

21

Adventures of a Disruptive Medical Physics Innovator

Thomas "Rock" Mackie

1   Introduction

When I graduated from high school, my ambition was to be a writer, a novelist if possible.

Unfortunately, I was objectively a terrible novelist likely because I had too little experience to inform my writing. To solve that problem, I took a series of jobs: bartender, waiter, construction worker, potash miner, meat plant worker, and pilot car driver—as an independent contractor. These jobs forced me to communicate effectively and confidently and, like President Biden now, except in extremely stressful times, most people could not tell that I had a stammer. During this period, my mother had been diagnosed with a "benign" tumour on her pituitary which we were told would grow and could eventually kill her. My father had retired from teaching in Saskatchewan, Canada, and was teaching in Alberta just across the border to supplement his pension, so staying at home to help my mother was important to me. Taking on some adult responsibilities likely created more confidence than going straight to university. However, the drinking age in Saskatchewan had just been lowered to 18 years old and I was also getting an education at partying in earnest, Ernest Hemingway style, so saint not I.

T. "Rock" Mackie (✉)
University of Wisconsin, Madison, WI, USA
e-mail: trmackie@wisc.edu

J. Van Dyk (ed.), *True Tales of Medical Physics*,
https://doi.org/10.1007/978-3-030-91724-1_21

My father nudged me into university. Two years after graduating from high school, he blithely informed me that he had forged my signature and had applied with my high school transcripts to the University of Saskatchewan. He showed me a letter stating that I was admitted. Of course, he said I could turn them down if I wanted—he was a believer in free will. The one thing about being a curious intellectual man-about-town, *bon vivant* failing novelist is that I had become curious about what it was like to attend a university. But I also had empirical evidence that the most intelligent, best looking girls in the bars and at the parties in a university town like Saskatoon were university students. I was doing what most American college-bound kids would abhor … I lived with my mother. My first year of university was a liberal mixed arts and science curriculum. I was leaning towards a philosophy major but I had an outstanding physics professor, Ray Skinner. Ray was a cosmologist and a communist. As an American, he couldn't get a job in the US, but Saskatchewan, being the first socialist jurisdiction in the Western Hemisphere, benefitted from his intelligence and great teaching. I decided to become a physics major at the end of my freshman year which meant that I still had 3½ years to go for a degree.

My high school physics was rusty, and I was still a part-time bartender and had not given up completely, good-time behavior; however, I soon realized, that getting high, makes calculus much harder. I also realized that, like bartending, knowing how to do something is different than being able to do it well. Luckily, at the end of my second year and because I was attending university, I met my future wife, Pam, moved in with her, and buckled down. Also, my father had quit teaching full-time and could better care for my mother. During the summer between my third and fourth year, a nuclear physics professor, Henry Caplan, heard that I had briefly been a miner and hired me to review a worker health and safety report that he was contracted to study. The proposed Cluff Lake uranium mine in northern Saskatchewan would have one of the highest uranium concentrations ever found—some of the ore assayed up to 50% uranium by mass. It was to be an open pit mine and the mining equipment would have to be shielded with lead and have filtered cabins to remove uranium dust and radon daughters (the decay products of radon that were themselves even more dangerous). I calculated that an atmospheric inversion would quickly cause incredible radon concentrations which could not be filtered out. The following year, 1978, Professor Caplan hired me as a night operator of a 500 MeV (MeV = million electron volts) high current electron accelerator.[1] Basically, I had three knobs

---

[1] That accelerator is still in service and is the electron injector for the Canadian Light Source in Saskatoon, Saskatchewan, the only synchrotron photon source in Canada. A synchrotron is a source

to turn. One would steer up and down, one right or left, and the third to turn the current down if I could not centre the beam. Eight hours of this in the middle of night was a much worse job than being a bartender, and I complained of being bored. One of Professor Caplan's post-docs, Dr. Man Kon Leong, asked me to find the error in a translated Russian treatise on relativistic coordinate transforms to pass the time (these transforms relate to the mathematical theory of relativity). That exercise took about 30 pages of work to find and prove. About half way through the summer, satisfied with my work for him already, Mon Kon asked me to accompany him to Uranium City in northern Saskatchewan to earn "big money" on a Canadian federal government project run by the Atomic Energy Control Board (AECB) to decontaminate the city from radon and radon daughters (when a radioactive isotope like radon decays it yields isotopes known as daughters). It would mean that my final half year of university would be interrupted, maybe permanently, but I would be employed full time as a professional physicist.

Uranium City was north of Cluff Lake near the shores of Lake Athabasca in the centre of a large established uranium mining zone. The first woman Chair of the AECB was a Saskatoon medical physicist who had been on the team that developed the Saskatoon cobalt-60 treatment unit in the early 1950s, Dr. Sylvia Fedoruk. The AECB had started other projects in Ontario when uranium mining and uranium refining company towns were found contaminated by tailings from operations.[2] A private civil engineering firm from Regina, Saskatchewan, Keith Consulting, landed the contract on a lucrative 15% cost-plus contract. The more the AECB spent, the more Keith Consulting made. Getting a salary raise was not hard. Keith Consulting's main reason for getting the contract was that they owned their own twin engine aircraft and the only way to get into Uranium City during the summer was by air, and so a good job perk was one weekend a month "down South". Using a private aircraft instead of the regularly scheduled airlines into Uranium City also helped their profits! Air transportation was supplemented in the winter by an ice road including a 50 km journey over a narrower part of Lake Athabasca. The Uranium City operation was closely monitored by Sylvia who came up often on a direct commercial flight from Saskatoon. Sylvia had hired Man Kon Leung and Man Kon hired me. Other than Sylvia,

---

of brilliant light that scientists use to gather information about structural and chemical properties of materials at the molecular level. Such facilities support a huge range of experiments with applications in engineering, health and medicine, cultural heritage, environmental science and many more.

[2] The mine in Uranium City was owned by Eldorado Mining and Smelting which was a "Crown Corporation" or a government owned or controlled corporation. The contamination was felt to be in all of the Eldorado company towns. The government regulator, the AECB, was paying for the clean-up which undoubtedly helped the balance sheet of Eldorado.

at the beginning, I was the only one who knew anything about radon and radon daughters.

We discovered early that there was no Eldorado uranium tailings problem, but that the sandy soil of the town was rich in radium and most of the modern houses that were well-built and well-sealed from the winter cold, had high concentrations. Older log homes or "prospector shacks" were fine. Radium-226 is found naturally within uranium ore and decays into radon-222, which has a three-day half-life. Being a noble gas, radon can travel a long way through dry sand and houses act as chimneys. We measured radon by passing air through a filter to absorb the radon daughters which are attached to dust particles while at the same time capturing radon, free of daughters, in a flask. The filters were put on top of a scintillator sheet and counted over three different intervals to determine the relative concentrations of each of the three radioactive daughter isotopes (make three measurements and solve for three unknowns). The flashes of light produced when a particle decayed was measured by a photomultiplier tube and the counts accumulated. The flasks with radon in them were coated with zinc sulfide, an efficient phosphor for the alpha particles given off by the decay of radon. Not only was Uranium City contaminated by nature, but also at Christmas time, when we flew out for time off, our crew took radon flasks and sampled indoor gas just before our return. We found that there was widespread contamination in Canadian prairie cities and towns. In a survey, the AECB instigated the following year, Winnipeg, which is built on a sandy former lake bottom, turned out to have 30% of its dwellings contaminated. Figure 1 is a photograph of me taking radon and radon daughter measurements in the company lab in Uranium City.

Pam and I spent 14 months in Uranium City which was at the height of a periodic uranium prospecting boom. The population of about 2000 was two-thirds men, many of them uranium miners and prospectors. There were two things that people did for entertainment … outdoor adventure and drinking. My wife and I have stories for a lifetime! One of the adventures was crossing the winter road to visit my father for a few hours with Pam and Man Kon. Also lured to the uranium boom in the North, my father was a substitute security guard (teaching is good training for humane policing) on the operation constructing the same mine at Cluff Lake that I had reviewed for Professor Kaplan. To get there from Uranium City required going south across Lake Athabasca on an ice road and then through huge sand dunes and virgin scrubby forest, uncountable rivers and small lakes for a total return trip of about 400 km. There were no towns, gas stations, emergency phones, or even people other than a couple of semi-truck drivers expected on the winter

Fig. 1 The author doing an experiment to test concrete sealants to prevent radon from entering homes. The experimental apparatus can be seen on the right; large paint cans sealed to concrete slabs that had been coated with different applications of sealant. Enough uranium ore, hand-harvested prospector style, in the countryside surrounding Uranium City, to create about 5×10^5 Bq/l of radon gas, was in the bottom can and then the gas collected in the top can was measured

road. We needed to take fuel in order to make the return trip. We put the gas in three 40 L containers, about double what we expected to use, and strapped them into the back of a covered pickup box. Past the sand dunes, which was the most jarring part of the trip, we thought we smelled gas. Looking in the box, we saw that one of the bungy cords had broken and a container had turned over spilling most of its gas into the enclosed truck box. Drop-by-drop the gas fell on the hot tailpipe of the truck and burst into steamy smoke. A stream of gas inside the tailpipe may have caused an explosion that, if not killing us or burning us badly, would have stranded us in the cold! We shoveled snow into the back of the truck carefully so as to not create a spark, then we swept out as much gas and snow as possible, then covered it with snow again. We set off reeking of gasoline doubting that we understood the vapor pressure physics of cold gasoline.

2   Becoming a Medical Physicist

In the summer of 1979, I returned to the University of Saskatchewan so that I could finish my degree. My good friend, Richard Crilly, who had just graduated with Honours Physics, took over my job in Uranium City. I could have stayed in the health physics field,[3] but Sylvia Fedoruk and her colleague, Bill Reid, told me that I should get a graduate degree in medical physics instead. My mother had recent cobalt-60 radiation therapy for her tumour; so, it was not hard to convince me. My undergraduate transcripts were not all that great because of my weak first two years of university and acceptance into graduate school was uncertain but Man Kon said that if I did well on my Physics Graduate Record Exam, I could likely find a program to let me in. He told me most of the questions came out of the Feynman Lectures on Physics and so I studied all three volumes (which I have used routinely throughout my career). It worked, and in 1980, I was accepted into the University of Alberta Physics Department in Edmonton for a Master's degree supervised by Dr. John Scrimger, a medical physicist at the affiliated Cross Cancer Institute. The early 1980s were a great time to be a medical physicist associated with the Cross. Professor Jerry Battista had a project to develop a 3D treatment planning system based on CT scans. (See Jerry Battista's Chap. 16.) He, Dr. Don Chapman, who was a radiation biologist, and Dr. Raul Urtasun, a radiation oncologist, had landed a provincial grant to study the feasibility of building a hospital-based heavy ion radiotherapy centre at the Cross. The only place in the world doing heavy ion therapy at that time was at the Bevalac at the University of California in Berkeley. Professor Cornelius Tobias who developed the Bevalac radiation therapy program graciously visited Edmonton. It was a very exciting period. The report said it would cost hundreds of millions of dollars, a pittance for oil rich Alberta, but in 1982 there was an oil bust and the government scrapped the project.[4]

I defended my Master's thesis in the summer of 1982 which was a study modeling the buildup in surface dose for high energy photon beams. At the Master's defense, the committee offered me two choices like an academic version of Jeopardy. I could pass with a Master's or go for a Ph.D. which would likely take a year or so and write another chapter testing the model as

---

[3] There is often confusion between health physics and medical physics. Health physics has to do with radiation protection whereas medical physics generally deals with patient-related physics activities.

[4] Dr. Jerry Battista, now emeritus professor of Medical Biophysics at the University of Western Ontario, speculates that it is likely that the MARIA report found its way to Chiba, Japan to be part of their heavy ion accelerator project. Dr. Niek Schreuder said that MARIA was also the inspiration for the University of Indiana Proton Center. (MARIA = Medical Accelerator Research Institute in Alberta).

a dose calculation engine for radiation therapy treatment planning. Thinking near completion and reckoning that my model was important, I took the challenge. At that time, I also sent a letter offering to work for the Atomic Energy of Canada Limited (AECL), a Canadian crown corporation which sold the leading treatment planning system called Theraplan. A few months later I got a form letter back from AECL saying that they evaluated my letter, but unfortunately, at this time, "we are not hiring". In fairness to AECL, the dose calculation method I was working on at the time would have taken about a day to compute. From Moore's Law[5] I knew that eventually computers would be fast enough and indeed much of my career has been about developing computational algorithms that were impractical to implement at the time.

I have told many graduate students two things: (1) show initiative, but even more important is (2) being in graduate school is the best part of your life; so, enjoy yourself. The Cross Cancer Institute was special. Soccer behind the institute was held weekly in all types of weather and competition was fierce with crude Canadian hockey-style checking versus foreign talent and skill. We often retired to the Power Plant, a graduate-student association owned pub. Our institute-wide Christmas parties were legendary, featuring music, dance and comedy talent reviews with Jerry Battista always the master of ceremonies. The Medical Physics Department some years featured skits parodying radiation oncologists, and another, a Moulin Rouge style-male kick line. We were very close and creative in other ways. Laboratory-grade ethanol somehow would find its way into the party punchbowl. This was not politically correct and borderline illegal, but working in the cancer field is stressful and administration was very understanding. (This would definitely not be allowed today!) We also gave back to the community in unusual ways. The Cross medical physicists were customarily responsible for marshalling the huge annual Edmonton Klondike Day parade floats with remuneration paid in beer and food at the local Molson brewery.[6] We had Halloween costumes fabricated in the cast and mould room in the radiotherapy department. We had a toga party à la the movie Animal House. Pam and I rented a small house and had many friends including Rick Crilly who also worked at the Cross and attended the University of Alberta for his graduate degrees. The extra chapter promised, turned into a volume and I defended my Ph.D. in

---

[5] Moore's Law conjectures that computer performance doubles every 18 to 24 months. It has held true for the last 50 years.

[6] Alberta had just legalized strip clubs and soon nearly every Edmonton bar and grill featured it thereby driving insane competition. The medical physicists took to doing cheap lunch buffets at one such establishment close by. It didn't take many visits for us to not pay attention to the surroundings and geek away as if it were just another university cafeteria.

the Summer of 1984. My graduate career was only four years! I thought at the time that I had made up for lost time, but I instantly missed the loss of camaraderie. I then went to work at the Allan Blair Cancer Centre in Regina with a plan of coming back some day to work at the Cross.

3   Working as a Medical Physicist

Going back to Saskatchewan after my Ph.D. turned out to be a godsend and a fantastic start to a professional career. I had offers from Dalhousie in Halifax, Nova Scotia, and the National Cancer Institute (NCI) Clinical Radiotherapy Center in Washington, but chose Regina mainly because Pam would be close to her parents and we also wanted to get married and have kids. Sylvia Fedoruk, who I knew from my Uranium City adventure and had mentored me to pursue medical physics, was the Chief Physicist for the whole province. She was my official boss and was very encouraging of her reports engaging in research.[7]

Peter Dickof, who officed next to me and who I shared coffee with daily, really grounded my theoretical approach to radiotherapy treatment planning and opened my mind to search for solutions to practical problems. He and Phil Morris, who had just left Regina and would later contribute to the Impac oncology information system (a computer data base for cancer patients), had written their own 2D treatment planning system and as well, developed a Moiré pattern contouring system attached to a radiotherapy simulator to determine surface contours of a patient. Phil had developed a record and verify system that was able to control a Siemens linear accelerator (linac) and water phantom system in order to obtain commissioning measurements under automatic control. These folks could not only program in Fortran but could code in assembler language and program logic controllers.

In 1986, Professor Herb Attix, who was the Chairman of the Department of Medical Physics, recruited me to the University of Wisconsin (UW). I have to backtrack a bit and describe my previous interactions with Herb. I had met him at my first American Association of Physicists in Medicine (AAPM) meeting in New Orleans in 1982. I had just walked into the convention

---

[7] Although the award for the best Canadian paper in medical physics is named the Sylvia Fedoruk Award and she worked on the first cobalt-60 unit, her medical physics career, celebrated as it was, is not why she is famous in Canada. She was one of Canada's best female athletes, especially in curling. She was a president of the Canadian Ladies Curling Association and inducted into the Canadian Curling Hall of Fame. She received the Order of Canada and was the Lieutenant Governor (Queen's representative) of Saskatchewan from 1988 to 1994. Sylvia Fedoruk Canadian Centre for Nuclear Innovation was opened at the University of Saskatchewan in 2012.

centre and asked a distinguished gentleman if he knew where the conference registration was. He said he was going there and we should walk together. He asked me about what I was working on and we talked for quite a while. He seemed interested in what I was doing and I later read Herb's work and came to know that he was "the" expert in a core part of my research. Similar to some Nordic countries, the University of Alberta had a policy of paying for external Ph.D. examiners to fly to Edmonton and be a full member of the dissertation committee. It was the first time Herb had been asked to do this and he had enjoyed the Canadian hospitality including a party at the Power Plant bar. Another vote of confidence for my recruitment came from UW Professor Paul DeLuca. He was about to take over from Herb as the Chair of Medical Physics and we agreed to a job interview at the AAPM meeting in Lexington, Kentucky. We chose to sit together at the Awards Luncheon so that we could talk informally. Towards the end of the luncheon, we heard my name called out. I was being asked to come up to the podium to receive the Farrington Daniels Award.[8] I may be unique in having a national award named after a distinguished professor of a university presented to me during a job interview for a position at the same university! Pam was pregnant and I was reluctant to leave Regina and at first, I turned down the offer. In spring 1987, I was given a much better offer from the UW (if you can, don't accept the first offer). At that time, Wisconsin's Department of Medical Physics was considered one of the best in the world and Pam said I would be foolish to turn it down and was willing to sacrifice being close distance-wise to her parents.

My position at Wisconsin was perfect for me! The Medical Physics Department was strong in diagnostic imaging science, which I wanted to do more of. In Regina, I was a clinical medical physicist who was encouraged to do research at a tiny institution. In Wisconsin, I was a researcher in a gigantic organization who was expected to do "some" clinical medical physics. Within a year of arriving, Dr. Tim Kinsella, who was the Chairman of the Department of Human Oncology,[9] asked me to develop a stereotactic radiosurgery (SRS) system, and pointed out a workshop to be held at Harvard's Joint Center for Radiotherapy on the subject. SRS is the treatment with small

---

[8] The AAPM award is named after a UW-Madison professor who first used thermoluminescence to dating crystals and who supervised the Ph.D. of John Cameron, the founder of the UW Dept of Medical Physics, who then applied it to radiation dosimetry.

[9] It is where human radiation oncology is clinically practiced and studied. There were two other departments at the UW-Madison with "oncology" in their name established before it. One in the Vet School and the other is a lab science program in the College of Agriculture and Life Sciences. In addition, medical oncology is within the Department of Medicine and surgical oncology within the Department of Surgery. The UW-Madison has 70 departments related to life sciences alone.

fields of radiation to the brain (now other body sites are also treated) from as many directions as possible. SRS was invented by Professor Lars Leksell at the Karolinska Institute in Stockholm, Sweden, who founded the Elekta Corporation which had developed a cobalt-60 based treatment unit called the GammaKnife. In North America, Professor Ervin Podgorsak at McGill University, as well as the Boston group, had begun using linear accelerators for SRS, which were not as mechanically precise as the GammaKnife but with sufficient quality assurance could more than suffice. I attended the Boston workshop and Dr. Wendell Lutz, the medical physicist responsible for their program, sent me the blueprints for a floor-mounted system which utilized a neurosurgical headframe with sharpened bolts that punctured through the scalp and into the skull, holding the patient immobilized. The frame also defined a coordinate system that could be adjusted to place the lesion (a tumour or a non-cancerous arteriovenous malformation) to be treated at the centre of the converging multiple beams called the isocentre. I had a duplicate system built at the University of Wisconsin's Physical Sciences Laboratory (PSL).[10] The workflow for use of the system had medical physicists involved all day and the degree of work involved meant that only a single treatment of radiosurgery could practically be delivered. It began with the patient wearing a bathing cap under the headframe and then getting a CT scan of the head with and without contrast agents. We would then find and record the coordinate system of the lesion in the scan and determine the size of a circular collimator aperture necessary to define the beam. The patient would then be brought down to the linac and placed on the linac couch with their headframe attached to the floor mount to mechanically plan the treatment. We set up the coordinate system found from the CT scan and if the linac and couch were perfectly aligned, this should place the patient's lesion at the isocentre (refer to the diagram of a linac in Fig. 2 in Art Boyer's Chap. 5). Instead of individual beams directed to the isocentre, the Boston technique had several converging beam arcs (meaning that the beam is on while the machine rotates about the patient) each directed to the isocentre with each arc corresponding to a different couch angle. Treating arcs instead of individual circular beams reduced the workload greatly. We marked the entrance of the arc beams by drawing their extent made visible by a light field shining through the specified circular beam collimator. The heuristic that we used to plan the treatment was that we did not want the arc paths to overlap at the surface of the brain.

---

[10] PSL may be the largest university fabrication shop in the world. They supplied huge parts for Fermilab, near Chicago, the Large Hadron Collider (LHC) at CERN (European Organization for Nuclear Research) in Geneva, Switzerland, and built all the systems for making the IceCube neutrino observatory at the South Pole.

Our goal was to spread the unwanted dose with the lowest dose at the brain surface. When this planning was done the patient could be released from the floor stand but would have to wear the neurosurgical headframe until the treatment was done. Our work had only begun.

The University of Wisconsin had a Theraplan treatment planning system from AECL, the same company I tried to work for. It was a 2D treatment planning system where we needed to reformat each of the arcs so that the contour of the patient's head could be digitized into the system. I was assigned a young biomedical engineer hospital employee, Mark Gehring, to calculate the head contours from the CT scan of the patient so that we could compute the dose from each of the arcs. His program produced a contour on paper and the contour path was traced using a digitizer into Theraplan. This allowed us to calculate accurately only the dose to the lesion and not the dose to the rest of the brain. It did not take Mark very long to be able to calculate a 3D rendition of the patient's head from the CT scan data and it became obvious that we should compute the dose ourselves. Within a few months, in 1989, Mark had a working treatment planning system and we treated our first patients with a program we called UW Stereo. His program also allowed us to eliminate the bathing cap and the mechanical simulation step taking up several conventional treatment slots from the linac and, most importantly, allowed the patient to relax between the end of the CT scanning procedure at the beginning of the day and the beginning of treatment at the end of the day.

While the UW SRS program and the UW Stereo treatment planning system were getting developed, my research group was starting to take shape. The first Ph.D. student I recruited myself was Tim Holmes who had a Master's degree from the UW and had been working at the Mayo Clinic in Rochester, Minnesota. I had just gotten my first NCI grant to extend our treatment planning code for radiotherapy optimization. We formulated an approach borrowed from single photon emission-computed tomography (SPECT) that was used to obtain 3D gamma camera images.

Optimization algorithms invariably produced non-uniform beam distributions. This is because there are normal tissues more sensitive to radiation that need to have a lower dose. The beams which are more directed toward these sensitive tissues can be reduced in intensity and beams which are not directed toward them can have higher energy fluence to make up the missing dose to the tumour. This had been pointed out to me by Karolinska Professor, Anders Brahme, and I credit him for beginning our search to find ways to practically deliver what would be called intensity-modulated radiation therapy (IMRT). In 1989, a Ph.D. student who was not in my group, Stewart Swerdloff, approached me to supervise a one-credit summer special topics course. I assigned him the problem of finding ways to modulate a beam to create the non-uniform beam distributions. Our favorite approach was to use technology not too dissimilar to a nineteenth century steam-powered knitting mill. As the beam arced around the patient, small radiation blocks would shuttle quickly across the narrow beam at the right time, blocking a portion of it. The speed of the radiation blocks allowed easy delivery of arc therapy so that the number of effective beams aimed at the patient would approach the number that Tim Holmes' SPECT-analogous optimization system was computing. We imagined that a patient could be treated slice-by-slice in the same way that a CT scanned the patient. Furthermore, Tim realized that the ideal geometry should be that of a CT scanner, a ring gantry. And, if on a ring gantry, why not throw a CT scanner on board so that the setup of the patient could be determined and adjusted if necessary? Tim named the technique "tomotherapy", which is Greek for slice therapy.

By 1990, our analysis of tomotherapy indicated that a slice-by-slice approach would require extraordinary care to immobilize the patient such that the slice junctions must match perfectly with respect to the patient. Small patient movements could cause overlap of the beam giving higher dose than computed, but even worse, could cause gaps where a thin slice of the tumour received too little dose.[11] In 1991, I read a paper by Dr. Willi Kalendar,[12] the leading CT scientist at Siemens, on spiral CT whereby spiral slices, analogous to a spiral cut Christmas ham, were acquired with the patient couch and CT gantry in continuous movement. He showed not only that there were some correctable issues of blurring, but also that the data acquisition could be much faster than a conventional CT scanner. Part way through reading the paper I realized that a small amount of blurring between the slices would be perfect for tomotherapy as it nearly eliminated the possibilities of hot or cold spots. GE Medical, then headquartered in Milwaukee, Wisconsin, had already jumped onto this approach for CT and called it "helical CT" so we called our approach to tomotherapy, "helical tomotherapy". Up to this time, we had not filed a patent and in 1991 we approached the Wisconsin Alumni Research Foundation (WARF), the UW's technology transfer agency, to file a magnum opus patent application describing iterative optimized treatment planning, our shuttle style radiation blocking system on a ring gantry that

---

11 The NOMOS Corporation, founded by Dr. Mark Carol, licensed our patent for a binary collimator to be added onto an existing linac and marketed a form of tomotherapy in 1995 that treated with IMRT two slices at once. (The binary collimator had a series of shielding blocks that could be either in the beam or out of the beam, such that part of the beam was effectively on or off.).

12 Dr. Willi Kalendar received his medical physics Ph.D. under Prof. Chuck Mistretta at the UW-Madison.

also had a CT scanner on board capable of CT image guidance and delivering IMRT with helical tomotherapy. We also submitted a paper in 1992 which took almost 18 months to get published in the journal Medical Physics because our approach was seemingly so outlandish to the reviewers.

In 1991, after treating hundreds of patients with our SRS system planned with UW Stereo, me being a socialist import from Saskatchewan, and since we had received similar help from Harvard, decided to hold a workshop to give the software away. We charged $500 for expenses to the two-day event that included a reception, food, coffee and 9-track magnetic tape media. Six centres including the Mayo Clinic attended. A few weeks later, we were told by most of the participants that they could not get permission to use the software. The US Food and Drug administration (FDA) had changed the rules some years before, and treatment planning systems were deemed to be regulated medical devices because they used images from systems that were FDA regulated. Nonplussed, we asked the UW Hospital to fund the effort to get FDA clearance telling them that: (1) for a contribution of $1 million we could be done in a year, (2) there was a market for the planning system, and (3) we were sure the hospital could be paid back many times over. The UW Hospital replied that it was not in their charter to be a business (amusing now considering that since that time they became a corporate entity with revenue in 2019 which was close to $4 billion). Adding to our woes, the hospital operated in a loss position in 1990 and some non-essential employees would be laid off. Two of the three programmers in my group, paid for in part by the hospital, were laid off including Mark Gehring. We, and especially Mark, who had put his heart and soul into UW Stereo, did not want to give up. We looked for a company who wanted to work with us to get FDA approval as well as convert UW Stereo into a general purpose 3D treatment planning system with a dose calculation method that I had worked on as a graduate student. ADAC, a company which had a small market share for a 2D treatment planning system, agreed to an "advance against royalties" deal – like an academic book deal! They would add some functionality like contouring digitizing support, a CT image repository backend and also handle the FDA submission. We started Geometrics Corporation[13] in 1992 (the time between former socialist and neophyte capitalist was, by necessity, short). The royalty to Geometrics was to be a bit over 30% (which is far under market rate, but we didn't know that). My project estimate was not very good. It took about three years, not one, and $3 million of ADAC's money, not $1 million, to get FDA approval (we joked later that academics should plan projects by

---

[13] The name Geometrics had the letters of the last name of the founders: Mark Gehring, Rock Mackie, Paul Reckwerdt with the last two letters going to Cam Sanders.

making cost and time estimates and then multiply by pi to get a more accurate estimate). Figure 2 shows a graphics presentation for UW Stereo from 1994. In Fall of 1996, ADAC said, "Great, you can start repaying us back the advance. We estimate, at the current sales rate, that might be a few years. How are you going to support yourselves and provide support for our sales efforts as specified in our contract?" There was also a clause in our contract with ADAC that gave them ownership of the software if our company went bankrupt because an advance against royalties is just another name for a loan with collateral provided by the intellectual property, in this case, the software code. They let us stew on this for a couple of weeks then came back and said, "We have a solution to your dilemma. How about if we buy you for $4.4 million and forgive your indebtedness to us?" Three of the four founders leapt on the deal. Cam Sanders thought it was too low and recommended we should go and find investors to tide us through until our five-year exclusivity agreement with ADAC was over. He was absolutely right! That is what we should have done (remember, don't take the first offer). Geometrics was likely worth $40 million at that moment and the Pinnacle-3 Radiotherapy Treatment Planning system went on to sell many hundreds of millions of dollars of our software including the number one market share position in radiation therapy treatment planning systems and, by the way, killing AECL's Theraplan product. 3D CT-based treatment planning was very disruptive to radiotherapy. For the first time, normal tissue could be avoided, and margins shrunk enough to almost eliminate morbid side effects and enable dose to be escalated. Pinnacle and other 3D planning systems led to the elimination

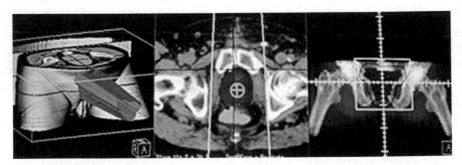

Fig. 2 UW Stereo (renamed Pinnacle-3) graphics presentation from 1994. The left panel is a 3D view of one treatment field for prostate therapy. The red lines indicate pseudo laser lines for positioning the patient. The central panel is a transverse CT slice with the prostate coloured red and the rectum blue. The right panel is a coronal view with the contour of the prostate field coloured red with a line surrounding it indicating the treatment field boundary. 3D treatment planning would eventually make the treatment margins much smaller than those shown here, thereby sparing normal tissue irradiation

of equipment called the radiotherapy simulator which had an x-ray source and an image intensifier system in a geometry that simulated the radiation treatment machine before CT scanners were developed. Geometrics had great software and disrupted the field, but we were lousy businessmen.

Throughout the Geometrics Corporation years, my UW group was working on two huge projects—tomotherapy and the BEAM-OMEGA project. I very much enjoyed the BEAM-OMEGA because I got to work with Dr. David Rogers and his fabulous colleagues at the National Research Council of Canada. (OMEGA = Ottawa-Madison Electron Gamma Algorithm) BEAM was a flexible "Monte Carlo" code using detailed models of radiation transport and random number sampling to simulate linear accelerators in order to gather information to better model treatment planning. OMEGA was an early Monte Carlo treatment planning system. OMEGA was too slow for photon beam calculations, but it was very useful for benchmarking radiation therapy treatment planning dose calculations. By 1995, we had the first BEAM-OMEGA workshop in Madison where the software was given away. I am very proud that the paper that described BEAM was for a long time one of the most-cited papers in the journal Medical Physics.

Beginning in 1992, with tomotherapy inventions filed, we made a concerted effort to find a corporate partner for the tomotherapy project. First, Varian said hell no, Second, Siemens courted us for a while before declining. Third, GE Healthcare said yes. GE's Global Radiotherapy leader, Per Johnson, who had spent time in the CT division, loved the idea. GE's CEO, Jack Welch, had a rule that GE would only stay in a business if they were number one or number two, and this seemed like the only hope for their radiotherapy business. That business was based in Buc, France, but as part of the deal, the Milwaukee headquarters wanted to be involved. In return for using the latest GE Discovery CT platform on which to build our tomotherapy system, we were to help engineer Discovery. From 1994 to 1997, we had our key people commuting to Milwaukee from Madison and, at the end, a lot of the technical knowledge on CT was absorbed by my group. It was a great experience in both how to and how not to do project management. One of the bad habits of GE was to move their engineers on and off huge manpower projects like pawns rather than letting them concentrate in small stable teams. In 1997, a GE Sagittaire linac killed a patient in the same manner as the AECL Therac linac accidents a decade before.[14] Jack Welch quietly arranged a sale to Varian and the UW never even received

---

[14] These accidents were caused by electrons that instead of hitting a tungsten target to create a high energy photon beam, went straight through to the patient because the target was missing resulting in extremely high dose to the patient.

a contract cancellation notice. GE went silent and quickly hired two of our staff members. I am of the opinion that had GE stayed the course, they could have had tomotherapy on the market by 2001 and may now be a dominant radiotherapy company.

4   TomoTherapy Inc

In summer 1997, my UW group had 20 people with many on the GE contract. Bills to the UW had not been paid and there was no certainty they ever would be. My employees again faced layoff notices. This is when the fire sale of Geometrics to ADAC paid off. Paul Reckwerdt, my senior academic scientist, and I reasoned that we could put some of our proceeds into another venture. In winter 1997, Paul and I founded TomoTherapy Inc. We were not even sure that the UW would support us, and we were building the prototype in the same UW Physical Sciences Lab that I had built the stereotactic floor stand a decade earlier. In spring 1998, I applied for a position at Harvard that was meant to lead the birth of a medical physics program that would include collaboration with MIT. I got a letter from Dr. Herman Suit, the Chairman of the Massachusetts General Hospital (MGH) Radiation Oncology Department, that they would offer me the position, but they were not interested in me bringing TomoTherapy Inc. along to Boston. Boston would have been too stuffy for prairie folk like Pam and I anyway. Paul DeLuca was still the Chairman of UW Medical Physics at that time and he along with Carl Gulbrandsen, the director of WARF, arranged for a "faculty retention package" of $1.1 million to be used for the UW tomotherapy project until TomoTherapy Inc could get an investment. This kept the bills paid and the staff hired until 1999 when we raised $3 million lead by John Neis, who was the managing partner of Madison's Venture Investors of Wisconsin and had just lost a close brother to cancer. The State of Wisconsin also contributed a loan of $300,000 to the company. Most of my group joined the company with Paul Reckwerdt as the interim CEO and I became the Chairman of the Board.

The company's first building was a 1400 m^2 former onion powder factory.[15] Throughout 1999, we were still building the UW prototype at PSL while we designed and constructed a bunker at our new building (there was nowhere at PSL we could safely test the machine). The company bunker was

---

[15] The minutes of the first staff meeting in the building has an agenda item concerning how we were going to control the plague of mice infesting our building. Onion powder turned out to be the catnip of mice as the smell never seemed to dissipate.

Fig. 3 Canadian customers inside the TomoTherapy company vault along with the first helical tomotherapy prototype based on a GE CT platform. Left to right are Jerry Battista, Colin Field, Gino Fallone, the author, Jake Van Dyk and David Murray (Tomotherapy's VP of Engineering)

built with double interleaved concrete walls of highway construction berm on top of a large floor slab that was built well outside of the walls to support its weight (see Fig. 3). Our building had a second story and so we needed a bunker roof made with prestressed concrete floor supports. It was entirely held up by gravity and would have mortified California earthquake inspectors. It had a huge maze entrance so that we did not need a door and we could wheel the finished TomoTherapy unit out in one piece for shipment. The first prototype was moved from PSL to the new bunker in 2000 for completion and testing. The first sales of our equipment in 2000 were based on the GE platform to Canadian friends including Jerry Battista formerly at the London Regional Cancer Centre and his colleague Jake Van Dyk (and the editor of this book) as well as Gino Fallone and Colin Field at the Cross (see Fig. 3). In 2001, the prototype with only a few hours of run time was moved to an old cobalt-60 vault, the smallest room in the UW Radiotherapy Center.[16] I

---

[16] We reasoned if it could go in there it could go anywhere; so, the product was actually designed to fit into this particular vault.

owe a lot to then Chairman of Human Oncology, Professor Minesh Mehta, who was always a big supporter of our work, for letting us occupy the vault and for promoting us to administrators.

The software to control the unit was not ready. We had some ambitious design requirements that included many first offs. The dose calculation needed to do dozens of dose calculations per iteration and there could be dozens of iterations. It was the first time that a computer cluster was used in medicine because it took 24 central processing units (CPUs) to do the optimized planning in 30 min to an hour. It would be the first time that a treatment plan would be displayed at a delivery console because we felt it important to show the therapists, who delivered the treatment, the plan they were delivering. It was the first system capable of taking a CT scan of the patient before treatment and compare it to the planning CT scan and the treatment plan. For the first time, IMRT completion code was written so that an interrupted treatment could be safely restarted and completed without replanning. The last requirement was very important because about 5% of the time our prototype failed to complete test treatments usually because of over-tight "trip specifications" on the controllers. A few things were easier as the treatment unit was designed to simplify the dose calculation model. For the first time we pledged to keep the service of the machine producing sufficiently close to the data generated in the factory so that the unit would not require commissioning measurements in the clinic which obviated beam commissioning software to perfect a beam model from the user's measurements (i.e., the treatment planning system was commissioned in the factory and only needed some checks in the clinic).

We did not want to repeat the major mistake of Geometrics by not hiring experienced business people. In 2000, TomoTherapy Inc hired an experienced CEO, John Barni, and Paul Reckwerdt stayed on as President and I stayed on as the Chairman of the Board. John had recently retired as the worldwide manager of Marconi (formerly Picker) CT business located in Cleveland. We hired other excellent people in the business functions of the company including regulatory and quality, operations, marketing, sales, and legal. Many of the managers came from Marconi or GE Medical. In the early days, we had offers to buy the company from Siemens for $60 million and CTI, the developer of the first PET/CT, for $110 million but we did not repeat our Geometrics mistake of taking early offers.

By the end of 2001, the software was ready, and we did our first CT scans of pet dogs from the UW Veterinary Radiotherapy program. The tomotherapy CT scanner used a 1 MeV beam from the linac and so perspicuity was between that of a therapy energy CT and a diagnostic CT (usually between 100 and 140 kV), but certainly good enough for setup verification. In February 2002, the FDA cleared our system and we started treating sinus tumours in dogs shortly thereafter. The treatments were amazingly successful. Conventionally, sinus tumours in dogs are treated with two lateral fields from each side of the snout and head which usually controlled the tumour, but the radiation blinded all of the dogs within a few weeks. Our tomotherapy treatments controlled the tumour and left all of the dogs sighted. Tomotherapy converted a 100% tragic side effect into a 0% complication. There was only some bleaching of the fur on the dog's snout as a side effect. On August 21, 2002, we treated our first patient, a woman who had back pain from spinal metastases. Figure 4 is a photograph of an early patient treated on the UW clinical tomotherapy machine posing with our clinical team. At the time, I was used to the great responsibility and patient intimacy from our SRS program. However, TomoTherapy Inc wanted me to stop my clinical work on the UW unit as they could be held responsible if I made a mistake. The UW transferred responsibility of the clinical tomotherapy program to other medical physicists. I didn't realize how much pressure I

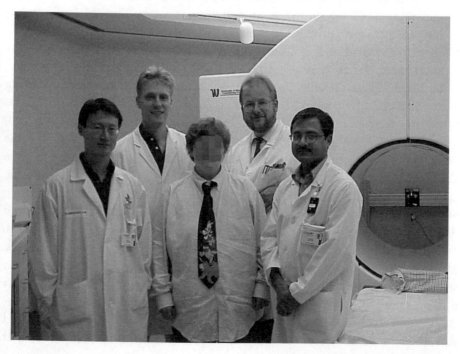

Fig. 4 One of the first patients treated on the first clinical TomoTherapy machine. L-R: Michael Lock MD, Jeff Kapatoes Ph.D., the patient, the author, Susanta Hui Ph.D. Notice that the unit bears the logo of my UW Tomotherapy Group

was under during this period until that responsibility ended. Today, I would never recommend anyone to be both the innovator and clinical pioneer and now it would not be allowed by all university conflict management policies.

At the end of 2002 and early 2003, the first GE platform based tomotherapy systems identical to the UW's were delivered to Edmonton, Alberta and London, Ontario in the capable hands of medical physicists Colin Field (Edmonton) and Tomas Kron (London). In parallel, led by ex-GE medical physicist David Murray, TomoTherapy had been developing its own system to advance beyond the GE Discovery platform. It used an early baggage CT scanner from Boston's Analogic Corporation. Analogic's founder, Dr. Bernie Gordon, also invented the analog to digital converter and so we also used their data acquisition system for the CT scanner detector. The megavoltage detector used on our systems was a refurbished single row xenon detector from an older Siemens CT scanner.[17] It was made with thin tungsten electrodes separating channels of high-pressure xenon gas. Really it was a tungsten detector because most of the photons detected interacted in the tungsten. Most of the electrons thereby set in motion escaped the thin tungsten electrodes and produced ionization in the gas. The system was completely immune to radiation damage. The "HiArt" version of tomotherapy was delivered in late 2003. I particularly remember the hard work of medical physicists Chet Ramsey, in Knoxville, Tennessee, and my close friend Rick Crilly, who at that time was in Rapid City, South Dakota.

In 2003, TomoTherapy moved into its own new three-story building with room for 250 people to work with dedicated manufacturing bunkers, a training bunker, and loading bays for parts and products. The next few years had explosive growth for TomoTherapy, doubling our sales in some years. We grew the service business with each sale because we wanted to have the fastest response time and at the beginning TomoTherapy had machine uptimes (fraction of time that the machine is available for clinical use) of 95% when it should have been 98% or higher. But putting service people close definitely made the customers and service engineers happy. We could hire excellent experienced service engineers because they did not have to travel. Many of our service engineers visited their centres every night to work on small upgrades and test the unit and within a few years, we had the best uptime in the industry, but it took us a lot longer than that to lose our reputation for poor uptimes. In 2005, John Barni retired and we hired Dr. Fred Robertson, a former CEO of Marquette Electronics who had led the sale of

---

[17] Xenon CT detectors were first used in the GE CT/M breast CT scanner introduced in 1975 which was also the first 3rd generation fan-beam CT scanner. (See Terry Peters' Chap. 10 regarding the description of four generations of CT scanners.).

Marquette to GE. Fred was a physician and so was also our de facto Chief Medical Officer. His goal was to prepare for a sale to a big company or to do an initial public offering (IPO).

By 2007, we had expanded to Europe and Asia and I became an "outside-looking" person for the company travelling extensively throughout the world visiting distributors and customers. It was a tremendously exciting few years for me and years of hard slogging for Pam who had our four teenagers to parent often alone. In May 2007, after a couple of profitable years and with a booming stock market, we did an IPO (Nasdaq: TOMO) and the value of the company on the opening stock day was $960 million and rose in the next five months to over $1.2 billion. At this point TomoTherapy Inc. had 600 employees and nearly $240 million in revenue.

In September 2007, when I was 53, I was diagnosed with advanced prostate cancer. There was aggressive stage tumours in both halves of my prostate and there was evidence of spread to my seminal vesicles and adjacent nodes. Surgery was not an option. There was only one technology I would trust … TomoTherapy. I was a pretty stubborn patient! I had long suspected that treating overweight people like me would be better with a "belly board" in the prone position (lying on my stomach) so that my stomach could sink into the carved-out cavity and use gravity to advantage. I also insisted on using a "rectal balloon" which I will leave to the imagination. I wanted to use modest "hypofractionation" (higher dose per treatment and fewer treatments) with extra dose to the magnetic resonance imaging (MRI) visible tumours in my prostate and simultaneous boosting at a lower total dose to the nodes and seminal vesicles. I did my own treatment planning. The treatment volume was nearly a 30 cm long field that would have required couch shifts on a Varian system, but not so on a TomoTherapy system. I was told several times by my anxious friends and colleagues that a doctor shouldn't treat themselves. My answer was that I was not a "real" doctor.

The sunny years turned gloomy for the company as well. In the spring of 2008, the company had to restate its public backlog forecast. We had "booked" several units to entrepreneurial chains of clinics without knowing where they would go or when they would ship in violation of our own accounting standards. Our accountants also discovered that the company was not profitable in service. This meant that the more we sold, the more it would impact our service business. The recession in the fall of 2008 took the shine off because hospitals were worried about the performance of their invested reserves (not-for-profits put profits into "reserve accounts") and the election win of President Obama made some hospital executives uncertain because they were unsure how far his promised healthcare reform would go.

Finally, Varian came out with Rapid Arc and their tremendous marketing power ("One Arc Is Enough") was landing orders that would have been ours a year before.

2009 was a terrible year for the company including deep layoffs. However, there was no downturn in Asia and, as the company expanded aggressively there, it meant more long-haul flights for me to exotic places. The company began recovering in 2010 and 2011. In Fall 2010 we had secret discussions of a "merger" with Accuray. We debated with Accuray management whether TomoTherapy would buy Accuray or Accuray would buy TomoTherapy with a merger of equals combination eliminated early. Accuray was braver than us because the stock market slump meant that the acquirer would have to reach better days while suffering loses with a meager cash balance. Accuray acquired TomoTherapy in May 2011.

The tomotherapy paradigm disrupted radiotherapy. It showed that with multi-field IMRT: (1) you rarely needed non-coplanar beams eliminating the need to rotate the couch improving geometrical accuracy and making the unit more compact, (2) you did not need dual photon energy equipped linacs but a lower energy beam was better, (3) there was no need for electron beams, and (4) finally you could visualize if the treatment was set up correctly instead of relying on absolute localization (as we had used for SRS with an invasive headframe) but needed only reliable immobilization. Recent systems like the View Ray MRIdian, Varian Halcyon, and Elekta Unity all follow the same coplanar, single energy, no electron beam, and pre-treatment image guidance paradigm of tomotherapy.

Throughout the TomoTherapy years, I had cut back my appointment at the UW. When the company was funded, I reduced to 75% appointment and when TomoTherapy went public, I cut back to 50%. In actuality, I still had a large group of students and still did teaching and mentoring but I at least escaped all of the time-consuming committee work that is part of academic life. I was lucky to have a very capable senior academic staff member, Dr. Michael Kissick, who had left plasma physics to help keep order. I also was now attracting highly motivated entrepreneurial graduate students who were self-starters and fully capable of great things with no micromanaging on my part. With the sale of TomoTherapy, I went back to being a full time academic but having worked in a narrow area for so long, I wanted to broaden my research and engage more with industry.[18]

---

[18] A paper I published in 2006 provides a more detailed version of the tomotherapy developments. It can be found in Physics in Medicine and Biology 51, 427–453 (2006).

5   The Morgridge Institute for Research and into Academic Retirement

In 2010, I became the inaugural director of the Medical Engineering group at the Morgridge Institute for Research. It is a private, not-for-profit research institute dedicated to supporting the UW-Madison and originally had a dual mission of groundbreaking basic science as well as applied science and technology transfer to industry. For the first time in my career, I had a large base budget in the style of a German "Herr Professor". The first order of business was to set up my group which became known as the "Fab Lab". In addition to high quality research space, we had a machine shop integral at the heart of the lab that included the first 3D printers on the UW campus. Our labs were underground as is befitting badgers[19] and medical physicists. When I started, I said that I would stay for five years and then retire. During that time, we worked on hundreds of interesting collaborative projects with more than 55 groups on campus with Fine Arts and the School of Music being some of the closest. We and our collaborators did everything from designing pianos to building apparatus to help determine the effects of gravity on plant growth, to designing a CT scanner for horses, to the feasibility of using a sub-critical assembly to produce fission isotopes. One of the most interesting projects was designing a lighting system for modern microscopy labs. These labs have to be kept in absolute darkness. If there are multiple instruments, setting up one instrument will prevent all others from collecting data. Our solution was to have high speed lighting synchronized with the frame acquisition of the microscopes. Fluorescent and multi-photon microscopes use a raster pattern like cathode ray television tubes. At the end of a line the laser and/or detectors are off and by having room-lighting diodes turn on at that instant and repeating at the end of every line, it is possible to have a well-lit room that is actually in the dark more than 95% of the time. I retired at the end of 2014 just after my 60th birthday and left the Fab Lab in the good hands of Professor Kevin Eliceiri. My retirement present was a symposium at Morgridge presented by friends and organized by UW Professor and very good friend, Robert Jeraj. Attendees giving talks included Anders Anhesjo, Jerry Battista, Thomas Bortfeld, David Jaffray, and Rick Crilly.

In retirement, I have founded several medical companies. HealthMyne was the first American "radiomics" company to extract features from medical images related to anatomy and morphology like size, texture, heterogeneity

---

[19] "Badgers" is the name of multiple sports teams at the University of Wisconsin.

and boundary complexity of segmented structures. HealthMyne's software is designed for both use of providers and researchers including pharmacology.

Two of the companies are spin-offs of the Morgridge Institute. Asto CT licensed our WARF-owned design for a CT scanner for horses. It has sold several systems that are capable of scanning horse legs or their heads and neck while they are standing and sleeping (horses can stand while sleeping). Horses are led onto the scanner with the scanner ring in the floor. The handler arranges two legs at a time to be inside the CT bore and the scanner rises out of the floor to scan the legs. They are given a light sedative to induce sleep. To scan heads and necks, the scanner ring rises up and rotates 90°. This time the horse sleeps while resting their head on a small stationary table while the scanner moves by them. OnLume licensed the imaging in "well-lit darkness" technology from WARF for use in fluorescent-guided surgery. Vascular surgery uses immunocyanine green fluorescent dye to image the vascular system. OnLume believes the technology is not in widespread use because of the dim conditions the surgeon has to maintain to use the technology and so instead of just a quality assurance tool, OnLume has a real-time image guidance tool. Existing systems also have jerky frame rates whereas OnLume presents enhanced fluorescent views to the surgeon superimposed on white light images at standard video rates. Systems are being used at major universities for clinical evaluation and research on the use of fluorescence of vascular, cancer and nerve image-guided surgery.

Leo Cancer Care is developing upright radiotherapy (i.e., the patient is in the vertical position rather than the usual horizontal position) and is owned by Asto CT. The synergy between Asto CT and Leo is that upright radiotherapy requires a CT scanner capable of scanning in treatment position. Leo's CT will be able to scan along multiple angles depending on the position of the patient including the standard recumbent position used in conventional radiotherapy. Leo Cancer Care was founded on research from Professor Paul Keall at the University of Sydney in Australia and acquired by Asto CT in 2019. Researchers at the MD Anderson Cancer Center, Harvard University and the Paul Scherrer Institute in Switzerland showed that non-coplanar beams are rarely needed with advanced proton delivery and that lung and liver cancer should be done upright. Leo also believes that breast cancer should be done in the forward-leaning position so that the breast position does not fall laterally over the chest wall forcing lung, and often the heart, to be in the field. Therefore, gravity is used to improve the breast position for radiotherapy, not worsen it. Some of our research indicates that the need for treating prostate cancer patients with a full bladder can be eliminated with upright radiotherapy—something that I can personally endorse from my experience with trying to be relaxed for the setup and delivery of prostate treatment, and having a full bladder, situations which are mutually exclusive. What has held back upright radiotherapy is the lack of a CT scanner that can scan upright. Ironically, the reliance on CT-based treatment planning that Pinnacle and other 3D treatment planning systems helped to engender, is holding upright radiotherapy back. Leo's solution is to use Asto CT's multi-axis CT scanning to do the CT simulation in the upright position. Leo is pioneering upright delivery first in proton and particle beam therapy because it can obviate the need in most cases for a large expensive rotating gantry. Our vision is that future radiotherapy centres will have multiple upright patient positioning systems as workhorse machines and, perhaps, one rotating gantry for special cases. Once again, we plan to disrupt the radiotherapy industry, this time by proving that it is typically better for people to be treated upright and there is limited need for expensive rotating gantries (see Fig. 5). We also think that scanning patients upright will be useful for low dose CT lung cancer screening, planning for spinal surgery as well as many diagnostic procedures.

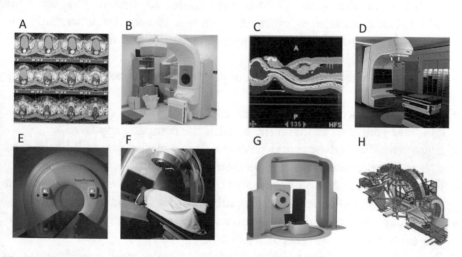

Fig. 5 Illustration of disruptions to radiotherapy in the past and a prediction for the future. Pinnacle shown here and other 3D CT-based treatment planning (a) disrupted planning using a planar x-ray equipped physical radiotherapy simulator (b). TomoTherapy and other intensity-modulated radiation therapy systems using multileaf collimators (MLC) (c) disrupted treatments with linacs not MLC equipped but instead equipped with a tray for hanging custom made blocks (d). TomoTherapy unit as an example of in-room image-guided radiotherapy (e) disrupted radiotherapy without image-guidance in the treatment room (f). The author predicts that upright proton radiotherapy (g) will disrupt proton and other particle beam radiotherapy with very expensive gantries (h)

6   Advice for Medical Physics Entrepreneurs

I often give advice to academic entrepreneurs. As a part-time job between January 2019 and June 2020 inclusively, I was the Chief Innovation Officer of UW Health, the largest medical provider in Southern Wisconsin and Northern Illinois. My responsibility was to help academic medical entrepreneurs start companies. Here is a compendium of advice I often give academics who are thinking of spinning off a company:

- Make sure you're tackling a big problem and that your solution is appropriate. Luckily, medical school faculty and academic medical providers are also customers of medical technology and have an applied practical viewpoint. They also know that inventions don't help patients until they can be developed, manufactured, and distributed in a highly regulated environment by the private sector.
- Your product should be efficacious, save money, or be easier to use (all three is best).
- See if your institution has an "incubator and/or accelerator" program. Our UW medical program is called the Isthmus Project. Understanding the basics of the medical industry is important even if you never actually start a company.
- Get your invention or software copyright protected by your technology transfer office.
- You need to disclose your potential conflict of interest to your institution and have your level of involvement with the company approved.
- Engage business advisors and when you raise money you will likely need to hire an experienced business person in your field.
- Get specialized advice on a quality management system and make sure you do all the development and testing using a strict regulatory framework.
- Get advice on marketing and reimbursement for your eventual product. If you cannot sell it, the company will fail.
- Be on the look-out for adjacent markets or make small modifications to your product that will open up other markets.
- You will need to provide product training and service support. Make sure to plan this into the "roll out" and account for this in the business plan as it is very costly.
- Hire people who can do the job that needs doing much better than you could. Make your other managers do the same. It is surprising how often managers are reluctant to hire people much more skilled than themselves.

- Be optimistic yet skeptical (especially of first offers[20]).

7   Final Thoughts

Whether you are a medical physicist who works in a hospital or works for industry the same principles apply. Show initiative and prepare yourself as well as possible. Learn enough of what your colleagues are doing so that you understand their perspective. Always be honest with yourself and your colleagues. Working in a clinic gives immediate self-satisfaction that your life is fulfilling and important but working in industry magnifies your skills by orders of magnitude. Being professional, hard-working, prepared, and pleasant to be with goes a long way to getting what you want done. And if you are smart and experienced and you are convinced of what your field should do, you should consider starting a company that is productively disruptive and ultimately help countless individual patients and mankind as a whole.

Acknowledgements I would like to thank my wife, Pam, for careful editing of the manuscript. I am the co-founder and chair of medical companies Asto CT, Health-Myne, Leo Cancer Care (owned by Asto CT) and OnLume. I am an investor and board member of Shine Medical Technologies. I am a board member of Cosylab. I am an investor in medical companies Image Mover, Oncora, Redox, and Stemina. I am a limited partner in medical venture firms, 30Ventures, HealthX, and Venture Investors of Wisconsin.

Thomas "Rock" Mackie is Professor Emeritus of Medical Physics at the University of Wisconsin-Madison. He has 40 years of experience in health and medical physics. In his academic career he had more than 180 peer-reviewed publications, more than 50 patented inventions and he supervised more than 40 Ph.D. students. He has co-founded six healthcare companies, three when he was also a professor and three since retiring in 2014. He is an investor and board member in several other companies. He is also a commissioner of the International Commission on Radiation Units and Measurements (ICRU) as well as the Vice-Chair of its board. Rock is also the chair of CARS, a not-for-profit company promoting the assessment of radiological sciences.

---

[20] As a rule of thumb, if you have time to investigate N items to choose from, then you should not choose any until you have sampled $\sqrt{N}$ of them. This advice applies to business and private life. For example, if you have time to look for 9 apartments don't settle on one until you have seen 3.

Awards

2019. Gold Medal—American Association of Therapeutic Radiology and Oncology (ASTRO) for outstanding contributions to the field of radiation oncology.

2019. John Mallard Award—International Organization of Medical Physicists (IOMP) for successfully applying innovation to clinical practice.

2014. Coolidge Award—American Association of Physicists in Medicine (AAPM)—for outstanding contributions to research, education and support of the mission of the AAPM.

2010. Honorary Member of European Society of Therapeutic Radiology and Oncology (ESTRO)—for significant contributions to radiation oncology.

2007. Ladies' Home Journal, Breakthroughs in Medicine Award—for significant contributions to medicine.

22

An Accidental Radiation Physicist

Radhe Mohan

1   Coming to America

As a medical physicist, I have had a rewarding and enjoyable career. Medical physics is a constantly evolving field offering immense opportunities to make contributions to health care, to conduct exciting, innovative research and grow as a scientist and as a human being. It has been my good fortune to have had the opportunity to work at the two top cancer centers in the world, 25 years at Memorial-Sloan Kettering Cancer Center (MSKCC) in New York and now, for approximately 20 years, at the University of Texas MD Anderson Cancer Center (MDACC) in Houston, Texas. In-between, I was at the Medical College of Virginia (now a part of Virginia Commonwealth University) in Richmond, VA for five years. I have been mentored by and have mentored and collaborated with some of the best scientists in the field at these institutions and elsewhere. Interacting with highly qualified individuals at all levels inspired me and exposed me to a diversity of ideas and opinions. Like the Brendan Graham song says, such interactions helped "raise me up to more than I can be."

At MDACC, my current place of work, we aspire to "Make Cancer History." Our mission is to eliminate cancer, a very inspiring and noble

R. Mohan (✉)
MD Anderson Cancer Center, Houston, TX, USA
e-mail: rmohan@mdanderson.org

goal. As medical radiation physicists, we are challenged and motivated to develop safer and more effective treatments. We strive to achieve this goal through collaboration, hard work and out-of-the-box thinking, all leading to the discovery of creative solutions to complex cancer problems.

I came to the US in 1965 with a passion for nuclear physics. My guess is that I was a part of a wave of scientists allowed in due to the nuclear arms race, the launch of Sputnik, and the missile crisis in 1962. Federal funding for research and jobs for physicists were plentiful. I was awarded a scholarship to attend the Duke University Ph.D. program in theoretical nuclear physics. By the time I received my Ph.D. in 1969, man had landed on the moon, the worries about Soviet dominance were receding, resources were being diverted to the war in Vietnam, Nixon was elected president, the funding for scientific research plummeted and jobs had disappeared. Some of my classmates went on to join the national labs operated by the Department of Energy. One became the scientific advisor to President Reagan, and another became the head of the Los Alamos National Lab.

I joined the Rutgers University nuclear physics lab as a post-doctoral fellow. During my second year there, I applied for faculty positions in physics. I had acquired decent software development expertise as a part of my Ph.D. and post-doctoral research, a skill that I used to develop a rudimentary word processing system. I sent out hundreds of applications and inquiries and received exactly the same number of polite rejections. Seeing the handwriting on the wall, I began taking courses in computer science.

Key Points Be flexible. Be prepared. Never give up.

2   Entering the Field

My officemate at Rutgers had applied for a medical physics post-doctoral position at MSKCC but was not interested in changing to an applied area and suggested I go for it. I interviewed with stalwarts in the field, including John Laughlin, the Chair of the Department of Medical Physics at MSKCC for over three decades. The department was struggling with developing software for radiation treatment planning. Fortuitously, my software development skills and computer science background, no matter how limited compared to today's standards, paid-off. (In hindsight, I was the one-eyed king among the software blind.) As luck would have it, instead of a post-doctoral fellowship, I was offered a junior faculty position ("Assistant Attending Physicist"). This turned out to be the best thing that could have happened. To develop software, not just for external beam radiation treatment planning but subsequently for many other applications, described in the paragraphs below, was a great opportunity. It helped me gain a good understanding of the underlying clinical medical physics in diverse areas, to conduct research and to translate the knowledge thus gained into clinical applications. This opportunity to broaden my knowledge and skills has served me well. In fact, it has produced an insatiable appetite to learn more and more.

Key points Expect the unexpected. Be adaptable.

3 Early days at Memorial Sloan-Kettering Cancer Center

3.1 External Beam Treatment Planning

There are two primary ways of delivering radiation to treat cancer. One applies radiation from external sources using multiple beams of high energy x-rays aimed at the tumour target to deliver therapeutic doses. The other is called brachytherapy, in which small radioactive sources are inserted into the tumour. I took over the development of the external beam radiation treatment planning system. Before we could use this system, we needed to calculate, as precisely as possible, the pattern of radiation dose to be delivered. This is done using data measured in water, which serves as a substitute for human tissue as we are mostly made of water. Let me share a story.

An experienced colleague and a novice (that's me) used a tank containing water, called a water phantom, with a water-proof radiation detector (a specially designed ionization chamber). The detector was positioned and controlled remotely by stepping motors at a series of positions (hundreds for each beam of radiation) in the water phantom, one at a time, to irradiate and measure dose deposited at each point. About halfway through the measurements one evening after the patient treatments had stopped, I got the hang of it and my colleague left. I finished the measurements at about 2 AM. Yet, what we accomplished in the whole evening of effort was only a small subset of the data required. I needed to empty the water tank and return the room to a condition suitable for patient treatments before the morning. Utilizing a motorized pump to empty the tank, I placed the end of the in-hose connected to the pump in the water tank and the out-hose into the sink, then turned the pump on. The force of the water exiting into the sink caused the hose to fly out of the sink and, before I could react, 35 gallons of water were on the treatment floor. I spent the rest of the night soaking water up with bed sheets and had the room ready just in time for the treatments to start.

This frustrating experience and the length of time required for the highly boring, repetitive work necessary to acquire a small amount of data led me to feel that there had to be a better way. I partnered with an electronics engineer to develop what may arguably be the first computer-controlled dose measurement system (see Fig. 1). It could automatically position the detector at each predetermined point, measure the dose and move to the next point until the entire radiation field was mapped. All this took a fraction of the time, while we got to enjoy coffee.

The field has advanced tremendously since. Not only have the dosimetry systems become highly advanced, but automation is used increasingly in all aspects of what medical physicists do, including quality assurance, treatment planning and treatment delivery. Artificial intelligence techniques are now being introduced to facilitate automation, ensure safety and quality, and improve the effectiveness of radiation treatments.

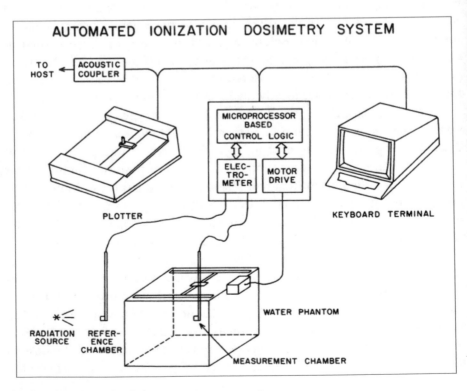

Fig. 1 A computerized dose measurement system

Key points

1. Necessity and boredom are the mother and father of invention.
2. Don't mindlessly accept the status quo. Be open to seeking solutions to old problems.

3.2   Memorial Dose Distribution Computation Service (MDDCS)

When I arrived at MSKCC in 1971, treatment planning was being done using input and output devices linked via modems to a commercial time-sharing computer service for the external beam, brachytherapy, and dose calculation systems we had developed. Considering the lack of commercial treatment planning systems, we had leveraged our in-house treatment planning systems to establish a non-profit, low-cost service to allow institutions and groups practicing radiation oncology throughout the US and Canada to plan treatments. This service started in 1971 and eventually had as many as 100 simultaneous users at any one time. Over 200 groups or institutions benefitted from it over its entire life. It was ended in 1996, by which time numerous advanced commercial treatment planning systems running on inexpensive high speed mini and microcomputer systems had become commonly available.

In 1975, John Laughlin, the chairman, accepted my recommendation to establish our own in-house computer network. Initially, the new system was based on a dual (for redundancy) Digital Equipment Corporation's DEC PDP series computer, which was eventually upgraded, to the DEC Vax series of computers. Until its demise, DEC was the second largest vendor of computer hardware and operating systems. At the initial phase of developing the network, I vividly recall the DEC sales representative convincing me that my treatment planning system would easily run on the baseline PDP 11/45 with its 8 kilobytes of memory and a few megabytes of disk capacity.

However, when the software installation specialist arrived, he laughed upon seeing my large stack of source code printout, saying that a fraction of the code will "blow the 8 KB memory." As one can imagine, I was panic stricken since the purchase of the computer was based on my recommendation. The DEC software specialist, realizing my predicament and the overzealousness of their representative to make the sale, agreed to provide an additional 8 KB of memory. That was not going to be enough either. In desperation, I went through an approximately 6-foot-high stack of DEC manuals end-to-end to learn how to "overlay" the software efficiently, component by component. I spent many a night in my office—in fact, I stored a folding hospital bed in it for occasional rest. (Later I heard from a colleague that John Laughlin had peeked into my office one day, seen the bed and questioned if "I was trying to save on the rent." He did not know me well then.) Eventually, I did manage to fit the entire treatment planning software into the 16 KB of memory. The thrill of success felt terrific. In fact, I became a local software guru for DEC in New York City, and they often referred their customers to me. Moreover, our department at MSKCC became a beta-test site for DEC equipment, and we enjoyed discounts on purchases.

Key points

1. Atychiphobia, fear of failure, can paralyze you or it can be a powerful motivator, depending on how one confronts obstacles.
2. Take risks, at least measured risks. There are no rewards without risks.

3.3   Distinction of Being the world's First Hackee—The Victim of the "414 Gang"

By the early 1980s, I was able to form a small team and establish the Computational Physics Service within the department, which provided treatment planning and dosimetry services to clinical physicists in our own institution, who were hard-wired into our computer system, as well to the outside institutions via the dial-in network.

One chain of events, which lasted for several months, is noteworthy. One morning in early June of 1983, our system manager, reviewing logins from the previous day, found some entries that were not recognizable as being among our users. Some new accounts had been created. He immediately deleted them, but they reappeared the next morning. That was alarming. We contacted DEC to find out how an unauthorized user could get into our system. They could not, or would not, explain (and never admitted that there was a backdoor way of getting into the system). This removal of accounts and their reappearance went on repeatedly for a few days. Moreover, the hackers deleted some accounting information in an attempt to hide their tracks. We contacted the New York City police, who indicated that this type of activity was not in their jurisdiction and they had better things to do. However, they suggested we contact the FBI. The FBI was even more dismissive. Feeling helpless, we sent an email to the intruders begging "Please note this is a hospital computer system. You may accidently damage or delete data and jeopardize patients' wellbeing. If you must use our computer system, please remain within your own account. We will not delete them." The hackers agreed, and the was problem solved!

A few months went by and the hackers continued using our system. One morning, I arrived at work early and was surprised to receive a phone call from a newspaper reporter from Milwaukee inquiring if there had been any intrusions into our computer system. It was followed by a deluge of calls to me and to the institution from numerous newspapers and magazines. I was ordered not speak to anyone unless authorized by MSKCC. They approved *The New York Times* and *Time Magazine* to interview me. Among the many questions the *New York Times* reporter asked was if any data were lost, to which I responded emphatically that no patient data were lost, only some user accounts were erased. Unfortunately, the headline in the *New York Times* the following morning was that patient data had been lost. As a result, the institution and I received many calls from concerned patients. One irate patient threatening to sue said that "I now know what happened to my records … you lost them." To which I replied "Madam, I am sorry for the loss of your records, but you were not even treated in our department."

I was also interviewed by Time magazine (see the story in http://content.time.com/time/subscriber/article/0,33009,949797,00.html). I recall that the photographer took six 36-frame rolls of photos of me in the computer room from every angle possible, but the one published was perhaps the saddest one they could find, with the caption "A somber Dr. Radhe Mohan at the keyboard of Memorial Sloan-Kettering machine," giving the impression that I was still devastated. Despite what the caption said, I was well past the whole affair.

Then the FBI came calling. What had happened was that the same gang of hackers had infiltrated a bank in Los Angeles and the Los Alamos nuclear weapons laboratory computers. The FBI wanted to know if we could give them information to identify the hackers. I told them we could do better. As you may recall, since we could not find a way to prevent the hackers from reentering our system, to protect our data from damage, we had made a deal with them to confine their activities to safe zones on the system, and that they were still dialing in. The FBI tapped the incoming lines to our system and were able to identify the hackers—the "414 gang," a name they had taken based on their area code.

Key point Believe that to every problem there is a solution.

4   CT Scanners and 3D Treatment Planning

The planning of radiation treatment up until the mid-1980s used to be based on two-dimensional x-ray projections of the anatomy, clearly a crude and suboptimal process. Our external beam treatment planning system, too, was two dimensional. CT (Computerized Tomography) scanners were introduced in the 1970s to create detailed cross-sectional images of the internal anatomy. Starting in the late 1970s, in partnership with one of the CT equipment vendors, we developed a CT-based 2D external beam planning system as an extension of the existing external beam treatment planning system. It required manual delineation of body and internal anatomy through the region of interest using crude devices, such as a "track ball," predecessor of the mouse. The system saw limited clinical use but became the foundation of the 3D CT-based external beam planning system.

As may be obvious, the human anatomy and tumours are three-dimensional. Facilitated by a contract from the US National Cancer Institute (NCI) in 1984, my small team and I embarked on the development of a 3D Conformal Radiation Therapy (3DCRT) planning system that utilized a stack of 2D CT images to create 3D images of the tumour target and the critical anatomy to be spared from excessive radiation exposure. The goal of this system was to use multiple, appropriately-shaped beams of high energy x-rays, aimed using the so-called "beam's eye view," to produce three-dimensional patterns of dose distributions in which high doses conformed to the shape of the tumour target and optimally spared normal critical tissues. Initially, I was reluctant to take this task on since I felt that my physician colleagues would not adopt it. I was concerned they would not be willing to take on the arduous task of drawing contours on each of a large stack of images for each patient. However, I am pleased to say that I was "persuaded" by my chairman to "just do it." I was incorrect in my assessment and the physicians, realizing the value of 3DCRT to patients, gradually adopted it, first for clinical trials and eventually for routine clinical practice, not just at MSKCC but worldwide. A lesson well learned. A story of our work appeared in the January 1987 issue of *National Geographic*, including a screen shot of 3D contours of a head and neck patient on the cover. (Interested readers can find the cover images and the related articles through a Google search for "National Geographic January 1987."). As a part of the development of this system under the NCI contract (which included three other institutions), I had the great fortune of working with Dr. Michael Goitein of Massachusetts General Hospital (MGH)—one of my heroes, a brilliant medical physicist, a broadly knowledgeable individual, and a thought leader willing to share his ideas and dispense advice without being condescending. I should mention there were parallel developments of 3DCRT systems at other institutions including MGH and the University of Michigan. Starting in the mid-1990s, intensity-modulated radiotherapy began replacing 3DCRT as the primary mode of radiation therapy of cancers (see below).

Key Points

- "If at first, the idea is not absurd, then there is no hope for it." - Albert Einstein
- "New ideas pass through three periods: (1) It can't be done; (2) It probably can be done, but it's not worth doing; (3) I knew it was a good idea all along!"—Arthur C. Clarke

5 Throwing Dice to Compute Radiation Dose

Simultaneously with the development of 3DCRT, I started to learn about the so called "Monte Carlo (MC)" technique to simulate, using computers, the transport of radiation (only photons and electrons at that time) in water and human tissues. We used MC to conduct research to improve our under-standing of the basic mechanisms of radiation transport and dose deposition and to develop algorithms for faster and more accurate computation of dose distributions in patients.

In the MC technique, random numbers are generated as if using a die with a virtually infinite number of sides. The name "Monte Carlo" comes from the casino in Europe, and the term is used because of the randomness of the approach. Particles of radiation are started from the source, one at a time, and traverse the beam-shaping devices and the patient's image, scat-tering and depositing radiation dose following the laws of physics. Millions of particles have to be transported to achieve sufficient accuracy. This number is typically a tiny fraction of the particles that actually bombard the patient. In this random process, in which particles, even if they have the same properties initially, travel random distances, scatter differently and deposit different doses at different points in the patient. The dose in any one volume element (voxel) is the accumulation of contribution from all particles. This phenomenon is akin to using a random poll involving a small sample of the population to predict the results of an election.

The semi-empirical systems for dose calculations make simplistic assump-tions and approximations that seriously compromise the accuracy of the predicted dose, which, in turn, may adversely affect outcomes of treatment and the validity of what we can learn from patients' response to treatments. Our goal in the 1980s, which proved to be naïve in the face of the slow speed of the computers available at that time, was to use MC in routine practice of radiation treatment planning. Only now, when computers are many thousands of times faster and the computations can be parallelized and distributed among a large number of CPUs, are MC techniques increasingly employed, though not in all applications.

Nevertheless, MC techniques proved to be invaluable to understand the consequences of the underlying assumptions and approximations of conventional methods of dose calculations. They were also critical for developing novel three-dimensional "pencil beam" dose calculation algorithms. Pencil beam dose distributions were generated with MC techniques and the algorithms integrated contributions of each of the pencils traversing the heterogenous anatomy and beam shaping devices.

An interesting episode related to our forays into the MC world was the acquisition near the end of 1980s of the MM50 Microtron treatment delivery machine from a Swedish manufacturer, Scanditronix. (A microtron is a type of particle accelerator in which electrons are accelerated in a linear accelerating structure but then bent by magnetic fields and re-entered into the same structure for further acceleration, multiple times until the appropriate energy is achieved.) A team of Scanditronix personnel visited us, accompanied by an internationally renowned medical physicist, to persuade us to purchase it. We were informed of its two attractive features: a multi-leaf collimator (see Fig. 2), a device to electromechanically create irregularly shaped x-ray beams, which could replace the use of heavy, cumbersome radiation-shielding blocks custom fabricated for each patient, and a 50 MV scanning photon "pencil" beam, the intensity of which could be modulated to shape dose distributions to a high degree. Based on my experience with the application of Monte Carlo techniques, I was skeptical about the claims of the narrowness of the scanning and offered to carry out MC simulations to determine its characteristics. However, before I could complete my calculations, the purchase agreement had been signed. The calculations, which were validated eventually with measurements, showed that the pencil beam was, in fact, more like a broom, not suitable for scanned intensity modulated applications. The multileaf collimator, on the other hand, was a highly innovative idea, but the Scanditronix version proved to be of limited value for routine practice. Being the first of its kind, it was not precise enough, and it required substantial maintenance. However, realizing its potential, we and many other institutions were able to persuade Varian Medical Systems and other vendors of treatment machines to develop multileaf collimator add-ons of their own.

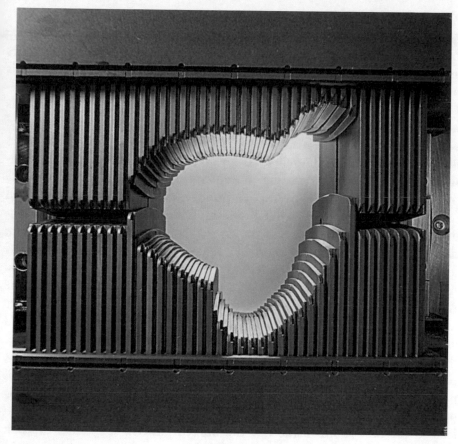

Fig. 2 A typical multi-leaf collimator. The leaves (vanes), driven by small stepping motors, can be moved in and out to shape the beam aperture to conform to the shape of the tumour target

Over the years, I have encountered surprising resistance to the use of MC both among the potential users as well as the vendors. In part, that may have been due to the insufficient speed of the computers and in part due to the natural resistance to change among us mortals. More recently, somewhat simplified versions of MC approaches that are an order of magnitude faster, yet are nearly as accurate, have been developed. While MC-based systems could have been adopted into clinical practice more than a decade ago, I am delighted to find that an increasing number of institutions have started using them, and vendors are now integrating them into their treatment planning systems.

Key point Recognize change as an opportunity for growth.

6   IMRT ("I'M Really Tired")

In the early 1990s, Dr. Thomas Bortfeld (now the Head of Radiation Physics at MGH), asked me if he could visit us. He was on his way back to his home country of Germany after completing his post-doctoral fellowship with Dr. Arthur Boyer at the MDACC (see Arthur Boyer's Chap. 5). I was his host and learned about an algorithm he had developed in which each of a set of multiple beams is subdivided into tiny "beamlets." The intensities of the beamlets are optimized by the algorithm to produce dose distributions that could conform to the tumour much more tightly compared to 3DCRT and spare normal tissues to a greater extent. This algorithm needed to be incorporated into a 3DCRT system, which MDACC lacked. Fortuitously, as noted above, we had developed our own 3DCRT system and, during the first week of Dr. Bortfeld's stay in New York, he and I worked together around the clock to "marry" his algorithm with our 3DCRT system. After his departure, I applied it to plan a few cases of prostate cancer and the results were astonishing. My physician colleagues were so impressed that they made statements like "Wow! I didn't know we could treat the prostate and at the same time spare the rectum." Some even went as far as to say, "With this kind of dose distributions with IMRT, who needs protons?" Initially, planning for such treatments was called "inverse planning" because, in this process, we start with a desired dose pattern and try to "invert" it using optimization techniques to compute the intensities of beamlets that would lead to the best approximation of the desired pattern. I, instead, called it "intensity-modulated radiotherapy" (IMRT), a name that has stuck.

Our leaders wanted to know how IMRT could be delivered, but to do so, we needed some help. The Chairmen of the Departments of Medical Physics and Radiation Oncology at MSKCC, Drs. Clifton Ling and Zvi Fuks, respectively, impressed with the potential of this new modality, persuaded Varian Medical Systems, the vendor of our treatment machines, to enable the leaves of the multi-leaf collimator to move under computer control while the radiation beam was on so that the positions and speed of the leaves could be varied to modulate intensities. (See Clif Ling's Chap. 9.) In the meantime, we (and others) developed the so-called "sliding window" dynamic multi-leaf collimator (DMLC) algorithms to deliver IMRT. We treated our first patient with IMRT in 1995, a prostate case (see Fig. 3).

While we were the leaders in the development and clinical implementation of DMLC-based IMRT, we were not alone in this field. The initial concepts of inverse planning and delivery of such treatments were actually envisioned and articulated by Dr. Anders Brahme. Moreover, a commercial

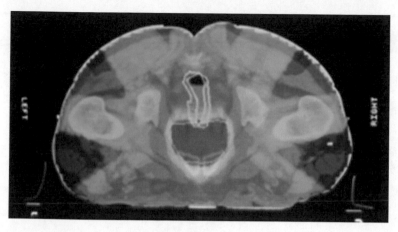

Fig. 3 The first patient treated with IMRT at Memorial Sloan Kettering Cancer Center—a prostate case. "With this kind of dose distributions with IMRT, who needs protons?"

vendor, NOMOS, developed a different way of delivering IMRT based on the principles of "Tomotherapy," conceptualized by Dr. T. Rockwell Mackie, using a rotating gantry and a slit DMLC (see Rock Mackie's Chap. 21). Since then, the field of IMRT has advanced considerably, leading to the improved optimality (i.e., clinical effectiveness) of dose distributions and greater efficiency of treatment delivery. However, conventional DMLC-based IMRT and its more recent incarnation, the volumetric modulated arc therapy (VMAT), have become the dominant modes of treatment. In VMAT, the gantry rotates and the DMLC leaves move continuously to deliver IMRT in just a few minutes compared to the static, multiple field IMRT, which may take as much as 20 min, sometimes even more.

In the early days of IMRT, the efforts required to plan treatments as well to perform quality assurance of each of a large number of DMLC beams were notoriously high. With regard to the treatment planning effort, each and every organ to be spared had to be explicitly delineated. In fact, one of the well-known medical physicists, in one of his presentations, redefined the acronym IMRT to mean "I'M Really Tired." There were IMRT sceptics, but many of us, who were passionate about its potential persisted and gradually developed hardware and software solutions for the obstacles. During the late 1990s and early 2000s, IMRT had become the most sought-after radiation treatment modality. So much so that it found its way into a tabloid (see Fig. 4). It did not name the source of information in the article but claimed that that if you have prostate cancer, you "are in luck," and suggested that for more information, call 1–800-TRYIMRT. Just out of curiosity, I called the

---

National Enquirer: 24 August 1999, Page 18, Article by G Carden *et al.*

Headline: *New prostate cancer treatment KOs tumors with fewer side effects*

The first lines of the article:

"The nearly one million men in the U.S. with prostate cancer are in luck. A new, super-precise radiation treatment system called IMRT promises a higher cure rate and far fewer side effects than traditional radiation treatment. ..."

Also included in the article: "For more information on IMRT call 1-800-TRYIMRT."

Fig. 4 Making IMRT a household word through a tabloid

number. The voice at the other end sounded familiar, and I asked, "Is this really you, Herb?" (Name changed to protect the guilty.) The source of the story turned out to be a well-known vendor.

My early efforts at MSKCC on IMRT were in collaboration with Drs. Chen Chui, C. Clifton Ling (see Chap. 9) and Qiuwen Wu. At the end of 1996, I left MSKCC to become the Director of Radiation Physics in the Department of Radiation Oncology at Medical College of Virginia (MCV), now Virginia Commonwealth University). Dr. Wu joined me there and we continued further development of IMRT and implemented it into a Phase I clinical trial of head and neck cancer patients. As a part of our research, we developed a novel IMRT technique to simultaneously deliver different levels of radiation doses to target volumes that require different doses. These volumes are the gross tumour volume (GTV), requiring the highest dose, and the clinical target volume (CTV), the region surrounding the GTV that has a lower density of cancer cells, requiring a lower dose. We called it "simultaneous integrated boost" or SIB.

In the early 2000s, I presented an idea at a Varian Users Conference about the "flattening filter-free (FFF)" mode of treatment delivery. High energy-x ray machines employ a flattening filter to flatten the x-ray beams, creating a uniform dose at a depth in the patient. Without such a filter, the dose in the middle of the beam would have too high an intensity. The need for flat beams was established in the 1960s, probably for ease of calculation of dose distributions and for shaping of fields. My point was that the tumour target volumes are not flat or cubes. Moreover, in this day-and-age of IMRT, why do we have to first flatten the beam and then unflatten it using IMRT and DMLC? The decades long fixation with flat fields did not seem rational. Omitting the flattening filter would lead to increased dose rates, therefore more efficient treatment delivery; a reduction in scattered x-rays in the beam, and therefore sharper beam boundaries; and, for higher energy x-rays, lower neutron contaminants, and, therefore, a reduction in the risk of secondary cancers. Soon thereafter, I accepted the position of Chairman of the Department of Radiation Physics at MDACC, where, in collaboration with my colleagues, Drs. Oleg Vassiliev and Uwe Titt, we continued to investigate the FFF mode. We presented our FFF findings again at a Varian Research Partners Symposium in 2006. This technology was introduced commercially by Varian in 2010 and soon afterwards by other vendors.

Key points

1. Collaboration pays off.
2. Question existing paradigms.

7   Core Values and Arrival at the UT MD Anderson Cancer Center

In 2001, I was invited to apply for the position of the Chair of the Department of Radiation Physics at the MDACC. Considering that most of my career had been devoted to research with relatively limited clinical experience, I felt that I had little chance in competition with several other applicants with a significantly greater clinical background. Nevertheless, I submitted a response and was interviewed by Drs. James Cox (then the Head of Division of Radiation Oncology), Kian Ang (Deputy Division Head), Margaret Kripke (Executive Vice President of Academic Affairs), John Mendelsohn (MDACC President) and others. Surprisingly, I was offered the position, which I happily accepted. Many years later, Dr. Kripke informed me that the key factor in selecting me was my appreciation and passion for certain principles and core values that I discussed with senior interviewers. It turned out that, just a couple of years prior when I was at MCV, I saw an interview on the Public Broadcasting System TV with Jack Welch, then the Chairman of General Electric, a rather controversial figure, who had built the company into among the most valuable in the world. He was known as "neutron Jack" for his harsh policy of firing the bottom 10% of his managers each year. During the interview, he emphasized the importance of core values and spelled out a list he had introduced at GE. They all made common sense, but until then I had never given any serious thought to core values. I was impressed by their potential and started to adapt them, create my own set and apply them in my small sphere of influence, partly for self-improvement and partly to create a friendly, collaborative, yet exciting and more productive work environment. It so happened that the MDACC leadership also explicitly believed in the core values of Caring, Integrity and Discovery. Many years later each of these were subdivided into three "sub-values." A sampling of my personal values, which, in many ways are similar to MDACC's values, and which I tried to adhere to and encourage my colleagues to embrace are:

1. Passion for excellence
2. Integrity, honesty and trustworthiness
3. Openness and free flow of communication
4. Caring and respect
5. Recognizing change as an opportunity for growth
6. Team spirit, collaborativeness and sharing
7. Rewarding progress, demanding commitment and accountability
8. Humility
9. Self-confidence and mindset for simplicity
10. Hatred for bureaucracy and resisting desire to control

Each of us needs to develop his or her own set of values and strive to live by them. I should mention that MDACC uses adherence to its values as a part of the annual performance appraisal of its employees.

8   Image-Guided and Adaptive Radiotherapy and Motion Management

The exquisite radiation dose patterns possible with IMRT that conform tightly to the tumour target and spare normal critical tissues become highly sensitive to anatomic changes over the protracted multi-fraction course of radiotherapy or to anatomic motion (e.g., due to breathing) during a single fraction. Such changes introduce uncertainty regarding the dose delivered to the patient and, therefore, may compromise treatment outcomes. For proton therapy (see next section), for intensity modulated proton therapy (IMPT) in particular, because of the sharp drop in dose at the end of the range of protons, such uncertainties may be even more damaging. To tackle the consequences of respiratory motion, Dr. Paul Keall, my colleague at Medical College of Virginia, defined the concept of 4-dimensional CT scanning (4DCT) in the late 1990s. He generated the first ever 4DCT image and laid the groundwork for evaluating the impact of respiratory motion on dose distributions. He also pioneered respiratory-gated treatments and the tracking of tumours with intensity-modulated radiation beams. A few years later, another one of my colleagues, Lei Dong, carried out pioneering research to assess the impact of inter-fractional anatomy changes on dose distribution. He developed and applied novel approaches to adapt radiation treatments to changes in anatomy and to add the contributions to dose from each of the treatment fractions to determine the accurate dose distribution patterns actually received by the patient.

Paul is now the Director of the Radiation Physics Laboratory at the University of Sydney in Australia, and Lei is Vice Chair of Radiation Oncology and Director of Medical Physics at the University of Pennsylvania in Philadelphia. I have been fortunate enough to have extraordinary young colleagues like Paul, Lei and many others who are succeeding in the field. It is matter of pride and joy for me that I was able to provide them guidance, resources and freedom to develop their ideas and talents to their fullest.

Key points

1. Mentoring is a reward in itself.
2. "We make a living by what we get, we make a life by what we give." Winston Churchill.

9   Proton Therapy

Initially, at the time of my arrival at the MDACC, I was of the view that radiotherapy can achieve all that is needed with photons and that protons and heavier ions are just esoteric, expensive pie-in-the-sky modalities not practical for large scale routine care. However, Dr. James Cox, the Head of the Division of Radiation Oncology at that time, was very interested in protons. He faced numerous obstacles, the primary one being financial. The institution was unwilling to invest well in excess of $100 million on a questionable venture. Through sheer determination, Dr. Cox negotiated with external investors to acquire the needed funding to start building the Proton Therapy Center of Houston (PTCH). We also needed a team of physicists with expertise in proton therapy to establish the facility. To compensate for my lack of exper-tise, I hired Dr. Alfred Smith, an experienced particle therapy physicist from MGH, to design the PTCH and its equipment and to manage its instal-lation in collaboration with Hitachi, the vendor. There were no commercial proton treatment planning systems at the time. So, leveraging my background in photon treatment planning, I personally took charge of the development of a proton treatment planning system in collaboration with Varian. The construction phase ended in 2006, and the treatments began. At about that time, Dr. Michael Gillin, whom I had recruited as the Deputy Chair and Chief of Clinical Radiation Physics, took charge of the clinical implementation of the PTCH. Initially, we treated patients with passively scattered proton therapy (PSPT), a 3DCRT analog of x-ray therapy, the only mode available to us then. In 2011, we started treating about 30–40% of proton patients with IMPT. Realizing the advantages of IMPT, MDACC is in the process of building another large proton therapy facility, which is entirely devoted to the delivery of IMPT.

In the beginning, there was considerable hype and excitement about proton therapy. Its possible limitations were overlooked by most in the field. It seemed to me, however, that, because of high gradients in their dose distributions, protons ought to be quite sensitive to changes in anatomy, which, if not properly accounted for, could adversely affect clinical outcomes. Furthermore, the biological effectiveness of protons relative to photons (RBE) was assumed to have a constant value of 1.1. This also did not seem to make sense considering that protons lose energy and slow down at an increasing rate as they penetrate the body. This causes them to ionize increasingly more densely, and, therefore, their RBE should increase with depth. However, my concerns were dismissed as being minor. Nevertheless, I persisted in investigating the biological effects of protons.

As I dug more into various issues related to proton therapy, I realized we should not take past assumptions about proton therapy for granted. While, in principle, proton therapy has great potential, to take advantage of this optimally, we needed to investigate the properties of protons in detail. In collaboration with MGH, I co-led two NCI P01 Program Project grants to investigate the physical, biological and clinical aspects of proton therapy. I am happy to note that the joint efforts by MGH and MDACC teams have resulted in the recognition of the impact of assumptions and approximations in the practice of proton therapy and in the development of numerous solutions to advance the state of the art.

An interesting part of the first P01 grant was a randomized trial to compare proton and photon therapy outcomes for lung cancer patients. There was considerable debate among my colleagues. Some felt that since "we already know that protons are better," it would be unethical to conduct such a trial and that no properly informed patient would give consent to being enrolled. They argued that the trial would be over in six months due to large differences in toxicity between the two arms. Parallels were drawn with a hilarious tongue-in-cheek publication of a fictitious randomized parachute jump trial (https://www.bmj.com/content/bmj/327/7429/1459.full.pdf), the objective of which was to determine whether parachutes are effective in preventing major trauma or death related to gravitational challenge. Proponents of our randomized trial, including me, argued that we are not certain if protons are better, due mainly to greater uncertainties in the dose distributions delivered and in their biological effectiveness. The trial was successfully completed. However, the results were negative in that protons were not proven to be superior. Nevertheless, the data generated by the trial has proven to be of enormous value in understanding the consequences of uncertainties and differences in response to protons vs. photons and in developing solutions and strategies to maximize their true potential.

Key points

1. Be curious and question everything. Ask questions no matter how stupid it makes you look.
2. Don't trust. Verify.

10   Radiotherapy and the Body's Immune System

A truly exciting aspect of the field of radiation therapy is that innovations never cease. The challenge, for someone as scatter-brained as I am, (euphemistically, I call myself a "tangential thinker") is to remain focused and not jump from one unfinished project to another. An example is that of the immunomodulatory effects of radiation therapy. Radiotherapy is both immunostimulant as well as immunosuppressive. It has been shown that certain types of cell death that elicit immune response stimulate the immune system to attack the tumour cells; whereas radiation to tissues outside the tumour kills the highly radiosensitive cells of the immune system (lymphocytes), often more than negating the stimulatory benefit of radiation.

In early 2017, a radiation oncologist colleague (Dr. Steven Lin) presented data of a large cohort of esophagus cancer patients treated with protons and photons showing the statistically significant advantage in survival for proton patients. The advantage was found to be associated with differences in radiation-induced lymphopenia (i.e., depletion of lymphocytes), which, in turn, was associated with differences in mean body dose between protons and photons. Immediately, a light bulb went on. The radiation "dose bath" that photon therapy is forced to deliver; or, conversely, the ability of protons to significantly reduce it, is an unrecognized potential benefit of proton therapy in terms of sparing the immune system. The low and intermediate doses outside the tumour may not express themselves as measurable injury to normal tissues, but they can kill highly radiosensitive lymphocytes that are circulating all over the body as well as are concentrated in immune structures, such as the spleen and lymph nodes. Once the underlying principles are understood, the immune response can be modelled, and the models translated into treatment plan optimization to maximally spare the immune system. This work has progressed rapidly since the initial findings, resulting in numerous publications and grant applications.

Key point "Do not follow where the path may lead. Go instead where there is no path and leave a trail."—Ralph Waldo Emerson

11 Radiotherapy in a FLASH

At the time of writing of this chapter, ultra-high dose rate radiotherapy, called FLASH radiotherapy, has become the rage. In contrast with the conventional low dose rate protracted radiotherapy, which requires a fractionated course of up to 40 (sometimes even more) treatments, with each daily fraction taking between 15 and 60 min, in FLASH radiotherapy, the entire treatment can be delivered in a fraction of a second. The question is whether FLASH is something real or just a flash in the pan. Around 2015, a medical physicist, Dr. Alejandro Mazal of Institut Curie, in Paris, France, presented results of a study conducted by Favaudon, et al. (https://www.ncbi.nlm.nih.gov/pub med/25031268) at his institution showing sparing of normal tissues at ultra-high dose rates. I was skeptical. Naively, I thought why should the dose rate matter? It is the dose deposited that determines the biological damage. Since Favaudon's work, many experiments have been carried out all over the world confirming the normal tissue sparing effect of FLASH and, equally importantly, showing that the response of tumours to FLASH and conventional low dose rates is about the same. The number of researchers involved in FLASH as well as the number of publications is increasing exponentially. The underlying mechanisms are not yet understood; however, multiple hypotheses are being offered. It turns out that the sparing effect of ultra-high dose rates was discovered in the 1960s and 70s for electron beams. Research activities remained on the back burner until Favaudon's efforts. The rekindling of interest in FLASH radiotherapy is being thought of as akin to "sleeping beauty awakened."

The interest in FLASH at the MDACC was sparked by Dr. Albert Koong, the Head of our Division of Radiation Oncology, who came to MDACC in 2017. He was very keen on establishing a FLASH research program here.

Over a glass of wine at a Division get together, he asked me if our Hitachi synchrotron-based proton therapy system could be used to achieve FLASH dose rates (>40 Gy/s, about 4000 times the dose rates that we have been using routinely in conventional treatments). I immediately did some back of the envelope calculations and found that, indeed, we could achieve more than sufficient dose rates for small animal studies without making any alterations to the system. This and other related activities led to the formation of our FLASH research group and multiple ongoing investigations. While most of the FLASH research around the world has involved electron beams, I believe that FLASH radiotherapy with protons will be the mainstay of clinical translation due to the ability of protons to reach clinically relevant depths.

A huge number of questions need to be answered. While I believe that FLASH is not just a passing fad, at the same time, it is not a cure all. Eventually, once we have a better understanding of the underlying mechanisms, it is likely that it will be among the important radiotherapy tools. The main advantages of FLASH seem to be the sparing of normal tissues without sparing the tumour, sparing of circulating immune cells, ultrashort treatment fractions and radiotherapy courses, and the elimination of concerns related to intra-fractional motion and inter-fractional anatomy changes.

The more we learn about FLASH, the more questions arise. Our team is contributing to understanding the basic mechanisms, to designing and conducting experiments to acquire in vivo and in vitro data, and to interpreting the results. The current dominant hypothesis for FLASH seems to be that, at extremely high dose rates, oxygen is depleted, making normal tissues hypoxic (i.e., low in oxygen content) and, therefore, resistant to radiation damage. Tumours are not spared, possibly because they are already low in oxygen. I have a different hypothesis: FLASH also spares cells of the immune system (T-lymphocytes) that infiltrate the tumour and kill tumour cells. Another hypothesis, that seems to be appropriate at least for radiation therapy with carbon ions, is that FLASH may actually generate oxygen within the tumour, which sensitizes tumours. The FLASH effect overall may be a combination of all of these factors.

Key points

1. At first blush, certain things don't seem possible, but, while skepticism is encouraged, do remain open to different ideas. You never know.
2. The more I learn, the more I realize how little I know.

12   Summary

During my long career, I have learned innumerable lessons, do's, and don'ts. Like most of us, I have had my share of disappointments, rejections, and successes. I have developed a sense of core values, though I sometimes find it hard to adhere to them. The field of radiation oncology physics has offered me opportunities to be creative and to find solutions that make a difference. I have contributed to improved treatments and outcomes of the most dreaded of all diseases. The field has also allowed me to meet and exchange ideas and collaborate with the top scientists and clinicians, to travel, and to give lectures on a wide range of topics at various institutions and conferences the world over. I hope my experiences will be helpful, especially to those who are contemplating entering this fascinating field.

While reading the series of anecdotes above, I hope you noticed that the life of a medical physicist is anything but boring. In fact, it can be downright exciting and fun. The pace of new discoveries in our field is accelerating. Based on the current state of knowledge, I imagine that the most cost- and clinically effective way to cure most types of cancer will be a combination of particle radiotherapy with protons or heavier ions, possibly at ultra-high dose rates, in combination with immunotherapy. I strongly believe this field is becoming more exciting, innovative, and important every day. The best is yet to come.

Dr. Radhe Mohan, Ph.D., FAAPM, FASTRO is a tenured Professor and the holder of the Larry and Pat McNeil Chair in the Department of Radiation Physics at the University of Texas MD Anderson Cancer Center, Houston, TX. After receiving his Ph.D. from Duke University and post-doctoral training at Rutgers University, Dr. Mohan started his medical physics career in 1971 at Memorial Sloan-Kettering Cancer Center in New York where he rose to the rank of Professor and Associate Chairman of the Department of Medical Physics. He left MSKCC in 1996 to become the Director of Radiation Physics at Medical College of Virginia. He joined MD Anderson as Chairman of the Department of Radiation Physics in January 2002, stepping down in October 2010 to focus on proton therapy research and clinical activities. Dr. Mohan's extensive experience includes pioneering contributions in many areas of radiation physics and oncology. Since 2005, his activities have been concentrated on physical, clinical, biological and immunomodulatory aspects of proton therapy and FLASH (ultrahigh dose rate) radiotherapy. He has nearly 500 publications including over 430 in peer-reviewed journals with an H-Index of 92. He was the Senior Physics Editor of The International Journal of Radiation Oncology, Biology, and Physics from 2002 through 2011.

Awards

2018. American Association of Physicists in Medicine (AAPM) Coolidge Gold Medal Award—the highest honor bestowed on an AAPM member in recognition of an eminent career in medical physics.

2013 American Society for Radiation Oncology (ASTRO) Gold Medal—the highest honor bestowed on members who have made outstanding contributions to the field of radiation oncology.

2010. Failla Memorial Award of the Radiological and Medical Physics Society of New York.

2004. Allan M. Cormack Gold Medal of the Association of Medical Physicists of India—for outstanding contribution to Medical Physics.

2003. American Association of Physicists in Medicine Edith H. Quimby Award for Lifetime Achievement in Medical Physics.

Epilogue

Airport Security

The head of the Clinical Physics group at the Princess Margaret Hospital in Toronto was Professor John Robert (Jack) Cunningham. He was world renowned for the computer programs that he developed for radiation treatment planning, a process that allows for the development of the optimized treatment technique for individual patients undergoing radiation treatment … it's all about giving the maximum dose to the tumour and minimizing the dose to healthy tissues. In the 1970s, he developed an agreement with the Commercial Products Division of the Atomic Energy of Canada Limited (AECL). They would commercialize his software and sell it to interested users around the world. As part of the sales process, a one-week training course was provided by AECL for the new users at their production site in Kanata, Ontario, Canada, just outside of Ottawa, Canada's capital city. The training course included a lot of hands-on use of the programs; however, one afternoon of the week was devoted to lectures on the theory and practical aspects of the software. Usually, Jack Cunningham flew from Toronto to Ottawa on the Tuesday of the course to provide these lectures for a full afternoon and then, after going out for dinner with the course participants, fly back to Toronto. After Jack had done this for a while, he asked me if I would consider doing some of these courses, especially when he was away on travels around the world. I readily agreed. Not only was this an opportunity to meet medical physicists from different parts of the globe but it also gave me some satisfaction in teaching some basic medical physics on issues related to treatment

J. Van Dyk (ed.), *True Tales of Medical Physics*, https://doi.org/10.1007/978-3-030-91724-1

planning for patients undergoing radiation treatment. I still have connections with medical physicists in various places around the world as a result of these training courses.

To expand on this story, I need to explain another aspect of activities of medical physicists working in radiation therapy. The radiation beams from radiation therapy machines need to be calibrated in such a way that the dose delivered to the patient is consistent with global standards so that all patients get exactly the same amount of radiation dose as indicated by the prescription. We use ionization chambers as our radiation detectors to determine dose. These are small cavities of air enclosed in a detector with plastic walls with positive and negative electrodes. These small ionization chambers are connected, usually via a long cable, to an electrometer—basically an electrical meter that measures the current or charge from the ionization chamber (see Art Boyer's Chap. 5 which includes a picture of an ionization chamber and an electrometer). These ionization chambers need to be calibrated at a national standards laboratory. In Canada, this laboratory is located at the National Research Council in Ottawa.

On one of the occasions that I was teaching the course at AECL in Kanata, one of our dosimetry systems from the Princess Margaret Hospital, had previously been sent to the National Research Council in Ottawa for calibration, and, because these are very sensitive detectors and electrometers, I was going to hand deliver it back to my hospital in Toronto. I picked up the system effectively as two components, one being the electrometer in its own case and the other being the ionization chamber on the end of a long cable which was transported in what looked like a briefcase. So, after the course and the dinner, I was taken to the airport with my two cases. I was somewhat anxious about what would happen at airport security with these electronic components. When I arrived at the security checkpoint, I placed the electrometer and the case with the ionization chamber with its cable on to the conveyor belt for the x-ray security check. It went through the machine and then I was allowed to walk through the metal detector. At this point, I was very concerned as to how they would deal with me and this electronics equipment. As I came to the other end of the conveyor belt, the security person said to me, "Do you work at the Princess Margaret Hospital?" When I said "yes", there was no further questioning, and I was let through. How she knew that this technology related to radiation therapy and that it was used by the Princess Margaret Hospital is still a mystery to me! However, it was reassuring that she did know and that more people are aware of medical physicists and their activities than I realized.

Border Patrol[1]

Professor Harold Johns was a world-renowned medical physicist. He developed one of the first cobalt-60 radiation therapy machines in the world while he was in Saskatoon, Saskatchewan, Canada. Before he developed that machine, which was first used in 1951, he was also involved with a radiation therapy machine known as a betatron. This was a cutting-edge machine at that time, which was able to accelerate electrons in a circular path in conjunction with rapidly changing magnetic fields. The electrons could be extracted directly for electron beam treatments or the electrons could be made to hit a target to generate high-energy x-rays also for cancer treatment. With his status and reputation, Harold Johns travelled quite regularly to the United States. On one occasion, Harold was coming back from the United States and crossing the border into Canada, probably at the south end of Saskatchewan. When the border customs officer looked at his passport, he looked up at Harold and said, "Do you have any betatrons to declare?" Of course, this took Harold totally by surprise since the general population does not know anything about betatrons or issues related to medical physics. It turns out that this customs officer had gone to the University of Saskatchewan and had taken some physics courses from Harold Johns and certainly knew about him and his work with betatrons. This is yet another example of people who are aware of medical physicists and what they do.

Quirks and Quarks[2]

I indicated earlier that the National Research Council in Ottawa, Canada is the national standards laboratory where the Canadian radiation detectors are calibrated to ensure consistency between all the institutions in Canada, as well as the rest of the world. In 1984, as a result of cost-cutting measures, the Canadian government, under the leadership of prime minister Brian Mulroney, decided that these laboratories for radiation measurements were

---

[1] I tell this story based on my memory as I heard it; however, I do not remember who told it to me. Attempting to establish its veracity, I contacted a couple of people who were quite close to Prof. Johns during that period of time. While they were able to relate other anecdotes, they were not able to confirm the truth of this story. If any reader is aware of this story, I would appreciate hearing from them. This is my disclosure seeing that I do want to honour the "*True Tales*" nature of this book.

[2] With input from Professor David W. Rogers of Carlton University, Ottawa, Canada and former Group Leader of the Ionizing Radiation Standards of the National Research Council, Ottawa, Canada.

to be closed. The entire group had been laid off. The Canadian Broadcasting Corporation (CBC) has an award-winning weekly science program called *Quirks and Quarks*. To quote from the Quirks and Quarks website, *"the program presents the people behind the latest discoveries in the physical and natural sciences, from the smallest sub-atomic particle to the largest objects in the sky and everything in between. The program also examines the political, social, environmental and ethical implications of new developments in science and technology. Quirks & Quarks is a program for people fascinated by the world above, below and around them. And you don't need a PhD to enjoy it."* People from the NRC were being interviewed on Quirks and Quarks along with Professor Jack Cunningham from the Princess Margaret Hospital in Toronto, in view of his international reputation and as an objective bystander not being directly involved in the NRC dosimetry labs. When Professor Cunningham was asked what the effects of closing the Canadian radiation standards laboratory would be, he provided a detailed and fairly complex analysis. At the end, the host of Quirks and Quarks, Jay Ingram, asked, "Dr Cunningham, does that mean more Canadians will die of cancer?" Professor Cunningham's response was "yes". As a result, within two days, Professor David Rogers, who then had a leadership position in the group laid off from the National Research Council (NRC), received interview requests from many major news outlets in the country, although he was not given permission to respond.

Later an article was published in Canada's national newspaper, The Globe and Mail, about Thomas Siddon, Canada's Minister for Science who played a major role in the cuts to the NRC. Three columns of the article were quite positive about his political career, but half-way down the last column, in a discussion of his cuts to the National Research Council, it states, *"Even worse, he also chose reductions that would hurt dying people, giving rise to the prediction that a unique NRC service to test equipment from cancer centres across the country would be wiped out"*. Not too much later, the radiation standards group at the NRC was reinstated with David Rogers as its leader.

There are two messages in this story. First, the work of medical physicists received national coverage by the CBC radio program (actually, Quirks and Quarks is also broadcast outside of Canada, thus it was really international coverage) as well as Canada's national newspaper and many other papers throughout Canada. Second, as the title of this book indicates, medical physics does save lives. Of course, the saving of lives goes well beyond dose calibration procedures, but they certainly are included in the total radiation therapy and diagnostic imaging activities.

Words of Wisdom

The Tales told in this book have demonstrated that medical physics covers a wide variety of topics, is a highly technical field, and requires significant background education, training and practical experience to provide accurate and safe diagnostics and treatments of patients. These award-winning medical physicists have aired some of their experiences, both positive and negative. Many of the chapters have provided "words of wisdom" based on real-life experiences, often expressed as "lessons learned.". They have also expressed their complete enjoyment of working in this field and how it has led to very satisfying careers. To encapsulate *all* the words of wisdom or lessons learned, and comments about career satisfaction here is impossible. Thus, I will leave it to the reader to enjoy the stories and to pick up the messages as the stories unfold.

The Final Word

The first three stories of this Epilogue indicate that indeed there are some people who know about medical physics, although these stories are fairly rare. Even today, the question still comes up, what is medical physics and what do medical physicists do? Having come this far in the book, you will now have a better appreciation for the world of medical physics and a clearer perception of the kinds of activities that occupy the time of medical physicists. While each of the stories told by these all-stars in medical physics only provides a small snippet of their experience, in composite they provide great insight into the world of medical physics … the net impact of which is that their work results in saving the lives of patients.

Index

Printed in the United States
by Baker & Taylor Publisher Services